THE
HANDY
TECHNOLOGY
ANSWER
BOOK

About the Authors

Naomi E. Balaban, a reference librarian for more than twenty-five years at the Carnegie Library of Pittsburgh, has extensive experience in the areas of science and consumer health. She edited, with James Bobick, *The Handy Science Answer Book, The Handy Anatomy Answer Book,* and *The Handy Biology Answer Book.* She has a background in linguistics and a master's degree in library science.

James Bobick, a retired librarian, was most recently Head of the Science and Technology Department at the Carnegie Library of Pittsburgh for sixteen years. He was also an adjunct professor in the School of Information Sciences at the University of Pittsburgh, teaching the science resources course for graduate students in library and information sciences. He is the author, with G. Lynn Berard, of *Science and Technology Resources: A Guide for Information Professionals and Researchers.* He has master's degrees in both biology and library science.

Also from Visible Ink Press

Please visit the "Handy" series website at www.handyanswers.com.

THE HANDY TECHNOLOGY ANSWER BOOK

Naomi Balaban and James Bobick

VISIBLE INK PRESS

Detroit

THE HANDY TECHNOLOGY ANSWER BOOK

Visible Ink Press®
43311 Joy Rd., #414
Canton, MI 48187-2075

Visible Ink Press is a registered trademark of Visible Ink Press LLC.

Most Visible Ink Press books are available at special quantity discounts when purchased in bulk by corporations, organizations, or groups. Customized printings, special imprints, messages, and excerpts can be produced to meet your needs. For more information, contact Special Markets Director, Visible Ink Press, www.visibleink.com, or 734-667-3211.

Managing Editor: Kevin S. Hile
Art Director: Mary Claire Krzewinski
Typesetting: Marco DiVita
Proofreaders: Larry Baker and Shoshana Hurwitz
Indexer: Larry Baker

Cover images: Google glass: Antonio Zugaldia; electric car: Ludovic Hirlimann; all other images: Shutterstock.

Library of Congress Cataloging–in–Publication Data

Names: Balaban, Naomi E., author. | Bobick, James E., author.
Title: The handy technology answer book / by Naomi Balaban and James E. Bobick.
Description: Canton, MI : Visible Ink Press, [2016] | Includes bibliographical references and index.
Identifiers: LCCN 2015029037 | ISBN 9781578595631 (pbk. : alk. paper)
Subjects: LCSH: Technology–Miscellanea.
Classification: LCC T47 .B2165 2016 | DDC 600–dc23
LC record available at http://lccn.loc.gov/2015029037

Printed in the United States of America

10 9 8 7 6 5 4 3 2 1

Contents

ACKNOWLEDGMENTS *ix*
PHOTO SOURCES *x*
INTRODUCTION *xi*

INTRODUCTION TO TECHNOLOGY ... 1

Introduction and Historical Background (1) ... Inventions and Patents (9) ... Societies, Publications, and Awards (14)

COMPUTERS ... 27

Introduction and Historical Background (27) ... Hardware and Software (30) ... Programming and Applications (40)

COMMUNICATIONS ... 49

Symbols, Writing, and Codes (49)

RADIO AND TELEVISION ... 61

TELECOMMUNICATIONS AND RECORDINGS ... 73

The Internet (80)

WEIGHTS, MEASURES, TIME, TOOLS, AND WEAPONS ... 91

Weights, Measures, and Measurements (91) ... Time (101) ... Tools and Machines (119) ... Weapons (127)

MINING, MATERIALS, AND MANUFACTURING ... 139

Mining (139) ... Coal (145) ... Minerals (147) ... Metals (156) ... Natural Materials (172) ... Manufacturing (174)

BUILDINGS, BRIDGES, AND OTHER STRUCTURES ... 193

Buildings and Building Parts (193) ... Roads, Bridges, and Tunnels (205) ... Miscellaneous Structures (216)

CARS, BOATS, PLANES, AND TRAINS ... 225

Cars and Motor Vehicles (225) ... Boats and Ships (241) ... Planes and Other Aircraft

(247) ... Military Vehicles (255) ... Trains and Trolleys (258)

ENERGY ... 263

Basic Concepts (263) ... Fossil Fuels (265) ... Nuclear Energy (278) ... Renewable and Alternative Energy Sources (288) ... Generation of Electricity (301) ... Measurement of Energy (303) ... Consumption and Conservation (305)

TECHNOLOGY AND THE ENVIRONMENT ... 315

Introduction, Historical Background, and Environmental Milestones (315) ... Land Use (322) ... Air Pollution (328) ... Global Climatic Change (337) ... Water Pollution (343) ... Solid and Hazardous Waste (349) ... Recycling and Conservation (358) ... Sustainability (366)

BIOTECHNOLOGY AND GENETIC ENGINEERING ... 369

Introduction and Historical Background (369) ... Methods and Techniques (376) ... Applications (384)

TECHNOLOGY AND MEDICINE ... 403

Introduction (403) ... Imaging Techniques (405) ... Medical Devices and Tools (408) ... Assistive Technologies and Prosthetic Devices (416) ... Medical and Surgical Procedures (419)

FURTHER READING   *425*
INDEX   *433*

Acknowledgments

We thank Roger Jänecke for his confidence in our ability to undertake and complete this project. We also thank page designer Mary Claire Krzewinski, typesetter Marco DiVita, indexer Larry Baker, and proofreader Shoshana Hurwitz. We thank managing editor Kevin Hile for his patience, great insights, and editorial skills, once again.

Finally, we thank our spouses, Carey and Sandi, for always being there. We owe you a lot!

Photo Sources

Buffalutheran (Wikicommons): p. 353.

Oliver Calder: p. 289.

Peter Campbell: p. 44.

Chemical Heritage Foundation: p. 373.

Michael Hicks: p. 33.

David Holt: p. 285.

Hu Totya: p. 2.

Intel Free Press: p. 30.

Steve Jurvetson: p. 382.

Mico Kaufman: p. 22.

Ehud Kenan: p. 84.

Koninklijke Nederlandse Akademie van Wetenschappen: p. 421.

Robert Lawton: p. 260.

NASA: pp. 250, 341, 347.

NASA/Kim Shiflett: p. 252.

Nimur (Wikicommons): p. 420.

Rfc1394 (Wikicommons): p. 120.

Runner1616 (Wikicommons): p. 71.

Shutterstock: pp. 6, 11, 13, 14, 19, 24, 38, 40, 46, 50, 52, 54, 65, 70, 74, 79, 82, 86, 89, 93, 99, 101, 103, 109, 113, 115, 117, 121, 127, 130, 133, 135, 137, 141, 143, 146, 149, 151, 154, 160, 163, 164, 166, 173 (top), 173 (bottom), 176, 178, 180, 182, 184, 187, 190, 195, 198, 199, 202, 204, 207, 212, 215, 217, 219, 220, 222, 229, 231, 233, 236, 240, 258, 266, 268, 272, 275, 279, 286, 292, 295, 298, 299, 302, 306, 310, 323, 326, 329, 331 (bottom), 334, 337, 344, 352, 356, 359, 362, 364, 371, 377, 379, 385, 388, 394, 405, 409, 412, 414, 417, 419.

Smithsonian Institution: p. 67.

Martin Streicher, Editor-in-Chief, LINUX-MAG.com: p. 36.

Tecno 567 (Wikicommons): p. 78.

Theoprakt (Wikicommons): p. 124.

U.S. Fish and Wildlife Service: p. 319.

U.S. National Archives and Records Administration: p. 331 (top).

U.S. Navy: p. 248.

Visible Ink Press: p. 57

Wellcome Images: p. 75.

Public Domain: pp. 16, 18, 28, 47, 62, 64, 106, 134, 169, 209, 226, 243, 245, 254, 256, 312, 391, 399.

Introduction

Technology is an essential part of our daily lives. The impact of technology on all aspects of society is unquestionable. Modern technology has revolutionized the way people live, work, and play. The use of technological tools, gadgets, and other resources can help us control and adapt to the environment as our society changes and evolves. It is difficult to imagine living without technology.

A number of fields have experienced significant improvements because of advances in technology. One of these fields is communication. Early communication was by physical letter. Later, from advances in Morse code to Skype, the ability to communicate with people has changed dramatically. Today, the Internet has made long distances almost irrelevant, allowing users to correspond with people all over the globe in an instant. Technology has increased our connectivity with cell phones and other devices, providing an always-on connection to the global communication network. Similar advances have occurred in other fields such as energy, transportation, manufacturing and productivity, computer and information science, biotechnology and genetic engineering, medicine and healthcare, and environmental technology. Technology will shape the future and it can be compatible with nature by developing clean energy, cleaner transportation, and low-energy housing. Technology is an integral part of the future.

The information in *The Handy Technology Answer Book* will appeal to those with a background in the field, as well as those seeking an introduction to technology in everyday life. The selection of chapter titles and the content within each chapter provides an interesting overview of the major divisions of technology for the general reader.

We are pleased to contribute another addition to the "Handy Answer" family of books. It is wonderful to see literary growth paralleling technological growth, from the parent of all of the titles in this series, *The Handy Science Answer Book* (first published in 1994) to this most recent offspring.

INTRODUCTION TO TECHNOLOGY

INTRODUCTION AND HISTORICAL BACKGROUND

What is the difference between science and technology?

Science and technology are related disciplines, but each has different goals. The basic goal of science is to acquire a fundamental knowledge of the natural world. Outcomes of scientific research are the theorems, laws, and equations that explain the natural world. It is often described as a pure science. Technology is the quest to solve problems in the natural world with the ultimate goal of improving humankind's control of their environment. Technology is, therefore, often described as applied science; applying the laws of science to specific problems. The distinction between science and technology blurs since many times researchers investigating a scientific problem will discover a practical application for the knowledge they acquire.

How do scientists and engineers differ?

Scientists and engineers both work to solve complex problems and seek answers to how things work in nature. Scientists focus on exploring and experimenting to make new discoveries, which are presented at conferences and published in peer-reviewed journals. Engineers apply the newest discoveries in science to develop, design, create, and build new products, new devices, and new systems, which is the technology used by society. Oftentimes, scientists and engineers collaborate to attain the final goal.

What are the main branches of engineering?

Traditionally, the four main branches of engineering are civil, mechanical, electrical, and chemical engineering. Other areas of engineering, many of which are subspecialties of the main branches of engineering, are aerospace, agricultural, biomedical, computer, environmental, fire protection, geotechnical, nuclear, petroleum, sanitary, and transportation.

Civil engineers design, construct, operate, and maintain construction projects, including roads, buildings, tunnels, airports, bridges, dams, and water supply and treatment systems.

Mechanical engineering is one of the most diverse branches of engineering. The focus is to design, develop, build, and test mechanical and thermal devices, including tools, engines, and machines. Some of the industries in which mechanical engineers tackle problems are automotive, aerospace, biotechnology, electronics, environmental control, and manufacturing.

Electric engineers design, develop, test, and supervise the manufacturing of electrical equipment, including electric motors, radar and navigation systems, communications systems, and power generation equipment. Related to electric engineers are electronics engineers, who design and develop electronic equipment such as broadcast and communications systems, including global positioning systems (GPS) and portable music players.

Chemical engineers focus on the production or use of chemicals, fuel, drugs, food, and other products. Chemical engineering encompasses applications from the basic sciences of chemistry, physics, biology, mathematics, and computing.

When was engineering first recognized as a profession?

The terms "engineer" and "engineering" were first used during the Middle Ages. However, engineering was not recognized as a formal profession until the nineteenth century when the Industrial Revolution led to an explosion of the various fields of engineering.

Who is considered to be the first engineer?

Engineers in the twenty-first century are defined by education and training. However, in previous centuries, there were individuals who applied knowledge to design structures consistent with engineering techniques. No one knows for sure, but one candidate for the first engineer is Imhotep (2650–2600 B.C.E.), who was the engineer responsible for the construction of the Step Pyramid in Egypt in 2250 B.C.E.

A statue of the Egyptian priest, chancellor, and architect Imhotep is on display at the Louvre in Paris, France. Imhotep is credited with designing the Step Pyramid.

What are some engineering discoveries that were made accidentally?

There are many engineering discoveries that were made accidentally. Some of the major ones include: 1) vulcanized rubber, 2) plastic, 3) synthetic dyes, 4) radioactivity, 5) rayon, 6) Velcro, 7) Teflon, 8) microwave ovens, 9) saccharin, and 10) cardiac pacemakers.

What is nanotechnology?

Nanotechnology is a relatively new field of science that focuses on the miniaturization of different technologies. It aims to understand matter at dimensions between 1 and 100 nanometers. Nanomaterials may be engineered or occur in nature. Some of the different types of nanomaterials, named for their individual shape and dimensions, are nanoparticles, nanotubes, and nanofilms. Nanoparticles are bits of material where all the dimensions are nanosized. Nanotubes are long, cylindrical strings of molecules whose diameter is nanosized. Nanofilms have a thickness that is nanosized, but the other dimensions may be larger. Researchers are developing ways to apply nanotechnology to a wide variety of fields, including transportation, sports, electronics, and medicine. Specific applications of nanotechnology include fabrics with added insulation without additional bulk. Other fabrics are treated with coatings to make them stain-proof. Nanorobots are being used in medicine to help diagnose and treat health problems. In the field of electronics, nanotechnology shrinks the size of many electronic products, such as silicon chips. Researchers in the food industry are investigating using nanotechnology to enhance the flavor of food. They are also searching for ways to introduce antibacterial nanostructures into food packaging.

When was the term "nanotechnology" first used?

Richard Feynman (1918–1988) hinted at the concept of nanotechnology at the American Physical Society meeting in 1959 by suggesting the possibility of doing things at the atomic level. However, Norio Taniguchi (1912–1999) first introduced the term "nanotechnology" in the paper "On the Basic Concept of 'Nanotechnology'" at the Proceedings of the International Conference on Production Engineering of the Japan Society of Precision Engineering in 1974. His definition of nanotechnology was "Nanotechnology mainly consists of the processing of separation, consolidation, and deformation of materials by one atom or one molecule."

How large is a nanometer?

A nanometer equals one-billionth of a meter. There are 25,400,000 nanometers in an inch. A sheet of paper is about 100,000 nanometers thick. As a comparison, a single-walled carbon nanotube, measuring one nanometer in diameter, is 100,000 times smaller than a single strand of human hair, which measures 100 micrometers or 100,000 nanometers in diameter. The DNA, in our genetic material, is in the 2.5-nanometer range, while red blood cells are approximately 2.5 micrometers (2,500 nanometers).

Where are the major high-technology centers in the United States?

There is no consensus as to the definition of a high-technology center in the United States. Different studies use different criteria to identify major technology centers. A report in *Forbes* ranked metropolitan areas based on the ratio of engineers per 1,000 employees. Although Los Angeles has the greatest total number of engineers (70,000), it does not have the greatest concentration of engineers. Not surprisingly, San Jose-Sunnyvale-Santa Clara, CA (includes Silicon Valley) has the greatest concentration of engineers.

U.S. High-Tech Centers

Metropolitan Area	Total Number of Engineers	Engineers per 1,000 Employees
San Jose-Sunnyvale-Santa Clara, CA	40,400	45.0
Houston-Sugar Land-Baytown, TX	59,070	22.4
Wichita, KS	5,870	20.9
Dayton, OH	7,650	20.8
San Diego-Carlsbad-San Marcos, CA	25,490	20.2
Greenville-Mauldin-Easley, SC	5,710	19.1
Albuquerque, NM	6,810	18.7
Boston-Cambridge-Quincy, MA-NH	43,340	17.5
Bakersfield-Delano, CA	4,680	17.1
Denver-Aurora-Broomfield, CO	20,910	17.0

Still another criterion to define a major technology center is based on the number of utility patents granted in a year. The following chart lists the top ten metropolitan areas based on number of utility patents granted:

Top 10 U.S. Metro Areas for Technology Based on Patents

Metropolitan Area	2000	2005	2010	2011	2012	2013
San Jose-Sunnyvale-Santa Clara, CA	5,812	6,502	10,074	10,256	11,517	12,899
San Francisco-Oakland-Fremont, CA	3,625	3,726	6,290	6,468	7,403	8,721
New York-Northern New Jersey-Long Island, NY-NJ-PA	5,689	3,959	6,383	6,252	6,944	7,886
Los Angeles-Long Beach-Santa Ana, CA	3,877	3,674	4,992	5,154	5,871	6,271
Boston-Cambridge-Quincy, MA-NH	2,972	2,768	4,330	4,537	4,986	5,610
Chicago-Joliet-Naperville, IL-IN-WI	3,101	2,246	2,933	3,033	3,503	3,766
Seattle-Tacoma-Bellevue, WA	1,141	1,518	4,052	3,597	4,094	4,364

Metropolitan Area	2000	2005	2010	2011	2012	2013
Minneapolis-St. Paul-Bloomington, MN-WI	2,244	1,989	2,827	3,113	3,085	3,445
San Diego-Carlsbad-San Marcos, CA	1,724	1,767	2,993	3,293	4,010	4,805
Detroit-Warren-Livonia, MI	2,120	2,018	2,222	2,253	2,671	3,000

Which metropolitan areas of the United States have experienced significant growth in the high-technology industries?

A report in *Forbes* ranked metropolitan areas based on the number of employees in traditional high-tech industries, such as software and engineering, along with the number of employees in STEM (science, technology, engineering, and mathematics) occupations in other industries. The top ten U.S. metropolitan areas were:

Top 10 U.S. Metro Areas by Employee Growth

Metropolitan Area	Tech Industry Employment Growth, 2001–2013	STEM Occupation Growth, 2001–2013
Nashville-Davidson-Murfreesboro-Franklin, TN	65.8%	12.3%
Raleigh-Cary, NC	54.7%	24.6%
Baltimore-Towson, MD	50.7%	19.6%
Indianapolis-Carmel, IN	50.4%	10.6%
Seattle-Tacoma-Bellevue, WA	45.5%	19.5%
San Antonio-New Braunfels, TX	45.1%	21.9%
Austin-Round Roc-San Marcos, TX	41.4%	17.1%
Salt Lake City, UT	38.0%	19.2%
San Francisco-Oakland-Fremont, CA	28.0%	5.5%
Houston-Sugarland-Baytown, TX	18.6%	24.1%

What are the major time periods of technology?

There are five major time periods in the history of technology. They are:

1. The Ancient World, 8000 B.C.E. to 300 C.E. and Middle Ages through 1599

2. The Age of Scientific Revolution, 1600–1790

3. The Industrial Revolution, 1791–1890

4. The Electrical Age, 1891–1934

5. The Atomic and Electronic Age, 1935 to the present

What were some of the technical advances made in each of the five major time periods of technology?

Technical progress in the Ancient World, 8000 B.C.E. to 300 C.E. and Middle Ages through 1599, was made in developing tools, materials, fixtures, and methods and procedures implemented by farmers, animal herders, cooks, tailors, and builders. Aqueducts, plumbing, the wheel, and sailing vessels are examples of technical advances in the earliest period of technology.

During the Age of Scientific Revolution, 1600–1790, most achievements were attributed to a relatively small number of individuals. Key instruments were invented that enabled an examination of the natural universe or to measure natural phenomena. Some of these instruments were the telescope, the microscope, the mercury barometer and thermometer, the spring balance, and the pendulum clock.

The principles that governed machines were thoroughly examined during the years of the Industrial Revolution, 1791–1890. Newly discovered principles were put in place, making it possible to build better machines and reduce principles to mathematical formulas. The use of interchangeable parts, steam engines, the electric telegraph, the transcontinental railroad, the phonograph, lightweight cameras, the typewriter, and news of the first automobiles were all accomplishments of the Industrial Revolution.

The most important invention of the Electrical Age, 1891–1934, was the automobile, which enjoyed widespread use by 1910. Other consumer items that were intro-

The Industrial Revolution gave civilization huge advances in the efficient production of goods, as well as some negative consequences, such as pollution.

duced and changed the everyday life of the average person during these years were safety razors, thermos bottles, electric blankets, cellophane, rayon, x-rays, and the zipper.

In the Atomic and Electronic Age, 1935 to the present, we are constantly witnessing a multitude of achievements and advances in science and engineering. A major advancement was the discovery of nuclear power and its use for both generating power and weaponry. Some examples of modern technological advances include turbojet engines, computers, genetic engineering, and the Internet.

What were some of the greatest engineering achievements of the twentieth century?

The National Academy of Engineering published a list of the twenty greatest achievements of the twentieth century in 2000. The achievements were selected based on their impact on society's quality of life. The absence of any one achievement would greatly change our daily lives—from our health benefits to how we travel, work, and spend our leisure time.

1. Electrification
2. Automobile
3. Airplane
4. Water supply and distribution
5. Electronics
6. Radio and television
7. Mechanization of agriculture
8. Computers
9. Telephone
10. Air conditioning and refrigeration
11. Highways
12. Spacecraft and space travel
13. Internet
14. Imaging
15. Household appliances
16. Health technologies and devices
17. Petroleum and petrochemical technologies
18. Laser and fiber optics
19. Nuclear technologies
20. High-performance materials

What are some of the greatest engineering challenges of the twenty-first century?

The National Academy of Engineering held a contest in 2015 soliciting ideas to solve fourteen Grand Challenges for Engineering in the twenty-first century. The goal of each challenge was to lead to a more sustainable, healthy, secure, and/or joyous world. The challenges were:

1. Make solar energy economical
2. Provide energy from fusion
3. Develop carbon sequestration methods
4. Manage the nitrogen cycle
5. Provide access to clean water

6. Restore and improve the urban infrastructure

7. Advance health informatics

8. Engineer better medicines

9. Reverse-engineer the brain

10. Prevent nuclear terror

11. Secure cyberspace

12. Enhance virtual reality

13. Advance personalized learning

14. Engineer the tools of scientific discovery

Who coined the terms "makers" and "maker movement"?

The terms "makers" and "maker movement" were coined by Dale Dougherty in 2005 while he was still at O'Reilly Media.

What is the maker movement?

The maker movement encompasses all modern inventors, designers, tinkerers, entrepreneurs, and do-it-yourselfers (DIYs). Technologies such as 3-D printers, 3-D scanners, laser cutters, and computer-aided design (CAD) software, which are now available to a larger audience through "maker spaces," have allowed individuals to design and manufacture new products themselves. Maker spaces are often shared collaborative settings in universities, public libraries, schools, and small companies that offer shared space and technologies. Some have claimed that the maker movement is the most recent industrial revolution and will have as great an impact on manufacturing as previous industrial revolutions.

When was the first Maker Faire held?

The first Maker Faire was held in 2006 in the San Francisco Bay area. Maker Faire is a gathering of tech people, crafters, educators, tinkerers, hobbyists, engineers, science clubs, authors, artists, students, and entrepreneurs who demonstrate and explore innovation and technologies in science, engineering, art, performance, and crafts.

INVENTIONS AND PATENTS

What is a patent?

A patent grants the property rights of an invention to the inventor. Once a patent is issued, it excludes others from making, using, or selling the invention in the United States. The U.S. Patent and Trademark Office issues three types of patents.

What are the three types of patents granted by the U.S. Patent Office?

The three types of patents granted by the U.S. Patent Office are:

1. Utility patents are granted to anyone who invents or discovers any new and useful process, machine, manufactured article, compositions of matter, or any new and useful improvement in any of the above.
2. Design patents are granted to anyone who invents a new, original, and ornamental design for an article of manufacture.
3. Plant patents are granted to anyone who has invented or discovered and asexually reproduced any distinct and new variety of plant.

How many patents have been issued by the U.S. Patent Office?

Over eight million patents have been granted by the U.S. Patent Office since its inception in 1790. In recent years, the number of patents issued on a yearly basis has risen dramatically due to the growth of various technologies and industries, such as electronics and pharmaceuticals. The following chart shows the numbers of patents of all types (utility, design, plant, and reissue) issued for selected years:

Year	Total Number of Patents Granted
1965	66,647
1970	67,964
1975	76,810
1980	66,170
1985	77,245
1990	99,077
1995	113,834
2000	175,979
2005	157,718
2010	244,341
2011	247,713
2012	276,788
2013	302,948

Who is the only U.S. president to receive a patent?

On May 22, 1849, twelve years before he became the sixteenth U.S. president, Abraham Lincoln (1809–1865) was granted U.S. Patent #6,469 for a device to help steamboats pass over shoals and sandbars. The device, never tested or manufactured, had a set of adjustable buoyancy chambers (made from metal and waterproof cloth) attached to the ship's sides below the waterline. Bellows could fill the chambers with air to float the vessel over the shoals and sand bars. It was the only patent ever held by a U.S. president.

Must all U.S. patent applications be accompanied by a drawing?

It is required by law, in most cases, that a drawing of the invention accompany the patent application. Composition of matter or some processes may be exempt, although in these cases the commissioner of patents and trademarks may require a drawing for a process when it is useful. The required drawing must show every feature of the invention specified in the claims and be in a format specified by the Patent and Trademark Office. Drawings must be in black and white unless waived by the deputy assistant commissioner for patents. Color drawings are permitted for plant patents where color is a distinctive characteristic.

Do the terms "patent pending" and "patent applied for" offer any protection?

"Patent pending" and "patent applied for" are used to inform the public that a patent application is on file in the U.S. Patent and Trademark Office (USPTO) for the item. According to the USPTO, individuals and companies using the terms falsely to deceive the public will be fined.

What is a provisional patent application?

The option of filing a provisional patent application has been available to inventors since June 8, 1995. It is a lower-cost, first-patent filing option in the United States. A provisional patent application does not require a formal patent claim or an oath or a declaration. It establishes an early effective filing date and allows the terms "patent pending" to be applied to the invention. A formal, nonprovisional patent application must be filed within twelve months for the earlier filing date to be effective.

Does the USPTO define what cannot be patented?

According to the USPTO, utility patents are provided for new, nonobvious, and useful processes, machines, articles of manufacture, and composition of matter. The following things cannot be patented:

- Laws of nature
- Physical phenomena
- Abstract ideas
- Inventions that are not useful (such as perpetual-motion machines) or are offensive to public morality
- Literary, dramatic, musical, and artistic works (which are able to be copyrighted)

What is a trademark?

A trademark protects a word, phrase, name, symbol, sound, or color that identifies and distinguishes the source of the goods or services of one party (individual or company) from those of another party.

What is the difference between the symbols ™ and ®?

The use of the ™ (trademark) or ᴿᴹ (service mark) designation may be governed by local, state, or foreign laws and the laws of a pertinent jurisdiction. The symbol ® may only be used when the mark is registered in the Patent and Trademark Office and may not be used before the registration is issued.

Why do companies choose not to patent trade secrets?

A trade secret is information a company chooses to keep secret so as not to divulge certain information to its competitors. Once a patent is filed, the information is in the public domain. Perhaps the most famous trade secret is the formula for Coca-Cola®.

How do trademarks "die" and become a generic name for a product?

When a trademark becomes so popular that the term is universally used as the generic name for the product, the trademark is considered "dead." Some ex-trademarks that are now in the public domain are zipper (sliding fasteners), linoleum, cellophane (cellulose sheets), aspirin (pain reliever), escalator (powered stairs), and thermos bottle (vacuum-insulated container). Oftentimes, the dictionary will define a word as a former trade-

---

Which individual was granted the most U.S. patents?

Thomas Edison (1847–1931) received 1,093 U.S. patents—the most for any single individual. Among his many inventions are the phonograph, magnetic iron ore separator, electrographic vote recorder, and motion pictures on celluloid film.

mark and now as a generic name. Companies will try to keep trademarks alive by either inserting the term "brand" between the mark and the generic name, e.g., Kleenex® brand tissues, or ensuring there is a generic term for the product, e.g., "correction fluid" for liquid Wite-Out®. Other examples of corporations that are trying to avoid a mark from becoming a generic name are Xerox® for photocopier and Google for doing an Internet search.

How do copyrights differ from patents and trademarks?

Copyrights are very different from patents and trademarks. A patent primarily prevents inventions, discoveries, etc., of useful processes, machines, etc., from being manufactured, used, or marketed without the permission of the patent holder. It is a grant of intellectual property rights to the inventor and heirs by the government. A trademark is basically a brand name. It is a word, name, or symbol to indicate origin or source of goods and to distinguish the products and services of one company from those of another. Copyrights protect the form of expression rather than the subject matter of the writings or artistic works. They protect the original work of authors and other creative people against copying and unauthorized public performance and recording.

How long does a patent, a trademark, or a copyright protect the holder?

Utility and plant patent applications filed on or after June 8, 1995, are granted for a term that begins on the date the patent is issued and usually ends twenty years from the date of filing the patent application. Maintenance fees are due three times during the life of the patent on the following schedule:

- three to three and a half years after the date of issue
- seven to seven and a half years after the date of issue
- eleven to eleven and a half years after the date of issue

The term of a design patent is fourteen years from the date of issue and is not subject to the payment of maintenance fees.

The term of a federal trademark is ten years, with ten-year renewal terms. It can be renewed if it remains currently in use in commerce, and if the registrant files an affidavit to that effect between the fifth and sixth year and within the year before the end of every ten-year period after the registration date.

The term of a copyright depends on several factors. In general, for a work that was created on or after January 1, 1978, copyright protection usually lasts for the life of the author plus an additional seventy years. Works that are published anonymously or under a pseudonym are usually granted copyright protection for a term of ninety-five years from the year of its first publication or a term of 120 years from the year of its creation, whichever expires first.

Which inventions are credited to Thomas Jefferson, even though he never received a patent?

Thomas Jefferson (1743–1826), the third U.S. president, is recognized as having invented the swivel chair, pedometer, shooting stick, a hemp-treating machine, and an improvement in the moldboard of a plow.

What is the National Inventors Hall of Fame?

The National Inventors Hall of Fame was founded in 1973 on the initiative of H. Hume Mathews (b. 1911), then chairman of the National Council of Patent Law Associations (now called the National Council of Intellectual Property Law Associations). The U.S. Patent and Trademark Office be-

President Thomas Jefferson was a jack of all trades, including inventor.

came a cosponsor the following year. A foundation, created in 1977, administers this prize, which honors inventors who conceived of and developed advances that contributed to the nation's welfare. Usually, the recognition is for a specific patent. Originally located in Akron, Ohio, the National Inventors Hall of Fame moved to the campus of the U.S. Patent and Trademark Office in Alexandria, Virginia, in 2008. On May 22, 2014, the newly renovated facility was opened. Each of the more than five hundred inductees is represented in the gallery of icons. Thomas A. Edison (1847–1931), the first inductee, was honored for his electric lamp, U.S. Patent #223,898, granted January 27, 1880. Some of the most recent inductees were William Bowerman (1911–1999), who received Patent #3,793,750 for the modern athletic shoe; Mildred Dresselhaus (1930–), who received Patent #7,465,871 for superlattice structure; and Willis Whitfield (1919–2012), who received Patent #3,158,457 for the clean room. Inductees have been responsible for such things as coaxial cable, crash test dummy, flexible flyer sled, windshield wiper, steam generator, transistor, hepatitis B vaccine, pH meter, frozen food, powered loom, wrinkle-free cotton, plasma display, and hundreds of other inventions.

Who was the first woman inducted into the National Inventors Hall of Fame?

The first woman inducted into the National Inventors Hall of Fame was Gertrude Belle Elion (1918–1999) in 1991. Elion's research at Burroughs Wellcome led to the development of drugs to combat leukemia, septic shock, and tissue rejection in patients undergoing kidney transplants. She shared U.S. Patent 2,884,667 with George Hitchings (1905–1998) for anti-leukemia drugs.

13

SOCIETIES, PUBLICATIONS, AND AWARDS

What was the first important scientific society in the United States?

The first significant scientific society in the United States was the American Philosophical Society, organized in 1743 in Philadelphia, Pennsylvania, by Benjamin Franklin (1706–1790). During colonial times, the quest to understand nature and seek information about the natural world was called natural philosophy.

What was the first national scientific society organized in the United States?

The first national scientific society organized in the United States was the American Association for the Advancement of Science (AAAS). It was established on September 20, 1848, in Philadelphia, Pennsylvania, for the purpose of "advancing science in every way." The first president of the AAAS was William Charles Redfield (1789–1857).

What was the first national science institute?

On March 3, 1863, President Abraham Lincoln signed a congressional charter creating the National Academy of Sciences, which stipulated that "the Academy shall, whenever called upon by any department of the government, investigate, examine, experiment, and report upon any subject of science of art, the actual expense of such investigations, examinations, experiments, and reports to be paid from appropriations which may be made for the purpose, but the Academy shall receive no compensation whatsoever for any services to the Government of the United States." The Academy's first president was Alexander Dallas Bache (1806–1867). The National Academy of Sciences, National Academy of Engineering, Institute of Medicine, and the National Research Council are known collectively as the National Academies. These private, nonprofit institutions serve as the country's preeminent sources of advice on science and technology, engineering, and medicine, often shaping the nation's policies.

U.S. President Woodrow Wilson requested the establishment of the National Research Council in 1916.

When was the National Research Council established?

The National Research Council was established in 1916 by the National Academy of

Sciences at the request of President Woodrow Wilson (1856–1924) "to bring into cooperation existing governmental, educational, industrial, and other research organizations with the object of encouraging the investigation of natural phenomena, the increased use of scientific research in the development of American industries, the employment of scientific methods in strengthening the national defense, and such other applications of science as will promote the national security and welfare." The National Research Council has six major divisions: (1) Behavioral and Social Sciences and Education, (2) Earth and Life Studies, (3) Engineering and Physical Sciences, (4) Policy and Global Affairs, (5) Transportation Research Board, and (6) the Gulf Research Program.

When was the National Academy of Engineering founded?

The National Academy of Engineering was founded in 1964. Its mission is to provide independent advice to the federal government on matters involving engineering and technology. Furthermore, it strives to promote a vibrant engineering profession.

When was the Institute of Medicine established?

The Institute of Medicine (IOM) was established in 1970 to help those in the government and the private sector make informed health decisions. The IOM conducts studies and research on health-related issues. Studies may be initiated as specific mandates from Congress or requested by federal agencies and independent organizations.

What was the first technical society?

John Smeaton (1724–1792) founded the Society of Civil Engineers in 1771. It was later renamed the Smeatonian Society of Civil Engineers. It was the first engineering society to be established anywhere in the world. It continues to meet, mostly for dinner meetings, but it also awards the prestigious Smeatonian Society of Civil Engineers Medal to one final-year student or young researcher in the United Kingdom each year. The recipient of the 2014 medal was Adam Stephenson.

What book is considered the most important and most influential scientific work?

Isaac Newton's 1687 book *Philosophiae Naturalis Principia Mathematica* (known most commonly as the abbreviated *Principia*), which he wrote in eighteen months, summarizes his work and covers almost every aspect of modern science. Newton introduced gravity as a universal force, explaining that the motion of a planet responds to gravitational forces in inverse proportion to the planet's mass. Newton was able to explain tides and the motion of planets, moons, and comets using gravity. He also showed that spinning bodies such as Earth are flattened at the poles. The first printing of *Principia* produced only five hundred copies. It was published originally at the expense of his friend, Edmond Halley (1656–1742), because the Royal Society had spent its entire budget on a history of fish.

Why was *De re Metallica*, published in 1556, an important book?

This sixteenth-century treatise on mining engineering remained the definitive work for two hundred years. The author, George Bauer (1495–1555), collected and mastered the expertise and knowledge of miners about metals and engineering that had not been recorded in any book. The book was published posthumously in 1556 under his Latinized name, Georgius Agricola. Latin scholars encountered great difficulty in translating the work because Agricola had coined many technical terms. The first English translation was published by Herbert Hoover (1874–1964) and his wife, Lou Henry Hoover (1874–1944), in 1912. Hoover was a mining engineer before his career in politics.

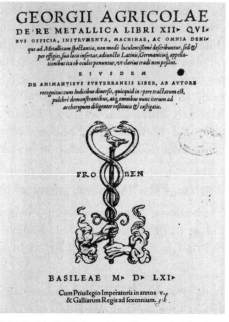

Published in 1556, *De re Metallica* was, for the time, an incredible repository of knowledge on engineering, mining, and metals.

What was the first scientific journal?

The first scientific journal was *Journal des Sçavans*, published and edited by Denis de Sallo (1626–1669). The first issue appeared on January 5, 1665. It contained reviews of books, obituaries of famous men, experimental findings in chemistry and physics, and other general-interest information. Publication was suspended following the thirteenth issue in March 1665. Although the official reason for the suspension of the publication was that de Sallo was not submitting his proofs for official approval prior to publication, there is speculation that the real reason for the suspension in publication was his criticism of the work of important people, papal policy, and the old orthodox views of science. It was reinstated in January 1666 and continued as a weekly publication until 1724. The journal was then published on a monthly basis until the French Revolution in 1792. It was published briefly in 1797 under the title *Journal des Savants*. It began regular publication again in 1816 under the auspices of the Institut de France, evolving as a general-interest publication.

What is gray literature?

Research scientists and engineers strive to be comprehensive in their quest for knowledge when investigating a topic. Therefore, they explore beyond traditional search resources such as journals and books. The information referred to as "grey literature" plays the important role of assisting the scholar in gaining a broader perspective on a topic and acts to fill in knowledge gaps. There are as many definitions of grey literature as there are forms of publications. Grey literature includes research summaries, topical

> ## What is the oldest continuously published scientific and technical journal?
>
> The *Philosophical Transactions* of the Royal Society of London, first published a few months after the first issue of the *Journal des Sçavans* on March 6, 1665, is the oldest, continuously published scientific and technical journal.

data sets, special industry and governmental publications, websites, newsletters, public industry reports, patents, and various other print and electronic formats. In fact, anything other than books and journals may be considered grey literature. One definition that is widely used and accepted was created by the Luxembourg Convention on Grey Literature held in 1997 and adopted by the Grey Literature Network Service (GLNS) at international conferences in 1997 and 2004. The GLNS definition of grey literature is: information produced on all levels of government, academics, business, and industry in electronic and print formats not controlled by commercial publishing, i.e., where publishing is not the primary activity of the producing body. Since the growth of the Internet, previously elusive materials considered grey literature are more easily retrieved, thereby making it feasible for more in-depth evaluative scholarly review processes and quality production.

What are technical reports?

Technical reports are documents that describe the progress or results of scientific or technical research and development. Most technical reports are usually produced in response to a specific request or research need and may serve as a report of accountability to a funding organization. They may report the results of new scientific and technical investigations or describe solutions to problems that will advance new technologies. Since they are geared towards a specialized audience, they are written in tight, concise language that focuses on pure content. The typical technical report is often confined to rather narrow limits, frequently describing progress on a particular project. For example, a technical report would not discuss how to design and build bridges of all types but may focus on a new way to design brackets that hold the bridge support cables or a new way to protect bridge surfaces from the effects of corrosion. Not all technical reports are immediately available to the general public because they may contain classified or proprietary information.

What are the three major categories of technical reports?

Technical reports usually fall into one of three categories: government-sponsored research reports, privately funded research reports, or academia-generated research reports. Privately funded research reports are produced as in-house technical reports by many corporations for proprietary use. These reports may have restricted use due to

17

pending patent applications and/or trade secrets. Technical reports are also generated as a result of the research in universities and other academic institutions. They are one method of reporting conditions in the field (e.g., engineers on a work site) and in the development of new software techniques, languages, and hardware component innovations (e.g., computer scientists). Research reports generated in the academic environment are a fast medium for communication and are not subject to the usual delays with peer review for scholarly publications. They make for an informal venue for sharing experimental approaches and hypotheses to be actively discussed and debated.

What is the National Technical Information Service (NTIS) Bibliographic Database?

The National Technical Information Service (NTIS) Bibliographic Database is the primary resource for accessing U.S. government-sponsored research and worldwide scientific, technical, and engineering information. NTIS is the central source for the sale of unclassified and publicly available information from research reports, journal articles, data files, computer programs, and audiovisual products from federal sources. The database covers 1964 to the present and is accessed via commercial vendors.

What is considered the first technical report?

Identifying the first technical report is difficult, but one of the earliest technical reports, and perhaps the beginning of technical writing, was prepared by Frontinus (ca. 35–ca. 103). He was the head administrator of the Roman aqueducts at the end of the first century C.E. In ancient times, engineers wrote detailed accounts of building projects for their rulers, complete with drawings. Frontinus had discovered faulty intake channels in the original works and is credited with writing a full account, known as *De Aquaeductu Urbis Romae* (*On the Aqueducts of Rome*), with his thoughts on improvements for the regulation of the water system.

What was the first technical report written in English?

Geoffrey Chaucer's (1343–1400) *Treatise on the Astrolabe* was written in 1391.

What are technical standards?

Standards are published documents that provide technical definitions and guidelines for designers and manufacturers to

A page from Geoffrey Chaucer's 1391 work, *Treatise on the Astrolabe* ensure the reliability of the materials, products, methods, and/or services people use every day. The purpose of developing and maintaining standards is to ensure that things work properly, safely, and efficiently. There are standards for such diverse items as toys, food, drugs, automobiles, jumbo jets, stairs, ladders, and nuclear reactors.

What are the various types of standards?

There are basically seven types of standards. They are:

- *Dimensional Standards,* which specify dimensions or sizes to achieve compatibility and interchangeability of component parts.
- *Materials Standards,* which specify the composition, quality, chemical, or mechanical property of materials such as alloys, fuels, and paints.
- *Performance Standards,* which set the minimum acceptable levels of efficiency and safety for a product or component.
- *Test Methods,* which recommend the tools, conditions, and procedures for comparing the quality or performance of materials and products.
- *Codes of Practice,* which specify procedures for the installation, operation, and maintenance for safety and uniform operation.
- *Terminology and Symbols,* which enable communication across disciplines through the use of standard nomenclature and graphic symbols.
- *Documentation Standards,* which spell out the specifications for layout, production, distribution, indexing, and bibliographic description of documents, generally known as classification standards.

What was one of the earliest documented standards?

One of the earliest documented standards was related to the measurement of length. In 1496, the distance between two marks on a bronze bar was decreed to be the Imperial Yard.

What was Eli Whitney's contribution to standards?

Eli Whitney (1765–1825) is universally recognized as the inventor of the cotton gin. However, he also played an important role in the development of standards. He is recognized as the individual responsible for the standardization of interchangeable parts. In 1798, he developed the idea of

Eli Whitney is most often remembered as the inventor of the cotton gin, a device that separates cotton from the seeds, saving considerable time and labor for farmers.

using machine-made components in the manufacture of rifles after receiving a contract from the federal government for the manufacture of ten thousand rifles for the U.S. Army. Whitney set up a musket factory in Hamden, Connecticut, which used interchangeable parts in the production operation.

How did the Baltimore fire in 1904 demonstrate the need for standards?

The need for standardization of interchangeable threaded components was dramatically demonstrated during the Baltimore, Maryland, fire in February 1904. A large number of fire engine companies went to Baltimore by special trains from many cities as far away as Philadelphia and New York to help control the fire. Once they reached Baltimore, they could not put their equipment to use in fighting the fire since every fire engine company had hoses with different-sized threads. The threads on the hoses did not match with one another or with the street hydrants in Baltimore. The firefighters had to helplessly stand back and watch the fire consume more than fifteen hundred buildings in an area of approximately seventy city blocks.

How do standards prevent disasters?

Standards affect all areas of our daily lives and play an important role in preventing disasters. At precisely 5:27 P.M., November 9, 1965, the electric power suddenly went off for an area of 40,000 square miles (103,600 square kilometers) of the northeastern United States. In over 10,000 elevators, more than a quarter of a million people hung suspended in midair, some as much as a tenth of a mile up. Not one elevator fell, and there were no injuries or fatalities. All were saved by the standard *Safety Code for Elevators and Escalators*, published by the American Society for Mechanical Engineers. Every elevator that is designed, manufactured, installed, or operated in the United States must conform to this standard. Of all those trapped that November day, probably fewer than twenty-five knew what saved their lives.

What is the difference between standards and specifications?

Standards are documents that provide technical definitions and guidelines for designers and manufacturers to ensure the reliability of the materials, products, methods, and/or services people use every day in a safe and efficient manner. Specifications may be thought of, basically, as documents to implement standards for a project. For example, a state

---

When was the Nobel Prize first awarded?

The Nobel Prize was established by Alfred Nobel (1833–1896) to recognize individuals whose achievements during the preceding year had conferred the greatest benefit to mankind. Five prizes were to be conferred each year in the areas of physics, chemistry, physiology or medicine, economic sciences, and peace. Although Nobel passed away in 1896, the first prizes were not awarded until 1901.

government agency that awards a contract for the construction of a new building may specify to its contractor the minimum level of illumination required in the various parts of the building. Such a statement is incorporated in a "specification." Both standards and specifications stipulate acceptable levels of dimensions, quality, performance, or some other aspect of materials, products, and processes. The scope of applicability of a standard is usually much larger and more significant than that for a specification.

What are codes?

Codes are mandated ordinances, regulations, or statutory requirements enforced by a government or its agencies for administering or controlling certain activities (such as building construction) or for promoting and protecting public health and safety. Enforcement may be at the local, state, national, or international level. Codes covering buildings are very extensive and one of the most important. Building codes are sets of regulations governing the design, construction, alteration, and maintenance of structures. They specify the minimum requirements to adequately safeguard the health, safety, and welfare of building tenants. Rather than create and maintain their own codes, most states and local jurisdictions adopt the model building codes maintained by the International Code Council (ICC). The three sections that deal with buildings are: 1) International Building Code (IBC), which applies to almost all types of new buildings; 2) International Residential Code (IRC), which applies to new one- and two-family dwellings and townhouses of not more than three stories in height; and 3) International Existing Building Code (IEBC), which applies to the alteration, repair, addition, or change in occupancy of existing structures. There are also codes specific to plumbing, mechanical, and electrical construction.

Is there a Nobel Prize for achievements in engineering?

There is no specific designation for engineering among the Nobel Prize fields. However, many recipients of the Nobel Prize have received the award for achievements that are technical in nature or for the basic science that has led to significant technical applications. For example, Charles H. Townes (1915–2015) received the 1964 Nobel Prize in Physics for the fundamental work that led to the construction of lasers. Similarly, the 2014 Nobel Prize in Physics was awarded to Isamu Akasaki (1929–), Hiroshi Amano (1960–), and Shuji Nakamura (1954–) for the invention of efficient blue light-emitting diodes. The 1993 Nobel Prize in Chemistry was awarded to Kary B. Mullis (1944–) for the invention of the polymerase chain reaction (PCR) method, which had a significant role in biotechnology.

How many awards does the National Academy of Engineering grant for achievements and innovations in engineering?

The National Academy of Engineering presents five different awards for achievements and innovations in engineering. They are:

1. The Charles Stark Draper (1901–1987) Prize is an annual award given for an accomplishment that has significantly impacted society by improving the quality of

life, providing the ability to live freely and comfortably. The 2015 prize was awarded to Isamu Akasaki (1929–), M. George Craford (1938–), Russell Dupuis (1947–), Nick Holonyak Jr. (1928–), and Shuji Nakamura (1954–) "for the invention, development, and commercialization of materials and processes for light-emitting diodes (LEDs)."

2. The Bernard M. Gordon (1927–) Prize is an annual award for innovation in engineering and technical education. The 2015 prize was awarded to Simon Pitts and Michael B. Silevitch of Northeastern University for developing an innovative method providing graduate engineers the skills to become effective engineering leaders.

3. The Fritz J. Russ (1920–2004) and Dolores H. Russ (1921–2008) Prize is awarded biennially for an achievement in bioengineering that improves the well-being and quality of life. The prize serves to encourage collaboration between the engineering and medical/biological professions. The 2015 prize was awarded to Graeme M. Clark (1935–), Erwin Hochmair (1940–), Ingeborg J. Hochmair-Desoyer (1953–), Michael M. Merzenich (1942–), and Blake S. Wilson (1948–) for the invention of cochlear implants.

4. The Founders Award is awarded annually to an outstanding member of the National Academy of Engineering for upholding the ideals and principles of the NAE. The 2014 award was presented to Robert A. Brown (1951–) for his commitment to diversity in engineering and leadership in transforming disciplines and institutions.

5. The Arthur M. Bueche (1920–1981) Award is awarded annually to an engineer dedicated to science and technology who has been active in determining U.S. science and technology policy and enhancing the relationship between industry, government, and universities. The 2014 award was presented to Siegfried S. Hecker (1943–) for contributions to nuclear science and engineering and nuclear diplomacy.

What is the highest honor for technological innovation in the United States?

The National Medal of Technology and Innovation (NMTI) is the highest honor for technological innovation in the United States. It was established by the Stevenson-Wydler Technology Innovation Act of 1980. The first medal was awarded in 1985 and subsequently has been awarded annually. It recognizes individuals or companies for their lasting contributions to America's competitiveness, standard of living, and quality of life through technological innovation along with strengthening

The National Medal of Technology and Innovation is the highest honor of its kind in the United States.

the nation's technological workforce. The 2012 laureates (awarded in 2014) of the National Medal of Technology and Innovation were:

- Charles W. Bachman (1924–) for fundamental inventions in database management, transaction processing, and software engineering

- Edith M. Flanigen (1929–) for innovations in the fields of silicate chemistry, the chemistry of zeolites, and molecular sieve materials

- Thomas J. Fogarty (1934–) for innovations in minimally invasive medical devices

- Eli Harari (1945–) of SanDisk Corporation for the invention and commercialization of flash storage technology

- Arthur Levinson (1950–) for pioneering contributions to the fields of biotechnology and personalized medicine, leading to the development of novel therapeutics for the treatment of cancer

- Cherry A. Murray (1952–) for contributions to the advancement of devices for telecommunications and leadership in the development of the science, technology, engineering, and mathematics (STEM) workforce in the United States

- Mary Shaw (1943–) for pioneering leadership in the development of innovative curricula in computer science

- Douglas Lowy (1942–) and John Schiller (1953–) for developing the virus-like particles and related technologies that led to the generation of effective vaccines that specifically targeted HPV (human papillomavirus) and related cancers

When was the National Medal of Science established by Congress?

The National Medal of Science was established in 1959 as a Presidential Award for "outstanding contributions to knowledge in the physical, biological, mathematical, or engineering sciences." In 1980, it was expanded to include the social and behavioral sciences. It may be awarded posthumously. The 2012 recipients (presented in 2014) of the National Medal of Science were:

- Bruce Alberts (1938–), Biological Sciences, for innovation in the field of DNA replication; promoting and improving science education and public policy

- Robert Axelrod (1943–), Behavioral and Social Sciences, for interdisciplinary work on the evolution of cooperation, complexity theory, and international security; use of social science models to explain biological phenomena

- May Berenbaum (1953–), Biological Sciences, for studies on chemical coevolution and the genetic basis of insect–plant interactions

- David Blackwell (1919–2010), Mathematics and Computer Sciences, for contributions to probability theory, mathematical statistics, information theory, mathematical logic, and Blackwell games and their impact on drug testing, computer communications, and manufacturing

- Alexandre J. Chorin (1938–), Mathematics and Computer Sciences, for new methods for realistic fluid-flow simulation used to model engines, aircraft wings, and heart valves
- Thomas Kailath (1935–), Engineering, for contributions to the fields of information and system science; translational science
- Judith P. Klinman (1941–), Chemistry, for discoveries of chemical and physical principles underlying enzyme catalysis; leadership among scientists
- Jerrold Meinwald (1927–), Chemistry, for the application of chemical principles to studies of plant and insect defense and communication; establishing chemical ecology as a core discipline for agriculture, forestry, medicine, and environmental science
- Burton Richter (1931–), Physical Sciences, for contributions to the development of electron accelerators; discoveries in elementary particle physics and contributions to energy policy
- Sean C. Solomon (1945–), Physical Sciences, for creative approaches to understanding the internal structure of the Earth, the Moon, and other terrestrial planets; leadership and inspiration of new generations of scientists

What is the Turing Award?

The Turing Award, considered the Nobel Prize in Computing, is awarded annually by the Association for Computing Machinery to an individual who has made a lasting contribution of major technical importance in the computer field. The award, named for British mathematician Alan M. Turing (1912–1954), was first presented in 1966. The Intel Corporation and Google Inc. provided financial support for the $250,000 prize that accompanies the award. In 2014, it was announced that the cash award provided by Google was raised to $1,000,000 effective with the 2014 award presented in June 2015. Recent winners of the Turing Award include:

A statue of mathematician Alan Turing is displayed at the University of Surrey in Guildford, England. Turing was famous for breaking the German's code during World War II with his Enigma Machine, a forerunner of the modern computer.

Year	Award Recipient
2009	Charles P. Thacker (1943–) for the design of the first personal computer—the Alto at Xerox Palo Alto Research Center (PARC)—tablet personal computers, and the Ethernet
2010	Leslie G. Valiant (1949–) for devising the theory of computation
2011	Judea Pearl (1936–) for the development of the calculus for probabilistic reasoning applied to artificial intelligence
2012	Shafi Goldwasser (1958–) and Silvio Micali (1954–) for research on the complexity theory as it is applied to the field of cryptography
2013	Leslie Lamport (1941–) for contributions to the theory and practice of distributed and concurrent systems
2014	Michael Stonebraker (1943–) for fundamental contributions to the concepts and practices underlying modern database systems

COMPUTERS

INTRODUCTION AND
HISTORICAL BACKGROUND

Who invented the computer?

Computers developed from calculating machines. One of the earliest mechanical devices for calculating, still widely used today, is the abacus—a frame carrying parallel rods on which beads or counters are strung. The abacus originated in Egypt in 2000 B.C.E.; it reached the Orient about a thousand years later and arrived in Europe in about the year 300 C.E. In 1617, John Napier (1550–1617) invented "Napier's Bones"—marked pieces of ivory for multiples of numbers. In the middle of the same century, Blaise Pascal (1623–1662) produced a simple mechanism for adding and subtracting. Multiplication by repeated addition was a feature of a stepped drum or wheel machine of 1694 invented by Gottfried Wilhelm Leibniz (1646–1716).

In 1823, English visionary Charles Babbage (1791–1871) persuaded the British government to finance an "analytical engine." This would have been a machine that could undertake any kind of calculation. It would have been driven by steam, but the most important innovation was that the entire program of operations was stored on a punched tape. Babbage's machine was not completed in his lifetime because the technology available to him was not sufficient to support his design. However, in 1991 a team led by Doron Swade (1946–) at London's Science Museum built the "analytical engine" (sometimes called the "difference engine") based on Babbage's work. Measuring 10 feet (3 meters) wide by 6.5 feet (2 meters) tall, it weighed 3 tons (2.7 tonnes) and could calculate equations down to 31 digits. The feat proved that Babbage was way ahead of his time, even though the device was impractical because one had to turn a crank hundreds of times in order to generate a single calculation. Modern computers use electrons, which travel at nearly the speed of light.

Based on the concepts of British mathematician Alan M. Turing (1912–1954), the earliest programmable electronic computer was the 1,500-valve "Colossus," formulated by Max Newman (1897–1984), built by T. H. Flowers (1905–1998), and used by the British government in 1943 to crack the German codes generated by the cipher machine "Enigma."

What was Howard Aiken's contribution to the development of the modern computer?

Howard Aiken (1900–1973) proposed and designed an automatic sequence-controlled calculator in 1936. He collaborated with engineers at the International Business Machine laboratory in Endicott, New

American physicist Howard Aiken developed what became the IBM Mark I computer.

York, to build the first large-scale digital calculator. Officially called the Automatic Sequence Controlled Calculator, it was known as the Harvard Mark I (Aiken was at Harvard at the time) or simply the Mark I. When it was installed at Harvard in 1944, it was 51 feet (15.5 meters) long and 8 feet (2.4 meters) high. It consisted of 72 rotating registers, thousands of relays, and 1,400 rotary switches and other components. There were 500 miles of wire and more than 3,000,000 soldered connections in the Mark I. It did not have a keyboard since data were entered using strips of punch-card paper. It was immediately used by the U.S. Navy to solve wartime calculation problems. It ran almost continuously during 1944 and 1945. Aiken was inducted into the National Inventors Hall of Fame in 2014 for his achievements in ushering in the digital age of computing.

What was MANIAC?

MANIAC (mathematical analyzer, numerator, integrator, and computer) was built at the Los Alamos Scientific Laboratory under the direction of Nicholas C. Metropolis (1915–1999) between 1948 and 1952. It was one of several different copies of the high-speed computer built by John von Neumann (1903–1957) for the Institute for Advanced Studies (IAS). It was constructed primarily for use in the development of atomic energy applications, specifically the hydrogen bomb.

It originated with the work on ENIAC (electronic numerical integrator and computer), the first fully operational, large-scale, electronic, digital computer. ENIAC was built at the Moore School of Electrical Engineering at the University of Pennsylvania between 1943 and 1946. Its builders, John Presper Eckert Jr. (1919–1995) and John William Mauchly (1907–1980), virtually launched the modern era of the computer with ENIAC.

What is an algorithm?

An algorithm is a set of clearly defined rules and instructions for the solution of a problem. It is not necessarily applied only in computers, but it can be a step-by-step procedure for solving any particular kind of problem. A nearly 4,000-year-old Babylonian banking calculation inscribed on a tablet is an algorithm, as is a computer program that consists of step-by-step procedures for solving a problem.

The term is derived from the name of Muhammad ibn Musa al-Khwarizmi (c. 780–c. 850), a Baghdad mathematician who introduced Hindu numerals (including 0) and decimal calculation to the West. When his treatise was translated into Latin in the twelfth century, the art of computation with Arabic (Hindu) numerals became known as algorism.

What is the difference between a "bit" and a "byte"?

A byte, a common unit of computer storage, holds the equivalent of a single character, such as a letter ("A"), a number ("2"), a symbol ("$"), a decimal point, or a space. It is usually equivalent to eight "data bits" and one "parity bit." A bit (a binary digit), the smallest unit of information in a digital computer, is equivalent to a single "0" or "1." The parity bit is used to check for errors in the bits making up the byte. Eight data bits per byte is the most common size used by computer manufacturers.

What units are used to measure computer storage space?

Computer storage space for hard drives and other storage media is calculated in base 2 using binary format with a byte as the basic unit. The common units of computer storage are:

Kilobyte (KB)	1,024 bytes
Megabyte (MB)	1,024 kilobytes or 1,048,576 bytes
Gigabyte (GB)	1,024 megabytes or 1,073,741,824 bytes
Terabyte (TB)	1,024 gigabytes or 1,099,511,627,776 bytes
Petabyte (PB)	1,024 terabytes or 1,125,899,906,800,000 bytes

How have computers made an impact on society?

Since the beginning of the modern computer age less than one hundred years ago, computers are now an integral part of our lives and society. Computers are used in business for accounting and analysis, customer support, ordering and reservation systems, human resources, risk management, quality control, and every other aspect of business operations. Computers are used in educational settings for individualized learning, student research, lesson plan and curriculum planning, grading, and distance learning. Computers are used as a tool for government agencies for compiling statistics, maintaining records, and collecting and analyzing data. Computers are embedded devices for automobiles, aircraft, and other machinery. Computers are used in the military for combat simulation, monitoring, encryption and decryption, intelligence gathering, and logistics. Computers are used in medical settings and hospitals to monitor patients, surgery, and other procedures. We use computers in our personal and professional lives for communication, entertainment, and learning. Our world is a technical world, and computers are a major tool in our daily lives.

HARDWARE AND SOFTWARE

What is the difference between computer hardware and software?

Computers have two major components: hardware and software. Hardware consists of all the physical devices needed to actually build and operate a computer. Examples of computer hardware are the central processing unit (CPU), hard drive, memory, modems, and external, peripheral devices such as the keyboard, monitor, printer, scanner, and other devices that can be physically touched. Software is an integral part of a computer and consists of the various computer programs that allow the user to interact with it and specify the tasks the computer performs. Without software, a computer is merely a collection of circuits and metal in a box, unable to perform even the most basic functions.

What is a silicon chip?

A silicon chip is an almost pure piece of silicon, usually less than one centimeter square and about half a millimeter thick. It contains hundreds of thousands of miniaturized electronic circuit components, mainly transistors, packed and interconnected in layers beneath the surface. These components can perform control, logic, and memory functions. There is a grid of thin, metallic strips on the surface of the chip; these wires are used for electrical connections to other devices. While silicon chips are essential to almost all computer operations, a myriad of other devices depend on them as well, including calculators, microwave ovens, automobiles, and VCRs.

Who invented the silicon chip?

The silicon chip was developed independently by two researchers: Jack Kilby (1923–2005) of Texas Instruments in 1958 and Robert Noyce (1927–1990) of Fairchild Semiconductor in 1959. They received the first Charles Stark Draper Prize in 1989 for the invention and development of the monolithic (i.e., formed from a single crys-

Robert Noyce co-created the microchip in 1959 and went on to found Intel Corporation in 1968.

tal) integrated circuit. Kilby was awarded the National Medal of Technology in 1990 and the Nobel Prize in Physics in 2000 for his discovery of the silicon chip. Noyce received the 1979 National Medal of Science (awarded in 1980). He was also a 1987 laureate of the National Medal of Technology and Innovation. Kilby and Noyce were both inducted into the National Inventors Hall of Fame—Kilby in 1982 and Noyce in 1983. Noyce cofounded Intel with Gordon Moore (1929–) in 1968.

What are the sizes of silicon chips?

Small silicon chips may be no more than 1/16th-inch square by 1/30th-inch thick and hold up to tens of thousands of transistors. Large chips, the size of a postage stamp, can contain hundreds of millions of transistors.

Chip Size	Number of Transistors per Chip
SSI—small-scale integration	Less than 100
MSI—medium-scale integration	Between 100 and 3,000
LSI—large-scale integration	Between 3,000 and 100,000
VLSI—very-large-scale integration	Between 100,000 and 1 million
ULSI—ultra-large-scale integration	More than 1 million

What is Moore's Law?

Gordon Moore (1929–), cofounder of Intel®, a top microchip manufacturer, observed in 1965 that the number of transistors per microchip—and hence a chip's processing power—would double about every year and a half. The press dubbed this Moore's Law. Despite claims that this ever-increasing trend cannot perpetuate, history has shown that microchip advances are, indeed, keeping pace with Moore's prediction. Moore received the National Medal of Technology in 1990 and was inducted into the National Inventors Hall of Fame in 2009.

How does the third generation Intel Core processor differ from the Intel 4004 processor?

When Intel launched the 4004 processor, the first programmable processor, in 1971, it began a new era in electronics. The Intel 4004 held 2,300 transistors, was 10 microns wide, and had a clock speed of 108 KHz. Forty years later, Intel launched the 3rd generation Intel Core processor, holding 1.4 billion transistors on a chip that is 22 nanometers wide. It has a clock speed of 2.9 GHz. These newest chips have also departed from the traditional flat transistors. Instead, the 3D Tri-Gate transistors are three dimensional. There is a pillar or fin on the surface of the chip with a gate on three sides of the fin, significantly increasing the speed and number of transistors.

Are any devices being developed to replace silicon chips?

When transistors were introduced in 1948, they demanded less power than fragile, high-temperature vacuum tubes; allowed electronic equipment to become smaller, faster, and more dependable; and generated less heat. These developments made computers much more economical and accessible; they also made portable radios practical. However, the smaller components were harder to wire together, and hand wiring was both expensive and error-prone.

In the early 1960s, circuits on silicon chips allowed manufacturers to build increased power, speed, and memory storage into smaller packages, which required less electricity to operate and generated even less heat. While through most of the 1970s manufacturers could count on doubling the components on a chip every year without increasing the size of the chip, the size limitations of silicon chips are becoming more restrictive. Though components continue to grow smaller, the same rate of shrinking cannot be maintained.

Researchers are investigating different materials to use in making circuit chips. Gallium arsenide is harder to handle in manufacturing, but it has the potential for greatly increased switching speed. Organic polymers are potentially cheaper to manufacture and could be used for liquid-crystal and other flat-screen displays, which need to have their electronic circuits spread over a wide area. Unfortunately, organic polymers do not allow electricity to pass through as well as the silicons do. Several researchers are working on hybrid chips, which could combine the benefits of organic polymers with those of silicon. Researchers are also in the initial stages of developing integrated optical chips, which would use light rather than electric current. Optical chips would generate little or no heat, would allow faster switching, and would be immune to electrical noise.

What is the central processing unit of a computer?

The central processing unit (CPU) of a computer is where almost all computing takes place in all computers, including mainframes, desktops, laptops, and servers. The CPU of almost every computer is contained on a single chip.

How is the speed of a CPU measured?

Separate from the "real-time clock" that keeps track of the time of day, the CPU clock sets the tempo for the processor and measures the transmission speed of electronic devices. The clock is used to synchronize data pulses between sender and receiver. A one megahertz clock manipulates a set number of bits one million times per second. In general, the higher the clock speed, the quicker data are processed. However, newer versions of software often require quicker computers just to maintain their overall processing speed.

The hertz is named in honor of Heinrich Hertz (1857–1894), who detected electromagnetic waves in 1883. One hertz is equal to the number of electromagnetic waves or cycles in a signal, which is one cycle per second.

What is the difference between RAM and ROM?

Random-access memory (RAM) is where programs and the systems that run the computer are stored until the CPU can access them. RAM may be read and altered by the user. In general, the more RAM, the faster the computer. RAM holds data only when the current is on to the computer. Newer computers have DDR (double data rate) memory chips. Read-only memory (ROM) is memory that can be read but not altered by the user. ROM stores information, such as operating programs, even when the computer is switched off.

What is a hard drive of a computer?

Hard disks, formerly called hard disk drives and more recently just hard drives, were invented in the 1950s. They are storage devices in desktop computers, laptops, servers, and mainframes. Hard disks use a magnetic recording surface to record, access, and erase data in much the same way as magnetic tape records, plays, and erases sound or images. A read/write head, suspended over a spinning disk, is directed by the central processing unit (CPU) to the sector where the requested data are stored, or where the data are to be recorded. A hard disk uses rigid aluminum disks coated with iron oxide to store data. Data are stored in files, which are named collections of bytes. The bytes could be anything from the ASCII codes for the characters of a text file to instructions for a software application to the records of a database to the pixel colors for an image. Current hard drive size ranges from several hundred gigabytes to four terabytes. Developers are working on designs for eight- and even ten-terabyte hard drives.

A hard disk rotates from 5,400 to over 10,000 revolutions per minute (rpm) and is constantly spinning (except in laptops, which conserve battery life by spinning the hard disk only when in use). An ultra-fast hard disk has a separate read/write head over each track on the disk so that no time is lost in positioning the head over the desired track; accessing the desired sector takes only milliseconds, the time it takes for the disk to spin to the sector.

Hard drive performance is measured by data rate and seek time. Data rate is the number of bytes per second that the hard drive can deliver to the CPU. Seek time is the amount of time that elapses from when the CPU requests a file to when the first byte of the file is delivered to the CPU.

Who invented the computer mouse?

A computer "mouse" is a handheld input device that when rolled across a flat surface causes a cursor to move in a corresponding way on a display screen. A prototype mouse was part of an input console demonstrated by Douglas C. Engelbart (1925–2013) in 1968 at the Fall Joint Computer Conference in San Francisco. Popularized in 1984 by the Macintosh from Apple Computer, the mouse was the

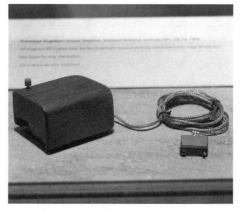

This prototype mouse that was invented by Douglas C. Engelbart in 1968 was made out of wood and had a simple button at the top.

33

result of fifteen years devoted to exploring ways to make communicating with computers simpler and more flexible. The physical appearance of the small box with the dangling, taillike wire suggested the name of "mouse." Engelbart received the Turing Award in 1997 and the National Medal of Technology and Innovation in 2000, and he was inducted into the National Inventors Hall of Fame in 1998 for his invention of the computer mouse.

In recent years, the mouse has evolved into other shapes and forms. One type is the wireless (or "tailless") mouse, which does not have a cord to connect to the computer. Wireless mice use radio signals or infrared to connect to the computer.

What is a USB port?

The Universal Serial Bus (USB) connectors first appeared on computers in the late 1990s. They have become the most widely used interface to attach peripherals, such as mice, printers and scanners, external storage drives, digital cameras, and other devices to a computer. Unlike older serial ports and parallel ports, USB ports are easy to reach and can easily be plugged in—even while the computer is in use.

What were floppy disks?

The first floppy disk drive was invented by Alan Shugart (1930–2006) in 1967 at IBM. Floppy disks, also called diskettes, were made of plastic film covered with a magnetic coating, which were enclosed in a non-removable, plastic protective envelope. Floppy disks varied in storage capacity from one hundred thousand bytes to more than two megabytes. The three common floppy disk (diskette) sizes varied widely in storage capacity.

Envelope size (inches)	Storage capacity
8	100,000–500,000 bytes
5.25	100 kilobytes–1.2 megabytes
3.5	400 kilobytes–more than 2 megabytes

Zip® disks were very similar to floppy disks, but the magnetic coating was of much higher quality. They were able to store up to 750 megabytes of data.

By the mid-1990s, floppy disks and Zip® disks had become obsolete as computer files and memory required larger storage, and computers were no longer being manufactured with floppy disk drives, although they can still be accessed by using an external floppy drive reader with a USB connection. The "save" button in the Microsoft Office toolbar is a picture of a floppy disk in recognition of the role floppy disks played in saving data.

What are some forms of portable storage media?

External hard drives are similar to the internal hard drive of the machine with storage capacity of up to and beyond 2 TB. Often using a USB port to connect, they provide an easy alternative for backup storage.

Compact discs (CDs) and DVDs are optical storage devices. There are three types of CDs and DVDs: read-only, write-once (CD-R and DVD-R), and rewritable (CD-RW and DVD-RW). While read-only CDs and DVDs are wonderful for prepackaged software, they do not permit a user to save their own material. CD-Rs store 700 MB of data and can be written once. DVD-Rs are similar to CD-Rs except they hold 4 to 28 times more data. Single-disc DVD-Rs are available that can store 4.7 GB of data or two hours of video. Read/write CDs use a different chemical compound that allows data to be recorded, erased, and rewritten.

Solid-state storage technology has no moving parts. One example is flash memory sticks. All cells are set to 0 on the memory chip before data are stored in a flash memory stick. When data are entered, electric charges are applied to certain cells. These charges pierce a thin layer of oxide and become trapped. The trapped charges become 1s. The binary code pattern of 0s and 1s is stored into the memory. Flash memory sticks are available with up to one terabyte of storage, providing large amounts of easily transportable storage.

Memory cards are used commonly in digital cameras, music players, smartphones, and tablets. Most computers (laptops and desktops) have ports to read memory cards, and they can be used to transfer files between devices. They are available in different physical sizes, speeds, and capacities. Standard-size memory cards measure $32 \times 24 \times 2.1$ millimeters and weigh about two grams. They are most frequently used in digital cameras. Mini memory cards measure $21.5 \times 20 \times 1.4$ millimeters and weigh about 0.8 grams. Designed for cell phones, they are no longer used frequently. The smallest memory cards, micro cards, measure $15 \times 11 \times 1$ millimeters and weigh 0.25 grams. They are used in cell phones, smartphones, and tablets. Adapters are available, so different sizes can be used in different-sized slots. They are available with capacities of two gigabtyes to two terabtyes. Speed is usually only of concern when saving photos. High-speed memory cards will save photos much more quickly—especially important for professional photographers.

What is the difference between operating system software and application software?

Operating system software tells the computer how to run the hardware resources and software applications. Application software provides the directions and instructions that allow computers to create documents and images, solve calculations, maintain files,

---

How does the amount of storage on a 64 GB flash drive compare to older storage media?

A single 64 GB flash drive can hold the same amount of data as four DVDs, 90 CDs, or 45,000 3.5-inch floppy disks.

and operate games. Examples of application software are word processing, databases, spreadsheets, graphics for desktop publishing, media, and games.

What does DOS stand for?

DOS stands for "disk operating system," a program that controlled the computer's transfer of data to and from a hard or floppy disk. Frequently, it was combined with the main operating system. The operating system was originally developed at Seattle Computer Products as SCP-DOS. When IBM decided to build a personal computer and needed an operating system, it chose the SCP-DOS after reaching an agreement with the Microsoft Corporation to produce the actual operating system. Under Microsoft, SCP-DOS became MS-DOS, which IBM referred to as PC-DOS (personal computer), which everyone eventually simply called DOS.

What are the tasks of an operating system?

An operating system is found in all computers and other electronic devices, such as cell phones. The operating system manages all the hardware and software resources of the computer. Operating systems manage data and devices, such as printers, in the computer. Operating systems have the ability to multitask, allowing the user to keep several different applications open at the same time. Popular operating systems for computers are Windows (Microsoft), OS X (Macintosh), and Linux.

Which computer was the first to have a graphical user interface?

The earliest operating systems were text-based, meaning they only allowed for text input. As computer usage changed, it became clear there was a need to develop a graphical

How did the Linux operating system get its name?

The name Linux is a combination of the first name of its principal programmer, Finland's Linus Torvalds (1970–), and the UNIX operating system. Linux (pronounced with a short "i") is an open-source computer-operating system that is comparable to more powerful, expansive, and usually costly UNIX systems, of which it resembles in form and function. Linux allows users to run an amalgam of reliable and hearty open-source software tools and interfaces, including powerful web utilities such as the popular Apache server, on their home computers. Anyone can download Linux for free or can obtain it on disk for only a marginal fee. Torvalds created the kernel—or heart of the system—"just for fun," and released it freely to the world, where other programmers helped further its development. The world, in turn, has embraced Linux and made Torvalds into a computer folk hero.

user interface (GUI) that would allow graphics, including icons and a drag-and-drop feature of text and icons. The Apple Macintosh personal computer was the first real operating system to have a GUI. Microsoft Windows followed, but instead of the operating systems having a GUI interface, it had a GUI interface on a text-based operating system. The user's mouse movements would be translated into text commands. Windows 95 was the first version to have a GUI.

What is the Android operating system?

The Android operating system, based on Linux, is used mostly for smartphones and tablets. It was developed by a startup company in 2003 with backing from Google. Google bought the company in 2005. Each version of Android is named after a dessert since, according to Google, the devices that use Android "make our lives so sweet." The current version, 5.0, is the Lollipop. Other versions have been named the Kit Kat, Jelly Bean, Ice Cream Sandwich, Honeycomb, Gingerbread, Froyo, Éclair, and Donut.

What does it mean to "boot" a computer?

Booting a computer is starting it, in the sense of turning control over to the operating system. The term comes from "bootstrap," because bootstraps allow an individual to pull on boots without help from anyone else. Some people prefer to think of the process in terms of using bootstraps to lift oneself off the ground, impossible in the physical sense, but a reasonable image for representing the process of searching for the operating system, loading it, and passing control to it. The commands to do this are embedded in a read-only memory (ROM) chip that is automatically executed when a microcomputer is turned on or reset. In mainframe or minicomputers, the process usually involves a great deal of operator input. A cold boot powers on the computer and passes control to the operating system; a warm boot resets the operating system without powering off the computer.

What is an expert system?

An expert system is a type of highly specialized software that analyzes complex problems in a particular field and recommends possible solutions based on information previously programmed into it. The person who develops an expert system first analyzes the behavior of a human expert in a given field, then inputs all the explicit rules resulting from their study into the system. Expert systems are used in equipment repair, insurance planning, training, medical diagnosis, and other areas.

What was the name of the personal computer introduced by Apple in the early 1980s?

Lisa was the name of the personal computer that Apple introduced. The forerunner of the Macintosh personal computer, Lisa had a graphical user interface and a mouse.

37

When did the first portable computer become available?

Perhaps the earliest portable computer was the Osborne 1, which was first available in April 1981. Created by Adam Osborne (1939–2003), the Osborne 1 is better described as a "luggable" computer than a portable computer. It weighed 24 pounds (10.9 kilograms), had two 5.25-inch floppy disk drives, a five-inch screen, 64K RAM, and a processor that ran on 4 Mhz, but it did not have a battery. It included a word processing program and a spreadsheet program. It was approximately the size of a sewing machine and fit under the seat of a commercial airliner. The cost was $1,795.

A 3-D printer replicates a wrench in this photograph. Such printers are getting more sophisticated, building such things as car parts and aircraft components.

What is 3-D printing?

Charles Hull (1939–) invented stereolithography, commonly referred to as 3-D printing, in 1984 as a way to build models in plastic layer by layer. The method uses UV light to cure and bond one polymer layer on top of the next layer. Although 3-D printing techniques were first used in research and development labs, it quickly became a tool to create models and prototypes in industry and manufacturing. It is now used to create automobile and aircraft components, artificial limbs, artwork, musical instruments, and endless other products. It is one of the products that has spurned the maker movement. Hull was awarded U.S Patent #4,575,330 in 1986 for his invention and was inducted into the National Inventors Hall of Fame in 2014.

What is a pixel?

A pixel (from the words pix, for picture, and element) is the smallest element on a video display screen. A screen contains thousands of pixels, each of which can be made up of one or more dots or a cluster of dots. On a simple monochrome screen, a pixel is one dot; the two colors of image and background are created when the pixel is switched either on or off. Some monochrome screen pixels can be energized to create different light intensities to allow a range of shades from light to dark. On color screens, three dot colors are included in each pixel—red, green, and blue. The simplest screens have just one dot of each color, but more elaborate screens have pixels with clusters of each color. These more elaborate displays can show a large number of colors and intensities. On color screens, black is created by leaving all three colors off; white by all three colors on; and a range of grays by equal intensities of all the colors.

The resolution of a computer monitor is expressed as the number of pixels on the horizontal axis and the number of pixels on the vertical axis. For example, a monitor described as 800 × 600 has 800 pixels on the horizontal axis and 600 pixels on the vertical axis. The higher the numbers, e.g., 1600 × 1200, the better the resolution.

What is the technology of a touch screen?

Touch screen technology relies on physical touch of the screen by the user using a finger or stylus for input. Instead of using a computer mouse to activate the cursor, merely touching the screen identifies the location and allows the user to modify the information on the screen. The basic underlying principle is that there is an electrical current running through a sensor on the screen. Touching the screen causes a voltage change, which indicates the location of the physical contact. Specialized hardware converts the voltage changes on the sensors into signals the computer can receive. Finally, software relates to the computer or other device, e.g., smartphone, what is happening on the sensor. The computer or device reacts to the inputted information accordingly.

What are the three types of sensors used in touch screens?

There are three types of sensors for touch screens: 1) resistive screens, 2) capacitive screens, and 3) surface acoustic wave screens. Resistive sensors are the most common in touch screens. The basic design is two thin sheets—a conductive and a resistive metallic layer—separated by a grid of plastic dots or spacers on a normal glass panel. Each sheet conducts electricity. There are four wires (hence the common name, 4-wire resistive screen) that measure the currents on the screen. The sheets contact each other at the spot where the user touches the screen. Other designs are 5-wire screens, which add a sheet. The additional sheet increases the durability of the screen since the user does not touch the screen that carries the current. An 8-wire screen has even greater durability since it has an extra set of wires.

A capacitive touch screen consists of a single, thin sheet on the glass panel. Touching the screen transfers the charge to the user, decreasing the charge on the capacitive layer. The decrease is measured and the computer or device can calculate the exact location of the physical contact. Capacitive sensors have a clearer screen than screens with resistive sensors. A disadvantage of capacitive sensors is they cannot sense physical contact with non-conductive objects, such as gloved fingers.

Touch screens that use wave interruption system sensors have two transducers—one for receiving and one for sending, which detect where the interference occurs. There are no metallic layers, and they offer the clearest resolution for displaying detailed graphics.

Who is considered the inventor of touch screen technology?

The technology for touch screens is traced to the digitizing tablet developed by George Samuel Hurst (1927–2010) in the 1970s. As early as 1971, scientists used a "touch sensor" to record data from graphs by placing the graph on the digitizing tablet and pressing the paper against the tablet with a stylus. Although not a transparent screen, it was the forerunner of the modern touch screen. The first commercial producer of digitizing tablets was Elographics (later known as Elo Touch Solutions). Elographics and Siemens Corporation developed a transparent version of the tablet on curved glass that could fit over a cathode ray tube (CRT) screen.

What devices or applications use touch screens?

Touch screens are commonly used for automated teller machines (ATMs), hand-held devices, including PDAs, tablets, and smartphones, information kiosks, and public computers, such as for ticket sales at airports, theaters, museums, and other public venues. They have been helpful for users with special needs that rely on assistive technology. The iPhone, introduced in 2007, uses only touch screen technology.

PROGRAMMING
AND APPLICATIONS

What is Hopper's rule?

Electricity travels 1 foot (0.3 meter) in a nanosecond (a billionth of a second). This is one of a number of rules compiled for the convenience of computer programmers; it is also considered to be a fundamental limitation on the possible speed of a computer—signals in an electrical circuit cannot move any faster.

What was the first major use for punched cards?

Punched cards were a way of programming, or giving instructions to, a machine. In 1801, Joseph Marie Jacquard (1752–1834) built a device that could do automated pattern weaving. Cards with holes were used to direct threads in the loom, creating predefined patterns in the cloth. The pattern was determined by the arrangement of holes in the cards, with wire hooks passing through the holes to grab and pull through specific threads to be woven into the cloth.

By the 1880s, Herman Hollerith (1860–1929) was using the idea of punched cards to give machines instructions. He built a punched card tabulator that processed the data gathered for the 1890 U.S. Census in six weeks (three times the speed of previous compilations). Metal pins in the ma-

Paper punched cards were an early way to input intructions into computers.

Who was the first programmer?

According to historical accounts, Lord Byron's (1788–1824) daughter, Augusta Ada Byron (1815–1852), the Countess of Lovelace, was the first person to write a computer program for Charles Babbage's "analytical engine." This machine was to work by means of punched cards that could store partial answers that could later be retrieved for additional operations and would then print the results. Her work with Babbage and the essays she wrote about the possibilities of the "engine" established her as a kind of founding parent of the art and science of programming. The programming language called "Ada" was named in her honor by the U.S. Department of Defense. In modern times, Commodore Grace Murray Hopper (1906–1992) of the U.S. Navy is acknowledged as one of the first programmers of the Mark I computer in 1944. Hopper received the National Medal of Technology and Innovation in 1991 in recognition of her accomplishments in developing computer programming languages.

chine's reader passed through holes punched in cards the size of dollar bills, momentarily closing electric circuits. Hollerith was inducted into the National Inventors Hall of Fame in 1990 for his invention of the punched card tabulator.

The resulting pulses advanced counters assigned to details such as income and family size. A sorter could also be programmed to pigeonhole cards according to pattern of holes, an important aid in analyzing census statistics. Later, Hollerith founded Tabulating Machines Co., which in 1924 became IBM. When IBM adopted the 80-column punched card (measuring 7 3/8 inches × 3 1/4 inches [18.7 centimeters × 8.25 centimeters] and 0.007 inches [0.17 millimeters] thick), the de facto industry standard was set, which endured for decades.

What are the generations of programming languages?

Computer scientists define the generations of computer programming languages in levels or steps of evolution.

A first-generation language (1GL) is called a "machine language," the set of instructions that the programmer writes for the processor to perform. It appears in binary form and is written in 0s and 1s.

A second-generation language is termed an "assembly" language because an assembler converts it into machine language for the processor.

A third-generation language is called a "high-level" programming language. Java and C++ are third-generation languages. A compiler then converts it into machine language, typically written like this:

```
if (chLetter ? 'B')
Console.WriteLine ("Usage: one argument");
return 1;//sample code
```

A fourth-generation language resembles plain language. Relational databases use this language. An example is the following:

FIND All Titles FROM books WHERE Title begins with "Handy"

A fifth-generation language uses a graphical design interface that allows a third- or fourth-generation compiler to translate the language. This is similar to HTML text editors because it allows drag and drop of icons and visual display of hierarchies.

Is assembly language the same thing as machine language?

While the two terms are often used interchangeably, assembly language is a more "user friendly" translation of machine language. Machine language is the collection of patterns of bits recognized by a central processing unit (CPU) as instructions. Each particular CPU design has its own machine language. The machine language of the CPU of a microcomputer generally includes about seventy-five instructions; the machine language of the CPU of a large mainframe computer may include hundreds of instructions. Each of these instructions is a pattern of 1s and 0s that tells the CPU to perform a specific operation.

Assembly language is a collection of symbolic, mnemonic names for each instruction in the machine language of its CPU. Like the machine language, the assembly language is tied to a particular CPU design. Programming in assembly language requires intimate familiarity with the CPU's architecture, and assembly language programs are difficult to maintain and require extensive documentation.

The computer language C, first developed in the late 1980s, is a high-level programming language that can be compiled into machine languages for almost all computers, from microcomputers to mainframes, because of its functional structure. It was the first series of programs that allowed a computer to use higher-level language programs and is the most widely used programming language for personal computer software development. C++ was first released in 1985 and is still used today.

Who invented the COBOL computer language?

COBOL (common business oriented language) is a prominent computer language designed specifically for commercial uses, created in 1960 by a team drawn from several computer makers and the Pentagon. The best-known individual associated with COBOL was then-Lieutenant Grace Murray Hopper (1906–1992), who made fundamental contributions to the U.S. Navy standardization of COBOL. COBOL excels at the most common kinds of data processing for business—simple arithmetic operations performed on huge files of data. The language endures because its syntax is very much like English and because a program written in COBOL for one kind of computer can run on many others without alteration.

Who invented the PASCAL computer language?

Niklaus Wirth (1934–), a Swiss computer programmer, created the PASCAL computer language in 1970. It was used in the academic setting as a teaching tool for computer scientists and programmers. Wirth received the Turing Award in 1984 for the development of PASCAL.

Which was the first widely used high-level programming language?

FORTRAN (FORmula TRANslator) was developed by IBM in the late 1950s. John Backus (1924–2007) was the head of the team that developed FORTRAN. Designed for scientific work containing mathematical formulas, FORTRAN allowed programmers to use algebraic expressions rather than cryptic assembly code. The FORTRAN compiler translated the algebraic expressions into machine-level code. By the late 1960s, FORTRAN was available on almost every computer, especially IBM machines, and utilized by many users. Backus received the National Medal of Science in 1975, the Turing Award in 1977, and the National Academy of Engineering Draper Prize in 1993 for his work in developing FORTRAN.

What was the first object-oriented computer language?

The first object-oriented computer language was Smalltalk, developed by Alan Kay (1940–) in the 1970s at Xerox's Palo Alto Research Center. Kay received the Turing Award in 2003 for his contributions to the development of object-oriented programming languages and personal computing. He received the Draper Prize in Engineering in 2004 for the development of the first practical networked personal computers.

Object-oriented languages are not a series of instructions but a collection of objects that contain both data and instructions. The objects are assigned to classes that perform specific tasks. Object-oriented languages include C++ and Java.

When was Java developed?

Java was released by Sun Microsystems in 1995. A team of developers headed by James Gosling (1955–) began working on a refinement of C++ that ultimately led to Java. Unlike other computer languages, which are either compiled or interpreted, Java compiles the source code into a format called bytecode. The bytecode is then executed by an interpreter. Java was adapted to the emerging World Wide Web and formed the basis of the Netscape Internet browser.

James Gosling headed the team that wrote the Java computer language.

What is the idea behind "open-source software"?

Open-source software is computer software where the code (the rules governing its operation) is available for users to modify. This is in contrast to proprietary code, where the software vendor veils the code so users cannot view and, hence, manipulate (or steal) it. The software termed open source is not necessarily free—that is, without charge; authors can charge for its use, usually only nominal fees. According to the Free Software Foundation, "free software" is a matter of liberty, not price. To understand the concept, you should think of "free" as in "free speech," not as in "free food." Free software is a matter of the users' freedom to run, copy, distribute, study, change, and improve the software. Despite this statement, most of it is available without charge.

Open-source software is usually protected under the notion of "copyleft," instead of "copyright," law. Copyleft does not mean releasing material to the public domain, nor does it mean near absolute prohibition from copying, like the federal copyright law. Instead, according to the Free Software Foundation, copyleft is a form of protection guaranteeing that whoever redistributes software, whether modified or not, "must pass along the freedom to further copy and share it." Open source has evolved into a movement of sharing, cooperation, and mutual innovation, ideas that many believe are necessary in today's cutthroat corporatization of software.

How did the term "glitch" originate?

A glitch is a sudden interruption or fracture in a chain of events, such as in commands to a processor. The stability may or may not be salvageable. The word is thought to have evolved from the German *glitschen*, meaning "to slip," or from the Yiddish *glitshen*, meaning "to slide or skid."

One small glitch can lead to a cascade of failure along a network. For instance, in 1997 a small Internet service provider in Virginia unintentionally provided incorrect

Where did the term "bug" originate?

The slang term "bug" is used to describe problems and errors occurring in computer programs. The term may have originated during the early 1940s at Harvard University, when computer pioneer Grace Murray Hopper (1906–1992) discovered that a dead moth had caused the breakdown of a machine on which she was working. When asked what she was doing while removing the corpse with tweezers, she replied, "I'm debugging the machine." The moth's carcass, taped to a page of notes, is preserved with the trouble log notebook at the Virginia Naval Museum.

router (a router is the method by which the network determines the next location for information) information to a backbone operator (a backbone is a major network thoroughfare in which local and regional networks patch into lengthy interconnections). Because many other Internet service providers rely on the backbone providers, the error echoed around the globe, causing temporary network failures.

What is a computer virus and how is it spread?

Taken from the obvious analogy with biological viruses, a computer "virus" is a name for a type of computer program that searches out uninfected computers, "infects" them by causing them to execute the virus, and then attempts to spread to other computers. A virus does two things: execute code on a computer and spread it to other computers.

The executed code can accomplish anything that a regular computer program can do: it can delete files, send emails, install programs, and copy information from one place to another. These actions can happen immediately or after some set delay. It is often not noticed that a virus has infected a computer because it will mimic the actual actions of the infected computer. By the time it is recognized that the computer is infected, much damage may have occurred.

Earliest computer viruses spread via physical media, such as floppy disks. Modern viruses propagate rapidly throughout the Internet. In May 2000, the "ILOVEYOU" virus made international headlines as it spread around the world in a single day, crashing millions of computers and costing approximately $5 billion in economic damages. Since then, criminals have learned to prevent their viruses from crashing computers, making detection of these viruses much more difficult. The highly advanced Conficker worm, first detected in 2008, operates silently, ensuring that the typical computer user won't realize that the virus is present.

When did the first computer virus spread via the Internet?

The first known case of large-scale damage caused by a computer virus spread via the Internet was in 1988. Robert Morris Jr. (1965–), a graduate student at Cornell Univer-

With just two digital paddles and a little white square representing a ball, Pong was a simple, but popular, tennis-style computer game that first appeared on store shelves in 1972.

sity, crafted a "worm" virus that infected thousands of computers, shutting down many of them and causing millions of dollars of damage.

What was the first computer game?

Despite the fact that computers were not invented for playing games, the idea that they could be used for games did not take long to emerge. Alan Turing (1912–1954) proposed a famous game called "Imitation Game" in 1950. In 1952, Rand Air Defense Lab in Santa Monica created the first military simulation games. In 1953, Arthur Samuel (1901–1990) created a checkers program on the new IBM 701. From these beginnings, computer games have become a multi-billion-dollar industry.

What was the first successful video-arcade game?

Pong, a simple electronic version of a tennis game, was the first successful video-arcade game. Although it was first marketed in 1972, Pong was actually invented fourteen years earlier in 1958 by William Higinbotham (1910–1994), who, at the time, headed instrumentation design at Brookhaven National Laboratory. Invented to amuse visitors touring the laboratory, the game was so popular that visitors would stand in line for hours to play it. Higinbotham dismantled the system two years later, and, considering it a trifle, did not patent it. In 1972, Atari released Pong, an arcade version of Higinbotham's game, and Magnavox released Odyssey, a version that could be played on home televisions.

How has computer gaming changed since the 1970s?

The computer game industry has grown to a multi-billion-dollar industry since the 1970s. The introduction of physically interactive computer games, such as Microsoft Xbox and Nintendo Wii, allowed game players to play virtual games, such as tennis, basketball, and baseball. Furthermore, games are no longer available just on consoles but are available on the Internet, where large groups of players can play simultaneously.

What was "The Turk"?

The Turk was the name for a famous chess-playing automaton. An automaton, such as a robot, is a mechanical figure constructed to act as if it moves by its own power. On a dare in 1770, a civil servant in the Vienna

The Turk was a mechanical figure that appeared to be able to play chess, though in reality a hidden person inside the machine moved the pieces.

Imperial Court named Wolfgang von Kempelen (1734–1804) created a chess-playing machine. This mustached, man-sized figure carved from wood wore a turban, trousers, and robe and sat behind a desk. In one hand it held a long Turkish pipe, implying that it had just finished a pre-game smoke, and its innards were filled with gears, pulleys, and cams. The machine seemed a keen chess player and dumbfounded onlookers by defeating all the best human chess players. It was a farce, however: its moves were surreptitiously made by a man hiding inside.

The Turk, so dubbed because of the outfit similar to traditional Turkish garb, is regarded as a forerunner to the industrial revolution because it created a commotion over devices that could complete complex tasks. Historians argue that it inspired people to invent other early devices such as the power loom and the telephone, and it even was a precursor to concepts such as artificial intelligence and computerization. Today, however, computer chess games are so sophisticated that they can defeat even the world's best chess masters. In May 1997, the "Deep Blue" chess computer defeated World Champion Garry Kasparov (1963–). "Deep Blue" was a 32-node IBM RS/6000 SP high-performance computer that used Power Two Super Chip processors (P2SC). Each node had a single microchannel card containing eight dedicated VLSI chess processors for a total of 256 processors working in tandem, allowing Deep Blue to calculate 100 billion to 200 billion chess moves within three minutes.

COMMUNICATIONS

SYMBOLS, WRITING, AND CODES

When was papyrus first used as a writing medium?

Papyrus is a plant (*Cyperus papyrus*) that grows in swamps and standing water. During ancient times it grew in the Nile Valley and Delta and along the Euphrates. It was used as a writing surface during the formative stages of civilization, but the origins of its first use are unknown. An unused roll was found in an Egyptian tomb from the First Dynasty (c. 3100 B.C.E). Papyrus was the main writing medium through the time period of the Roman Empire, but it was replaced by less expensive parchment in the third century C.E.

What changes occurred in the paper manufacturing process that have contributed to the loss of historical writings?

Most commercial cellulose paper manufactured within the last century is acidic. The acid makes paper brittle and eventually causes it to crumble with only minor use. The problem comes from two features of modern paper: the paper manufacturing process results in cellulose fibers that are very short, and acid is introduced (or not removed by purification) during manufacture. Acid in the presence of moisture degrades the fibers, and the acidic hydrolysis reaction repeatedly splits the cellulose chains into smaller fragments. The reaction itself produces acid, accelerating the degradation. Ironically, the older the paper, the longer it lasts. Paper manufactured up until about the mid-nineteenth century was made from cotton and linen. These early papers had very long fibers, the key to their longevity. Today's newsprint paper is the weakest paper; it is unpurified and it has the shortest fibers. Consequently, newspapers generally fade and yellow within a few months.

Acidic paper can be de-acidified. Books, for example, can be dipped in or sprayed with an alkaline solution. However, this process does not reverse brittleness. Once the damage is done to paper fiber, it is irreversible. Paper can crumble in as little as fifty years, which has placed many older manuscripts in danger.

The most durable modern papers are alkaline, where chalk is added to neutralize acid. Books that are marked in the front matter with the symbol for infinity set inside a circle ∞ generally indicate that the paper was specially prepared to meet the requirements of American National Standard Institute/National Information Standards Organization standard Z39.48-1992 (revised 2009) entitled Permanence of Paper for Publication and Documents in Libraries and Archives.

What was a Linotype machine?

The Linotype was the brand name for a printing device invented by Ottmar Mergenthaler (1854–1899) in 1886 and first

The Linotype machine was the method for setting type in newspapers and other printed materials from the late nineteenth century through the 1960s.

used to produce the *New York Tribune*. The text line was cast with molten lead when the operator keyed in and finished a line of text. Linotype typesetters were ubiquitous in the newspaper industry (and were even used on the battlefields of World War I) until about 1960, after which they were replaced with photocomposition. Mergenthaler was inducted into the National Inventors Hall of Fame in 1982 for his invention of the Linotype machine.

What is the standard phonetic alphabet?

Also known as the International Radiotelephone Spelling Alphabet or NATO Phonetic Alphabet, this is a system using the twenty-six letters of the English alphabet and then assigning each letter a word. In this way, people communicating via radio or telephone may spell out certain words or phrases and be assured they are understood regardless of language differences or problems with audio transmissions. The letters are assigned as shown below.

Letter	Phonetic Equivalent
A	Alpha
B	Bravo
C	Charlie
D	Delta
E	Echo

Letter	Phonetic Equivalent
F	Foxtrot
G	Golf
H	Hotel
I	India
J	Juliett
K	Kilo
L	Lima
M	Mike
N	November
O	Oscar
P	Papa
Q	Quebec
R	Romeo
S	Sierra
T	Tango
U	Uniform
V	Victor
W	Whiskey
X	X-ray
Y	Yankee
Z	Zulu

Who invented the Braille alphabet?

The Braille system, used by the blind to read and write, consists of combinations of raised dots that form characters corresponding to the letters of the alphabet, punctuation marks, and common words such as "and" and "the." Louis Braille (1809–1852), blind himself since the age of three, began working on developing a practical alphabet for the blind shortly after he started a school for the blind in Paris. He experimented with a communication method called night writing, which the French army used for nighttime battlefield missives. With the assistance of an army officer, Captain Charles Barbier (1767–1841), Braille pared down the method's twelve-dot configurations to a six-dot one and devised a code of sixty-three characters. The system was not widely accepted for several years; even Braille's own Paris school did not adopt the system until 1854, two years after his death. In 1916, the United States sanctioned Braille's original system of raised dots, and in 1932 a modification called "Standard English Braille, Grade 2" was adopted throughout the English-speaking world. The revised version changed the letter-by-letter codes into common letter combinations, such as "ow," "ing," and "ment," making reading and writing a faster activity. Grade 2 includes the 26 letters of the alphabet, punctuation, and contractions for letter combinations. There are Braille codes for every language and non-verbal languages, such as math and music.

Before Braille's system, one of the few effective alphabets for the blind was devised by another Frenchman, Valentin Haüy (1745–1822), who was the first to emboss paper to help

51

the blind read. Haüy's letters in relief were actually a punched alphabet, and imitators immediately began to copy and improve on his system. Another letter-by-letter system of nine basic characters was devised by Dr. William Moon (1818–1894) in 1847, but it is less versatile in its applications.

What is the Morse Code?

The success of any electrical communication system lies in its coding interpretation, for only series of electric impulses can be transmitted from one end of the system to the other. These impulses must be "translated" from and into words, numbers, etc. This problem plagued early telegraphy until 1835 when American painter-turned-scientist Samuel F. B. Morse (1791–1872), with the help of Alfred Vail (1807–1859), devised a code composed of dots and dashes to represent letters, numbers, and punctuation. Telegraphy uses an electromagnet—a device that becomes magnetic

Samuel Morse (shown here) is remembered for creating the code that bears his name, but people frequently forget he patented the code with co-creator Alfred Vail.

when activated and raps against a metal contact. A series of short, electrical impulses repeatedly can make and break this magnetism, resulting in a tapped-out message.

Having secured a patent on the code in 1837, Morse and Vail established a communications company on May 24, 1844, sending the first long-distance telegraphed message, "What hath God wrought," from Morse in Washington, D.C., to Vail in Baltimore, Maryland. This was the same year that Morse took out a patent on telegraphy. Morse was inducted into the National Inventors Hall of Fame in 1975 for his contributions to telegraphy.

The International Morse Code (shown below) uses sound or a flashing light to send messages. The dot is a very short sound or flash; a dash equals three dots. The pauses between sounds or flashes should equal one dot. An interval of the length of one dash is left between letters; an interval of two dashes is left between words.

A . -	J . - - -	S . . .
B - . . .	K - . -	T -
C - . - .	L . - . .	U . . -
D - . .	M - -	V . . . -
E .	N - .	W . - -
F . . - .	O - - -	X - . . -

G - - .   P . - - .   Y - . - -

H . . . .   Q - - . -   Z - - . .

I . .   R . - .

1 . - - - -   6 - . . . .   Period . - . - . -

2 . . - - -   7 - - . . .   Comma - - . . - -

3 . . . - -   8 - - - . .

4 . . . . -   9 - - - - .

5 . . . .   0 - - - - -

How does ASCII code work?

ASCII (pronounced "ASK-eee") is the abbreviation for American Standard Code for Information Interchange. ASCII is a coding system that uses 128 different combinations of a group of seven bits to form common keyboard characters, including upper- and lowercase A to Z, numbers 0 to 9, and special symbols such as "!" or "@" or "#." ASCII assigns each of these characters a number between 0 and 127. These assignments are graphed on a vertical line (00, 10, 20, through 70) and a horizontal line (00, 01, 02 to 09, 0A, 0B, 0C to 0F). Each character is then arranged with the vertical row position first followed by the horizontal row. For example, the name John would be coded as follows:

J54A

o570

h568

n56E

ASCII was created in 1963. The U.S. government and the American National Standards Institute (ANSI) officially adopted it in 1968. ASCII is unsuitable for most non-English languages and for complex computer applications because it is limited to only 128 different combinations of characters. More robust codes based on 16 and 24 bit, such as Unicode—which supports 65,356 different characters—are gradually becoming integrated into the newer operating systems and software applications.

What were Enigma and Purple in World War II?

Enigma and Purple were the electric rotor cipher machines of the Germans and Japanese, respectively. The Enigma machine, used by the Nazis, was invented in the 1920s and was the best-known cipher machine in history. One of the greatest triumphs in the history of cryptanalysis was the Polish and British solution of the German Enigma ciphers. This played a major role in the Allies' conduct of World War II.

In 1939, the Japanese introduced a new cipher machine adapted from Enigma. Code-named Purple by U.S. cryptanalysts, the new machine used telephone stepping switches instead of rotors. U.S. cryptanalysts were able to solve this new system as well.

Cryptography—the art of sending messages in such a way that the real meaning is hidden from everyone but the sender and the recipient—is done in two ways: code and cipher. A code is like a dictionary in which all the words and phrases are replaced by code words or code numbers. A code book is used to read the code. A cipher works with single letters, rather than complete words or phrases. There are two kinds of ciphers: transposition and substitution. In a transposition cipher, the letters of the ordinary message (or plain text) are jumbled to form the cipher text. In substitution, the plain letters can be replaced by other letters, numbers, or symbols.

Which code was never deciphered during World War II?

The secret code of the Navajo Code Talkers was never deciphered by the enemy during World War II. Twenty-nine Navajo tribe members in the U.S. Marine Corps at the beginning of World War II went to San Diego and developed a code based on their native language. The code consisted of three parts: the Navajo word, its translation, and its military meaning. For instance, the Navajo word "ha-ih-des-ee" translates to "watchful." The military meaning is "alert." The original 29 Navajo grew to more than 400 by the end of the war, and the code book increased from 274 original words to 508 words by war's end.

What are the 10-codes?

Almost as many different codes exist as agencies using codes in radio transmission. The following are officially suggested by the Associated Public Safety Communications Officers (APSCO):

Two members of the Navajo Code Talkers are honored in a Veterans' Day Parade in New York City in 2012.

Ten-1	Cannot understand your message.
Ten-2	Your signal is good.
Ten-3	Stop transmitting.
Ten-4	Message received ("O.K.").
Ten-5	Relay information to _____.
Ten-6	Station is busy.
Ten-7	Out of service.
Ten-8	In service.
Ten-9	Repeat last message.
Ten-10	Negative ("no").
Ten-11	_____ in service.
Ten-12	Stand by.
Ten-13	Report _____ conditions.
Ten-14	Information.
Ten-15	Message delivered.
Ten-16	Reply to message.
Ten-17	En route.
Ten-18	Urgent.
Ten-19	Contact _____.
Ten-20	Unit location.
Ten-21	Call _____ by telephone.
Ten-22	Cancel last message.
Ten-23	Arrived at scene.
Ten-24	Assignment completed.
Ten-25	Meet _____.
Ten-26	Estimated time of arrival is _____.
Ten-27	Request for information on license.
Ten-28	Request vehicle registration information.
Ten-29	Check records.
Ten-30	Use caution.
Ten-31	Pick up.
Ten-32	Units requested.
Ten-33	Emergency! Officer needs help.
Ten-34	Correct time.

How was a palindromic square thought to be used as a secret code?

A palindrome is a sequence of characters (words, letters, or a combination of them) that reads the same from left to right and right to left. Examples include the girl's name Hannah and the year 2002. A "recurrent" palindrome occurs where reading from right to left and left to right creates separate meanings such as the words "trap" and "part." A palindromic square is a complex sequence of letters forming a square that is the same when read from all variations of left to right and top to bottom. Some historians believe that the following intricate square written on a Roman wall in England was a code message left by an early Christian while eluding persecution.

55

SATOR

AREPO

TENET

OPERA

ROTAS

What do the lines in a UPC bar code mean?

The North American Universal Product Code (UPC), or bar code, is a product description code designed to be read by a computerized scanner or cash register. It consists of twelve numbers in groups of "0"s (dark strips) and "1"s (white strips). A bar will be thin if it has only one strip or thicker if there are two or more strips side by side.

The first number describes the type of product. Most products begin with a "0"; exceptions are variable-weight products such as meat and vegetables (2), health-care products (3), bulk-discounted goods (4), and coupons (5). Since it might be misread as a bar, the number 1 is not used.

The next five numbers describe the product's manufacturer. The five numbers after that describe the product itself, including its color, weight, size, and other distinguishing characteristics. The code does not include the price of the item. When the identifying code is read, the information is sent to the store's computer database, which checks it against a price list and returns the price to the cash register.

The last number is a check digit, which tells the scanner if there is an error in the other numbers. The preceding numbers, when added, multiplied, and subtracted in a certain way, will equal this number. If they do not, a mistake exists somewhere.

In 2005 the European Article Number (EAN) code was introduced. This system uses thirteen numbers instead of twelve. These are now the only codes allowed for point-of-sale trade items, such as at the grocery store.

Who was responsible for designing the Universal Product Code (UPC)?

George Laurer (1925–) modified a design originally proposed by Norman Joseph Woodland (1921–2012) and is generally credited with the development of the UPC. Laurer's system was a pattern of stripes that would be readable even if it was poorly printed. The

When was the first retail item sold by scanning a Universal Product Code (UPC)?

The first retail item sold by scanning a UPC was a multipack of chewing gum at Marsh Supermarket in Troy, Ohio, on June 26, 1974.

pattern of stripes was adopted by a consortium of grocery companies in 1973, when cashiers were still punching in all prices by hand. Within a decade, the UPC and optical scanner brought supermarkets into the digital age. Laurer received no royalties for this invention and IBM, where he was employed as an engineer, did not patent it. Woodland received the National Medal of Technology and Innovation in 1992 for his contributions to bar code technology. He was inducted into the National Inventors Hall of Fame in 2011 for inventing the first optically scanned bar code.

An example of a QR code

The QR code, named for "quick response," was developed by Masahiro Hara and a team of researchers at Denso Wave Incorporated in Japan in 1994. It was developed because bar codes were limited to twenty alphanumeric characters and could only be read in one direction. The QR code is a two-dimensional code. It can be read across and up and down, greatly increasing the amount of information the code can contain. QR codes began to be used by manufacturing and industry for tracking and production purposes. The specifications of the code are available to the public, and it has grown to be used in a wide range of applications from airline tickets to business cards to advertising. QR codes can be read using dedicated readers or with smartphones and tablets. Hara and his team of developers received the European Patent Office's Popular Prize in 2014.

When were price look-up codes first used by supermarkets?

The stickers with a four-digit number found on produce are Price Look-Up (PLU) codes. These codes were first used by supermarkets in 1990 to make check-out and inventory control easier, faster, and more accurate. These codes are four digits used to identify bulk produce, including fruits, dried fruit, vegetables, herbs, nuts, and flavorings. There are over thirteen hundred universal PLU codes assigned to produce. The PLU codes for conventionally grown produce fall in the 3000 and 4000 ranges. Different varieties and different sizes of produce have unique codes. For example, a small D'Anjou pear is 4025, and a large D'Anjou pear is 4416. Organically grown produce is identified with a "9" in front of the four-digit code. Genetically engineered produce is identified with an "8" in front of the four-digit code.

How can a grocery store customer become involved in data mining?

Most grocery stores offer a free card that allows instant price reductions, which functions like a plastic, reusable coupon. The only catch is that the customer must sign up for the card and reveal demographic information, such as age and gender. The particulars of each sale are stored in a database each time the customer uses the card, including facts such as day and time of purchase and items purchased. The grocery store then uses "data mining" to better target its sales campaigns and position its in-store displays. Data mining, as its name implies, is a computer and statistical technology where computers sift through extensive data to extract patterns, identify relationships, and allow for predictions. For instance, if patterns show that diapers are purchased before 3 P.M. on weekends by men aged 26 to 35, and most wine is purchased after 3 P.M. by men aged 46 to 55, then the store can move diapers to a prominent, high-traffic location during the day and place items close by that men of the same age group also purchase. Then, after 3 P.M., the store can move the diapers out and the wine into the same prominent location, along with items that men in the 46-to-55 age range also purchase.

What is "phishing"?

"Phishing" is a technique used by cybercriminals to trick users into giving out their personal information. Phishing email or text messages will usually look like they are from legitimate businesses—especially banks or credit card companies or other financial institutions or retail establishments. They often will include threatening statements requiring individuals to update personal information, often by clicking on links, or their accounts will be closed. The links will often lead to sites where the criminals can capture personal information passwords, or other identity information and/or infect computers with viruses. Never respond to emails that request sensitive personal information.

When was the Standard Book Numbering (SBN) system first implemented?

The Standard Book Numbering (SBN) system was first implemented in 1967 in the United Kingdom. It was devised when W. H. Smith, the largest book retailer in Great Britain, moved to a computerized warehouse and wanted a standard numbering system for all the books it carried. In 1968, the International Organization for Standardization (ISO) Technical Committee on Documentation (TC 46) set up a working group to investigate the possibility of expanding the British SBN for international use. The ISO standard 2108, the International Standard Book Number (ISBN), was approved in 1970 and most recently revised in 2005.

What does the code that follows the letters ISBN mean?

The ISBN is an ordering and identifying code for book products. It forms a unique number to identify that particular item. The first number of the series relates to the language the book is published in; for example, the zero is designated for the English language. The second set of numbers identifies the publisher, and the last set of numbers identifies the particular item. The very last number is a "check number," which mathematically makes certain that the previous numbers have been entered correctly. In 2007 ISBNs were expanded to thirteen digits. The thirteen digits are divided into five parts:

1. The prefix 978, which indicates the book publishing industry.
2. A group or country identifier, which identifies a national or geographic group of publishers.
3. A publisher identifier, which identifies the publisher within the group.
4. A title identifier, which identifies a particular title or edition.
5. A check digit, which validates the ISBN.

RADIO AND TELEVISION

Who invented radio?

Guglielmo Marconi (1874–1937), of Bologna, Italy, was the first to prove that radio signals could be sent over long distances. Radio is the radiation and detection of signals propagated through space as electromagnetic waves to convey information. It was first called wireless telegraphy because it duplicated the effect of telegraphy without using wires. On December 12, 1901, Marconi successfully sent Morse code signals from Newfoundland to England. Marconi was awarded the Nobel Prize in physics in 1909 for his contribution to development of wireless telegraphy. In 1975, he was inducted into the National Inventors Hall of Fame for the development of wireless telegraphy and radio.

In 1906, American inventor Lee de Forest (1873–1961) built what he called "the Audion," which became the basis for the radio-amplifying vacuum tube. This device made voice radio practical because it magnified the weak signals without distorting them. The next year, de Forest began regular radio broadcasts from Manhattan, New York. As there were still no home radio receivers, de Forest's only audiences were ship wireless operators in New York City Harbor. De Forest was inducted into the National Inventors Hall of Fame in 1977 for his invention of the Audion.

What was the first radio broadcasting station?

The identity of the "first" broadcasting station is a matter of debate since some pioneer AM broadcast stations developed from experimental operations begun before the institution of formal licensing practices. According to records of the Department of Commerce, which then supervised radio station WBZ in Springfield, Massachusetts, received the first regular broadcasting license on September 15, 1921. However, credit for the first radio broadcasting station has customarily gone to Westinghouse station KDKA in Pittsburgh for its broadcast of the Harding-Cox presidential election returns on November 2, 1920. Unlike most other earlier radio transmissions, KDKA used electron tube technology to generate the transmitted signal and hence to have what could be described as broadcast quality. It was the first corporate-sponsored radio station and the first to have a well-defined commercial purpose—it was not a hobby or a publicity stunt. It was the first broadcast station to be licensed on a frequency outside the amateur bands. Altogether, it was the direct ancestor of modern broadcasting.

How are the call letters beginning with "K" or "W" assigned to radio stations?

These beginning call letters are assigned on a geographical basis. For the majority of radio stations located east of the Mississippi River, their call letters begin with the letter "W"; if the stations are west of the Mississippi, their first call letter is "K." There are exceptions to this rule. Stations founded before this rule went into effect kept their old letters. So, for example,

In 1901 Italian electrical engineer Guglielmo Marconi proved that radio signals could transmit messages over long distances. He won the Nobel Prize in Physics in 1909 for his work.

KDKA in Pittsburgh has retained the first letter "K"; likewise, some western pioneer stations have retained the letter "W." Since many AM licensees also operate FM and TV stations, a common practice is to use the AM call letters followed by "-FM" or "-TV."

Why do FM radio stations have a limited broadcast range?

Usually radio waves higher in frequency than approximately 50 to 60 megahertz are not reflected by the Earth's ionosphere and are lost in space. Television, FM radio, and high-frequency communications systems are therefore limited to approximately line-of-sight ranges. The line-of-sight distance depends on the terrain and antenna height but is usually limited to somewhere between 50 and 100 miles (80 and 161 kilometers). FM (frequency-modulation) radio uses a wider band than AM (amplitude-modulation) radio to give broadcasts high fidelity, especially noticeable in music—crystal clarity to high frequencies and rich resonance to base notes, all with a minimum of static and distortion. Invented by Edwin Howard Armstrong (1890–1954) in 1933, FM receivers became available in 1939. Armstrong was inducted into the National Inventors Hall of Fame in 1980 for his invention of FM radio.

Why do AM stations have a wider broadcast range at night?

The variation of broadcast range is caused by the nature of the ionosphere of the Earth. The ionosphere consists of several different layers of rarefied gases in the upper atmosphere that have become conductive through the bombardment of the atoms of the atmosphere by solar radiation, by electrons and protons emitted by the sun, and by cosmic rays. One of these layers reflects AM radio signals, enabling AM broadcasts to be received by radios that are great distances from the transmitting antenna. With the coming of night, the ionosphere layers partially dissipate and become an excellent reflector of the short waveband AM radio waves. This causes distant AM stations to be heard more clearly at night.

What is HD Radio® Technology?

HD Radio® Technology upgrades traditional analog AM and FM radio broadcasts to digital broadcasts. In 2002, the Federal Communications Commission (FCC) authorized iBiquity Corporation to begin digital broadcasting. Radio stations are able to broadcast in analog and digital simultaneously. The digital format is transmitted as a continuous digital data stream, in which the digital signal is compressed using iBiquity's compression technology. It is broadcast by a transmitter designed specifically for HD Radio broadcasting. Radio listeners with an HD Radio receiver hear high-quality digital broadcasts without static and hiss. In addition, broadcasters can create extra channels, known as HD2, HD3, and HD4, adjacent to the original channel with other listening options such as sports, talk, and other music options. HD Radio® technology can also display text, such as song titles and album covers.

How does satellite digital radio differ from traditional AM and FM radio?

Unlike traditional AM and FM radio stations, satellite radio is broadcast via satellites placed in orbit over the Earth similar to satellite TV. Although the FCC assigned certain frequencies in the 2.3-GHz microwave S-band in 1992, commercial satellite radio service did not begin until 2001 (XM Satellite Radio) and 2002 (Sirius Satellite Radio). XM Radio launched two geostationary Earth orbits in 2001. It launched two newer satellites in 2005 and 2006. The ground station transmits a signal to the two GEO satellites, which bounce the signals back to radio receivers on the ground. Ground transmitters and repeaters provide service where the satellite signal is blocked. There are more than 170 channels of digital radio. Satellite radio displays the song title, artist, and genre of music on the radio.

Sirius Radio also uses a system of satellites, but instead of GEO satellites, it uses a system of three SS/L-1300 satellites that form an inclined elliptical satellite constella-

---

When was the first commercial broadcast with HD Radio® technology?

The first radio station to broadcast with HD Radio® technology was WDMK-FM, KISS 102.7 FM in Detroit on January 7, 2003.

tion. The elliptical orbit path of each satellite positions them over the United States for sixteen hours per day, with at least one satellite over the country at all times. In 2007 Sirius Radio merged with XM Radio to form SiriusXM Radio. Unlike traditional AM and FM radio stations that rely on advertising revenue for their operations and is free to listeners, satellite radio is only available to subscribers for a monthly fee.

How do submerged submarines communicate?

Using frequencies from very high to extremely low, submarines can communicate by radio when submerged if certain conditions are met and depending on whether or not avoiding detection is important. Submarines seldom transmit on long-range, high-radio frequencies if detection is important, as in war. However, Super (SHF), Ultra (UHF), or Very High Frequency (VHF) two-way links with cooperating aircraft, surface ships, via satellite, or with the shore are fairly safe with high data rate, though they all require that the boat show an antenna above the water or send a buoy to the surface.

Which actress received a patent for a system that reduced the risk of detection of radio-controlled torpedoes?

Hedy Lamarr (1914–2000) received U.S. Patent #2,292,387 for frequency shifting or hopping as a way to reduce the risk of detection or jamming of radio-controlled torpedoes in 1942. Since radio signals were broadcast on a specific frequency, they could easily be jammed by enemy transmitters. Lamarr, along with composer George Antheil (1900–1959), devised a system of rapidly switching frequencies at split-second intervals in seemingly random order. When senders and receivers switched ("hopped") frequencies at the same time, the signal was clear. Lamarr and Antheil were inducted into the National Inventors Hall of Fame in 2014 for their secret communication system. Al-

Actress Hedy Lamarr (shown here in 1948) was also an inventor.

though the technology was never used during World War II, it was used during the Cuban Missile Crisis in 1962. Frequency shifting is used in cellular telephones and Bluetooth systems, enabling computers to communicate with peripheral devices.

What technology is being developed that could potentially replace UPC codes and/or magnetic strips to identify and track information?

Radio frequency identification systems (RFID) are gaining popularity to identify and track information. There are two parts to an RFID system: a tag or a card that can store and modify information and a transmitter with an antenna to communicate the information. There are passive and active RFID tags. Passive tags do not have a power supply. The reading signal induces the power to transmit the response. They tend to be small and lightweight but can only be read from distances of a few inches to a few yards. Passive tags are often attached to merchandise to reduce store loss. They are often used in "smart cards" for transit systems. Active RFID tags have their own battery to supply power so they can initiate communication with the reader. The signal is much stronger and may be read over greater distances. Active RFID tags have been used to track cattle and on shipping containers.

Who invented RFID technology?

The earliest development of RFID technology dates back to World War II. Once radar was invented in 1935, German, Japanese, British, and American planes were detectable while they were still miles away. However, it was impossible to determine whether a plane was an enemy aircraft or friendly aircraft. The Germans discovered that if pilots rolled their planes on their approach to the airfield, it would change the radio signal reflected back, thus alerting the radar crew on the ground that they were German planes. This was the first passive RFID system. The British devised the first active RFID system by installing a transmitter on each plane. When the transmitter received signals from radio stations on the ground, friendly aircraft would broadcast a signal back identifying it as friendly aircraft.

Uses of radio frequency communications continued to grow after World War II as a way to identify objects remotely. It developed as a technique to detect shoplifters in retail locations. The first patent (U.S. Patent #3,713,148A, entitled Transponder Apparatus and System) for an active RFID tag with rewritable memory was awarded to Mario W. Cardullo (1935–) on January

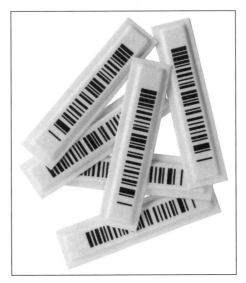

RFID tags like these are attached to items in stores and help track product information and sales.

23, 1973. Charles Walton (1921–2011), often referred to as the father of RFID technology, received U.S. Patent #3,752,960 in August 1973 for a passive transponder or "key card" used to unlock a door without using a traditional key.

What are some uses of RFID tags?

RFID tags are used in a wide variety of applications, including:

- automatic fare payment systems for transit systems
- automatic toll payments for highways and bridges
- student ID cards
- U.S. passports
- tracking cattle
- identification chips placed beneath the skin of pets to help identify lost pets and return them to their owners
- tracking goods from shipment to inventory
- smart cards for locks

Where was the first electronic highway toll collection system implemented?

The earliest experiments with electronic toll collection systems were for buses on the Golden Gate Bridge and with the Port Authority of New York and New Jersey in the 1970s. Transponders were the size of a brick and cost over $1,000, and the antennas to read the transponders were mounted in the pavement. Experimentation continued through the 1970s with designs for a windshield-mounted transponder being developed in Sweden, Italy, the United States, and Canada. TransCore, formerly called Amtech, pioneered the TollTag system in Texas based on RFID technology development in the Pentagon's National Laboratory in Los Alamos, New Mexico. The first electronic toll collection system was implemented on the Dallas North Tollway in 1989. The original transponders were read-only, passive, backscatter 0.92 GHz transponders with no battery. Many are still in use on the Texas toll road.

Who was the founder of television?

The idea of television (or "seeing by electricity," as it was called in 1880) was offered by several people over the years, and several individuals contributed a multitude of partial inventions. For example, in 1897 Ferdinand Braun (1850–1918) constructed the first cathode ray oscilloscope, a fundamental component to all television receivers. In 1907, Boris Rosing (1869–1933) proposed using Braun's tube to receive images, and in the following year, Alan Campbell-Swinton (1863–1930) likewise suggested using the tube, now called the cathode-ray tube, for both transmission and receiving. The man most frequently called the father of television, however, was a Russian-born American named Vladimir K. Zworykin (1889–1982). A former pupil of Rosing's, he produced a practical method of amplifying the electron beam so that the light/dark pattern would

Vladimir Zworykin demonstrates a cathode ray television set to Westinghouse employee Mildred Birt in this 1934 photo.

produce a good image. In 1923, he patented the iconoscope (which would become the television camera), and in 1924 he patented the kinescope (television tube). Both inventions rely on streams of electrons for both scanning and creating the image on a fluorescent screen. By 1938, after adding new and more sensitive photo cells, Zworykin demonstrated his first practical model. Zworykin was inducted into the National Inventors Hall of Fame in 1977 for his invention of the early cathode ray tube.

Another "father" of television is American Philo T. Farnsworth (1906–1971). He was the first person to propose that pictures could be televised electronically. He came up with the basic design for an apparatus in 1922 and discussed his ideas with one of his high school teachers. He documented his ideas one year before Zworykin, and this was critical in settling a patent dispute between Farnsworth and his competitor at the Radio Corporation of America. Farnsworth eventually licensed his television patents to the growing industry and let others refine and develop his basic inventions. Farnsworth was inducted into the National Inventors Hall of Fame in 1984 for his invention and contributions to the development of television.

During the early twentieth century, others worked on different approaches to television. The best known is John Logie Baird (1888–1946), who in 1936 used a mechanized scanning device to transmit the first recognizable picture of a human face. Limitations in his designs made any further improvements in the picture quality impossible.

How does rain affect television reception from a satellite?

The incoming microwave signals are absorbed by rain and moisture, and severe rainstorms can reduce signals by as much as 10 decibels (reduction by a factor of 10). If the installation cannot cope with this level of signal reduction, the picture may be momentarily lost. Even quite moderate rainfall can reduce signals enough to give noisy reception on some receivers. Another problem associated with rain is an increase in noise due to its inherent noise temperature. Any body above the temperature of absolute zero (0°K or –459°F or –273°C) has an inherent noise temperature generated by the release of wave packets from the body's molecular agitation (heat). These wave packets have a wide range of frequencies, some of which will be within the required bandwidth for satellite reception. The warm earth has a high noise temperature, and consequently, rain does as well.

What name is used for a satellite dish that picks up TV broadcasts?

"Earth station" is the term used for the complete satellite receiving or transmitting station. It includes the antenna, the electronics, and all associated equipment necessary to receive or transmit satellite signals. It can range from a simple, inexpensive, receive-only Earth station that can be purchased by the individual consumer to elaborate, two-way communications stations that offer commercial access to the satellite's capacity. Signals are captured and focused by the antenna into a feedhorn and low-noise amplifier. These are relayed by cable to a down converter and then into the satellite receiver/modulator.

Satellite television became widely available in the late 1970s when cable television stations, equipped with satellite dishes, received signals and sent them to their subscribers by coaxial cable.

Who built the first home satellite dish?

Henry Taylor Howard (1932–2002) built the first home satellite dish in 1976. Howard was an electrical engineering professor at Stanford. He built a large, dish-shaped antenna system in his backyard that picked up broadcasts that cable television providers were broadcasting via satellite for distribution to their subscribers. When he sent a check for $100 to Home Box Office (HBO) for its programming, HBO returned his check with a comment that it only dealt with large cable providers and not with individuals. Howard wrote "The Howard Terminal Manual" so other engineers could build similar systems. Shortly afterward, HBO and other cable providers began to scramble their broadcasts.

What is the difference between "analog" and "digital"?

The word "analog" is derived from "analogous," meaning something that corresponds to something else. A picture of a mountain is a representation of the mountain. If the picture is on traditional color film, it is called an analog picture and is characterized by different colors and range of color on the paper. However, if the photographer took the picture with a digital camera, the image is digital: it is stored as a series of numbers in the camera's memory. Its colors are discrete. In communications and computers, a range of continuous variables characterizes analog. Digital, in contrast, is characterized by measurements that happen at discrete intervals. Digital representations are, therefore, more precise. Common examples of analog media include records (music) and VCR tapes (movies). Counterpart examples of digital media include CD-ROM discs and DVD movies.

How does digital television differ from analog television?

Digital television is characterized by a different type of signal, format, and aspect ratio from analog television. Digital signals can transmit more information than analog signals. Standard digital television has 480 horizontal scan lines and 640 vertical scan lines, which produce a better quality of picture and sound than analog television. High-definition television formats are 720p, 1080p, and 1080i. The "i" and "p" stand for "interlaced" and "progressive," respectively. In the interlaced format, half of the scan lines

---

When did television stations in the United States stop analog broadcasting?

Full-power television stations in the United States stopped analog broadcasting, switching to only digital broadcasting, on June 13, 2009. At that time, television viewers with an analog television who did not subscribe to cable or satellite services required a digital converter box in order to continue receiving television broadcasts. A digital converter box converts the digital signals that are broadcast by the television stations to analog signals, which can then be displayed and viewed on an analog television set.

(that is, every other line or thirty lines) are updated every sixtieth of a second. In the progressive format, the full picture updates every sixtieth of a second, providing a sharper image with less flicker. Finally, there is a different aspect ratio in the HD formats of digital television. Analog television has a 4:3 aspect ratio, meaning the screen is four units wide and three units high. The aspect ratio for HD digital television is 16:9, meaning the screen is sixteen units wide and nine units high.

How do flat-panel displays differ from traditional screens?

Flat-panel screens are different because they do not use the cathode-ray tube. Cathode-ray tube monitors—the monitors that until recently were nearly omnipresent around the world—work by bombarding a phosphorescent screen with a ray of electrons. The electrons illuminate "phosphors" on the screen into the reds, greens, and blues that form the picture. In contrast, flat-panel displays use a grid of electrodes, crystals, or vinyl polymers to create the small dots that make up the picture. Flat-panel screens are not a new idea: LCD (short for liquid crystal display) watches and calculators going back decades have relied on crystal-based, flat-panel displays. The newest flat-panel screens found on laptop computers use plasma display panels (PDPs). They can be very wide—over 3 feet (1 meter)—but only a few centimeters thick. A PDP screen is made of a layer of picture elements in the three primary colors. Electrodes from a grid behind it produce a charge that creates ultraviolet rays. These rays illuminate the various picture elements and form the picture.

How does OLED technology differ from LCD technology?

Organic light-emitting diodes (OLED) are thin panels made from organic materials that emit light when electricity is applied through them. The components of an OLED are:

- Substrate to support the OLED
- An anode layer, which removes electrons (adds electron "holes") when a current flows through the device

69

- Organic layers consisting of organic (carbon-based) polymers
- Conducting layer which transports the electron "holes" from the anode
- Emissive layer where the light is made when electrons are transported from the cathode
- Cathode, which injects electrons when a current flows through the device

Unlike LCD displays, OLEDs emit light so they do not require a backlight. This allows them to be very thin. An OLED device is usually just 100 to 500 nanometers thick—about 200 times smaller than a single human hair. A major use of OLED technology is for television.

OLED (organic light-emitting diodes) technology makes the latest advance in televisions—curved screens—possible.

The advantages of OLED versus LCD technology include better image quality, quick response time, and reduced consumption of electricity. The first working OLED displays were developed in 1987 by Ching W. Tang (1947–) and Steven Van Slyke at Eastman Kodak.

What is the Dolby noise reduction system?

The magnetic action of a tape produces a background hiss—a drawback in sound reproduction on tape. A noise reduction system known as Dolby—named after Ray M. Dolby (1933–2013), its American inventor—is widely used to deal with the hiss. In quiet passages, electronic circuits automatically boost the signals before they reach the recording head, drowning out the hiss. On playback, the signals are reduced to their correct levels. The hiss is reduced at the same time, becoming inaudible. Dolby received the National Medal of Technology and Innovation in 1997 and was inducted into the National Inventors Hall of Fame in 2004 for inventing technologies to improve sound quality in recordings.

Who developed the first video tape recorder?

The first video tape recorder (VTR) was developed by Charles Ginsburg (1920–1992). Ginsburg led the research team at Ampex Corporation that developed a machine that consisted of recording heads rotating at high speed that applied high-frequency signals on magnetic tape to record live television images. The Ampex VRX-1000 (later renamed the Mark IV) was introduced in April 1956. Ampex won an Emmy Award for the invention of the VTR in 1957. Ginsburg was inducted into the National Inventors Hall of Fame in 1990 for the development of the VTR, which led to significant technological advances in broadcasting and television program production.

What was the first commercial broadcast replayed using a video tape recording?

The first commercial broadcast using the Ampex VRX-1000 was the CBS delayed broadcast of *Douglas Edwards and the News* on November 30, 1956, in Angeles. Although the VRX-1000 cost $50,000, network television recognized the potential of the technology to revolutionize television broadcasting. It was no longer necessary to rely on live broadcasts. Not only could recorded programs be edited, but with four time zones in the United States, the same show could be aired at different times throughout the country.

When did the first video cassette recorder for home use become available?

The first video recorder for home use was released by Sony in 1964. Ampex and RCA

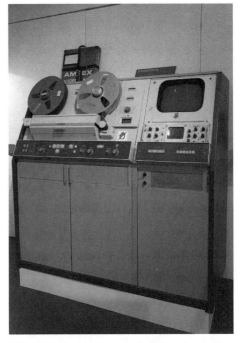

This 1960 video tape recorder (called an Ampex VR-2000) is on exhibit at the National Czech Technical Museum in Prague.

released models in 1965. While these were geared to the home consumer and less expensive than the original Mark IV (VRX-1000), they were still not accessible to the average consumer. Schools, businesses, and wealthy consumers were the primary consumers of these early video recorders. It was not until the 1970s that video cassette recorders began to penetrate the consumer market. Sony introduced Betamax in 1975, and JVC introduced VHS in 1976. Home use VCRs were popular through the 1980s and 1990s. By the early twenty-first century, DVDs became the dominant technology, leading to the demise of magnetic tape.

What is digital video recording (DVR)?

A digital video recorder (DVR) consists of a hard drive connected to the outside world through various jacks (similar to the ones used to hook up other equipment, such as a cable box). The television signal is received by the built-in tuner through either antenna, cable, or satellite. Signals from antenna or cable go into an MPEG-2 encoder, which converts the data from analog to digital. The signal is then sent to the hard drive for storage and then to the MPEG-2 decoder, which converts the signal back from digital to analog and sends it to the television for viewing. Depending on the system, it is possible to record up to three hundred hours of programming. Unlike VCR devices, a DVR is tapeless. It is also possible to begin watching a recording prior to the completion

71

of the recording. Some systems have dual tuners that permit the recording of different programs on different channels at the same time. It is also possible to access a system remotely. TiVo introduced the first DVR in March 1999.

When was the first commercial communications satellite used?

In 1960 *Echo 1,* NASA's first communications satellite, was launched. Two years later, on July 10, 1962, the first commercially funded satellite, *Telstar 1* (developed and funded by American Telephone & Telegraph [AT&T] Bell Laboratories), was launched into low Earth orbit. It was also the first true communications satellite to relay not only data and voice but television as well. The first broadcast, which was relayed from the United States to England, showed an American flag flapping in the breeze. The first commercial satellite was *Early Bird*, which went into regular service on June 10, 1965, with 240 telephone circuits. *Early Bird* was the first satellite launched for Intelsat (International Telecommunications Satellite Organization). Still in existence, the system is owned by member nations—each nation's contribution to the operating funds is based on its share of the system's annual traffic.

What is a geosynchronous orbit?

A satellite in a geosynchronous orbit will appear to be stationary in its location above the Earth. In truth, it is orbiting at the same speed as the Earth's rotation at 22,300 miles (35,800 kilometers) above the Earth, giving the impression that it is in a fixed location. *Early Bird* was the first geosynchronous satellite. Groups of three or more satellites are in geosynchronous orbit to provide global coverage for communications.

When was the first "live via satellite" television broadcast between the United States and Europe?

The first trans-Atlantic live via satellite television broadcast was July 23, 1962. During the historic broadcast, Walter Cronkite (1916–2009) announced, "Good evening, Europe. This is the North American continent live via AT&T *Telstar*, July 23, 1962, 3:00 P.M. Eastern Daylight Time." The split screen had an image of the Statue of Liberty and the Eiffel

Tower. The short program included images of baseball games, Niagara Falls, the World's Fair in Seattle, and a press conference with President John F. Kennedy (1917–1963). It ended abruptly after nearly 20 minutes when the *Telstar* satellite passed out of the range of the AT&T ground station in Andover, Maine. Since *Telstar* was a non-geostationary satellite, the broadcast between the North American continent and Europe was only possible during the short period when *Telstar* was visible to the ground stations on both sides of the Atlantic. Images from Europe to the United States were broadcast during *Telstar*'s next orbit, three hours later, when the ground stations on both sides of the Atlantic were again visible from the satellite.

Alexander Graham Bell patented the telephone in 1876.

Who invented the telephone?

Alexander Graham Bell (1847–1922) was the first person to receive a patent for the telephone. He filed for a patent on February 14, 1876, and received U.S. Patent #174,465 on March 7, 1876. The telephone consisted of a transmitter and receiver—each a thin disk in front of an electromagnet. Bell's telephone was based on the concept that a continuous, varying electric current could transmit and reconvert continuously varying sound waves. While Bell received the first patent, other inventors had also worked on inventing the telephone. Bell was inducted into the National Inventors Hall of Fame in 1974 for his contributions to improve telegraphy and the development of the of telephone. Elisha Gray (1835–1901) filed a caveat (an announcement of an invention) on the same day that Bell filed for his patent. According to the records at the patent office, Bell was the fifth entry and Gray was the thirty-ninth entry on February 14, 1876. Gray was inducted into the National Inventors Hall of Fame in 2007 for his improvements to the technology of telegraphy. Another person to design a talking telegraph was Antonio Meucci (1808– 1889). He filed a caveat for a talking telegraph or telephone on December 28, 1871. He renewed the caveat in 1872 and 1873. The U.S. House of Representatives passed House Resolution 269 on June 11, 2002, honoring Meucci's contributions and work in the development of the telephone.

What was the first telephone transmission?

The first sentence transmitted over the telephone was from Alexander Graham Bell (1847–1922) to his assistant, Thomas Watson (1854–1934). Watson heard Bell's request, "Watson, come here, I want you," while in an adjoining room.

When did caller ID first become available to the general public?

Caller ID first became available to the general public in April 1987 in New Jersey. In the mid-1980s, telephone companies implemented a new system to route calls based on the international standard Signaling System 7 (SS7). Information about the call setup, termination, and other data was handled on a different data circuit. Using the new SS7 system, it was possible to include the calling party's phone number and name to the telephone network's central office. New Jersey Bell was the first telephone company to realize the data could be sold to phone system subscribers. Since the telephones in use at the time did not have the capability to display the caller ID information, a special caller ID unit was necessary to see the calling party's information. The box was attached to the phone and for a monthly service fee to the local telephone company, the customer could see the name and number of the calling party. The first caller ID unit was sold at a Sears store in Jersey City, New Jersey, in April 1987.

How does VoIP differ from traditional telephone calls?

Voice over Internet Protocol (VoIP) is a technology that allows users to make voice calls using a broadband Internet connection. VoIP services convert the human voice into a digital signal, which can travel over the Internet. If calls are made to a traditional telephone, the signal is converted to a regular telephone signal before it reaches the destination. VoIP calls can also be made from one computer to another computer. Depending on the VoIP service provider, calls may be limited to people with the same service or to anyone with a telephone number, including local, long distance, mobile, and international phone numbers. Some service providers work over a computer or a special VoIP phone, while others work using a traditional phone connected to a VoIP adapter. An advantage of VoIP calling is that subscribers may not need to subscribe to and pay for a traditional telephone line. VoIP calls made from a computer with a broadband connection do not incur charges beyond those for the broadband service. A disadvantage of VoIP calling is that calling may not be available during a power outage. Also, most VoIP services do not connect directly to 9-1-1 emergency calling.

How does a fax machine work?

Telefacsimile (also telefax, facsimile, or fax) transmits graphic and textual information from one location to another through telephone lines. A transmitting machine uses either a digital or analog scanner to convert the black-and-white representations

Alexander Bain, inventor of the fax machine

of the image into electrical signals that are transmitted through the telephone lines to a designated receiving machine. The receiving unit converts the transmission back to an image of the original and prints it. In its broadest definition, a facsimile terminal is simply a copier equipped to transmit and receive graphics images.

The fax was invented by Alexander Bain (1811–1877) of Scotland in 1842. His crude device, along with scanning systems invented by Frederick Bakewell (1800–1869) in 1848, evolved into several modern versions. In 1924, faxes were first used to transmit wire photos from Cleveland to New York, a boon to the newspaper industry.

How does a fiber-optic cable work?

A fiber-optic cable is composed of many very thin strands of coated glass or plastic fibers that transmit light through the process of "cladding," in which total internal reflection of light is achieved by using material that has a lower refractive index. Once light enters the fiber, the cladding layer inside it prevents light loss as the beam of light zigzags inside the glass core. Glass fibers can transmit messages or images by directing beams of light inside itself over very short or very long distances up to 13,000 miles (20,917 kilometers) without significant distortion. The pattern of light waves forms a code that carries a message. At the receiving end, the light beams are converted back into an electric current and decoded. Uses include telecommunications; medical fiber-optic viewers, such as endoscopes and fiberscopes, to see internal organs; fiber-optic message devices in aircraft and space vehicles; and fiber-optic connections in automotive lighting systems.

Fiber-optic cables have greater "bandwidth": they can carry much more data than metal cables or wires. Because fiber optics is based on light beams, the transmissions are more impervious to electrical noise and can also be carried greater distances before fading. Fiber-optic cables are much thinner than metal wires; the glass core of a single fiber is less than one-tenth the thickness of a human hair. Fiber-optic cable delivers data in digital code instead of an analog signal, the delivery method of metal cables; computers are structured for digital, so there is a natural symbiosis.

When were the first mobile telephones developed?

Mobile phones were first developed by AT&T in the 1940s. The earliest type of mobile phone was permanently attached to an automobile and was powered by the car's battery. It also had an antenna that had to be mounted outside the vehicle. The second type of mobile phone was a transportable, or bag, cellular phone. It had its own battery pack that allowed owners to detach it from the car and carry it in a pouch. However, most bag phones weighed about 5 pounds (2.25 kilograms) and were not very practical when used this way. The first hand-held portable cellular phone was developed in the early 1970s. It used 14 large-scale integration chips, weighed close to 2 pounds (0.9 kilograms), and talk time was measured in minutes.

When was the first cell phone call made using a portable, hand-held phone?

Martin Cooper (1928–), an employee of Motorola, first demonstrated the use of a portable cell phone on April 3, 1973, when he called Joel S. Engel (1936–) of AT&T Bell Labs. Engel received the National Medal of Technology and Innovation in 1994 for his contributions to the development of cellular phones. Cooper and Engel both received the Draper Prize for Engineering for their contributions developing the world's first cellular telephone systems and networks.

The first cellular network, Advanced Mobile Phone System (AMPS), was tested in July 1978 in the Chicago suburbs. The system consisted of ten cells, each of which was approximately one mile across. Most of the phones in the network were car phones and not portable phones.

Who were the first subscribers to Mobile Telephone Service?

Mobile Telephone Service (MTS) began on June 17, 1946, in St. Louis, Missouri. The first subscribers were Monsanto Chemical Company and a building contractor named Henry Perkinson. Even though all the phones had to operate from cars or trucks and there were only six channels, the service was very popular with waiting lists of individuals wishing to subscribe. MTS was expanded and offered in twenty-five cities within a short time.

When did cell phones begin using digital technology in America?

Digital cell phones (2G technology) were introduced in 1995 and called PCS (personal communications services) phones. They are digital because they convert the speech into a chain of numbers, making the transmission much clearer, and enhances security by making eavesdropping more difficult. It also allows for computer integration. Users can send and receive email and browse the Web. They are cellular because antennas on metal towers—visible on the horizon of nearly every reasonably developed location in America—send and receive all calls within a geographical area. Transmissions are carried over microwave radiation. Wireless companies can use the same frequency over again in different cells (except for cells that are nearby) because each cell is small and therefore requires only a limited range signal. When users move across cells, the signal is "handed off" to the next cell tower.

How does a smartphone differ from a cell phone?

There is no precise definition of a smartphone, but generally, it is a cell phone with built-in applications and Internet access. The earliest smartphones essentially combined features of a cell phone with those of a personal digital assistant (PDA). IBM designed the first smartphone in 1992 called the Simon personal communicator. Today, smartphones are generally equipped with an operating system that allows them to send emails (often syncing with a computer), browse the Web, view spreadsheets and documents, send and receive text messages, operate as an MP3 player, play games, and have a camera for taking digital pictures or videos. More than cell phones, smartphones are comparable to private miniature computers.

What is 4G LTE?

The designation 4G stands for fourth-generation technology for data access over cellular networks. Data includes everything except phone calls and simple text messages (SMS). It includes everything else that is done on smartphones and tablets, such as accessing the Internet and web browsing, checking email, downloading apps, watching videos, and downloading pictures. It is essential for Internet use when a hard-wired or Wi-Fi network is not available. The advantage of 4G over 3G technology is the speed. It is almost as fast to download and upload using 4G technology as on home and office computers using Wi-Fi. Many companies offer a 4G network, but they are not all the same.

LTE stands for Long Term Evolution. It was developed by the 3rd Generation Partnership Project (3GPP), a group focused on telecommunications standard development. Most technical experts consider 4G LTE to be the fastest version of 4G and the one that adheres most closely to the official technical standard.

When was the first text message sent?

Neil Papworth (1969–) sent the first text message to Richard Jarvis at Vodafone in December 1992. Since mobile phones did not have keyboards in 1992, he typed the message on a computer keyboard. The message was "Merry Christmas." According to recent data, 1.91 trillion texts were sent in 2013.

How are compact discs (CDs) made?

The master disc for a CD is an optically flat glass disc coated with a resist. The resist is a chemical that is impervious to an etchant that dissolves glass. The master is placed on a turntable. The digital signal to be recorded is fed to the laser, turning the laser off and on in response to the binary on-off signal. When the laser is on, it burns away a small amount of the resist on the disc. While the disc turns, the recording head moves across the disc, leaving a spiral track of elongated "burns" in the resist surface. After the recording is complete, the glass master is placed in the chemical etchant bath. This developing removes the glass only where the resist is burned away. The spiral track now contains

Which cell phone was the first to have a full keyboard?

The Nokia 9000i Communicator was the first to have a full keyboard. It was first manufactured in 1997.

a series of small pits of varying length and constant depth. To play a recorded CD, a laser beam scans the three miles (five kilometers) of playing track and converts the "pits" and "lands" of the CD into binary codes. A photodiode converts these into a coded string of electrical impulses. In October 1982, the first CDs were marketed; they were invented by the Philips Company (the Netherlands) and Sony (Japan) in 1978.

What is the life span of CD-R and CD-RW discs?

The life span of unrecorded CD-R or CD-RW discs is estimated at five to ten years. CD-RW discs can be rewritten approximately one thousand times, according to industry experts. The life span of recorded CD-Rs and CD-RW discs is influenced by the quality of the manufacturing process, the actual properties of the discs, and how they are handled and stored. Manufacturers have claimed that CD-R discs have a potential life span of fifty to two hundred years, and CD-RW discs have a potential life span of twenty to one hundred years. The actual life spans may be closer to the low end of the span. One of the main problems is that the aluminum substratum on which the data are recorded is vulnerable to oxidation.

How does a DVD differ from a CD?

A DVD is much larger than a CD. A standard DVD can hold about seven times more data than a CD holds. The additional storage capacity provides enough storage for a full-

Wearing a headset that surrounds one's field of vision, this young woman enters a virtual reality world that can seem quite real.

length movie. DVD discs and players first became available in November 1996 in Japan and March 1997 in the United States.

What is virtual reality?

Virtual reality combines state-of-the-art imaging with computer technology to allow users to experience a simulated environment as reality. Several different technologies are integrated into a virtual reality system, including holography, which uses lasers to create three-dimensional images, liquid crystal displays, high-definition television, and multimedia techniques that combine various types of displays in a single computer terminal.

THE INTERNET

What is the Internet?

The Internet is the world's largest computer network. It links computer terminals together via wires or telephone lines in a web of networks and shared software. With the proper equipment, an individual can access vast amounts of information and search databases on various computers connected to the Internet or communicate with someone located anywhere in the world as long as he or she has the proper equipment.

Originally created in the late 1960s by the U.S. Department of Defense Advanced Research Projects Agency (DARPA) to share information with other researchers, the Internet expanded immensely when scientists and academics using the network discovered its great value. Despite its origin, however, the Internet is not owned or funded by the U.S. government or any other organization or institution. A group of volunteers, the Internet Society, addresses such issues as daily operations and technical standards.

Who coined the term "information superhighway"?

A term originally coined by U.S. senator (and future vice president) Al Gore (1948–), the information superhighway was envisioned as a high-speed electronic communications network to enhance education in America in the twenty-first century. Its goal was to help all citizens, regardless of their income level, connecting all users to one another and providing every type of electronic service possible, including shopping, electronic banking, education, medical diagnosis, video conferencing, and game playing. As the World Wide Web grew, it became the "information superhighway."

Who invented the World Wide Web?

Tim Berners-Lee (1955–) is considered the creator of the World Wide Web (WWW). The WWW is a massive collection of interlinked hypertext documents that travel over the Internet and are viewed through a browser. The Internet is a global network of computers developed in the 1960s and 1970s by the U.S. Department of Defense's Advanced

Research Project Agency (hence the term "Arpanet"). The idea of the Internet was to provide redundancy of communications in case of a catastrophic event (like a nuclear blast), which might destroy a single connection or computer but not the entire network. The browser is used to translate the hypertext, usually written in Hypertext Markup Language (HTML), so it is human-readable on a computer screen. Along with Gutenberg's invention of the printing press, the inception of the WWW in 1990 and 1991, when Berners-Lee released the tools and protocols onto the Internet, is considered one of humanity's greatest communications achievements. He received the Draper Prize for Engineering for developing the World Wide Web in 2007.

How does dial-up Internet service differ from broadband Internet service?

Dial-up Internet service was the most commonly used way most individuals accessed the Internet during the 1990s. Dial-up services consisted of a modem that transmitted network information over the telephone lines. Transmission speeds using the V.90 modem were less than 56 kbps bandwidth. In addition to having low transmission speeds, dial-up users could not use the telephone to make or receive calls and the computer to access the Internet simultaneously.

The major difference between dial-up service and broadband service is that broadband allows users to access the Internet at significantly higher speeds. In general, broadband transmission speeds are at least 256 kbps. In addition, broadband service allows transmission of voice and data channels over the same wires or cables, allowing traditional telephone usage simultaneously with Internet access.

What are the different broadband technologies?

Broadband Internet service is available in several different technologies. Not all technologies are available in all geographic areas. Internet access is often offered with other services, such as traditional telephone service and television. Common broadband technologies are Digital Subscriber Line (DSL), cable modem, fiber optic, wireless, and satellite.

DSL uses traditional copper telephone wires to transmit both voice telephone calls and Internet data. Typically, speeds are quicker to download from the Internet to a computer than to upload from a computer to the Internet. It is also known as high-speed Internet access.

Cable modem allows cable companies to provide broadband service using the same coaxial cables that deliver picture and sound to television sets. Cable modem transmission speeds are similar to DSL speeds.

Fiber-optic technology converts electrical signals to light, which is sent through glass fibers. Fiber transmission speeds are much higher than DSL or cable modem speeds. Fiber can also provide Voice over Internet Protocol and video-on-demand services.

Wireless fidelity (Wi-Fi) connects end-user devices to a local Internet service using short-range wireless technology. Fixed wireless technology is a good option in areas that are sparsely populated and do not have other broadband technologies. Wi-Fi is popular in many public settings as a way to access the Internet.

Satellite is generally used in remote areas or sparsely populated areas. Satellite broadband users must have a satellite dish, a satellite Internet modem, and a clear line of sight to the provider's satellite. Transmission speeds may be slower than DSL service and is disrupted in poor weather.

What is Wi-Fi?

Wi-Fi is a wireless local area network. The Wi-Fi Alliance certifies that network devices comply with the appropriate standard, IEEE 802.11. A Wi-Fi hotspot is the

A Wi-Fi router box creates localized access to the Internet and can network computers together within a limited range.

geographic boundary covered by a Wi-Fi access point. One research study reported there were 5.69 million public hotspots in 2015 and projected there would be over 13 million public wifi hotspots in 2020.

What is the client/server principle and how does it apply to the Internet?

The client/server principle refers to the two components of a centralized computer network: client and server machines. Clients request information and servers send them the requested information. For example, when an individual uses his computer to look at a Web page, his computer acts as the client, and the computer hosting the web page is the server. Browsers enable the connection between clients and servers. During the 1990s, Netscape, an outgrowth of the early browser Mosaic, and Internet Explorer were the dominant browsers. Eventually, Microsoft's Internet Explorer became the dominant browser. Other frequently used browsers are Mozilla Firefox, Safari, and Google Chrome.

What is a Uniform Resource Locator, or URL?

A Uniform Resource Locator (URL) can be thought of as the "address" for a given computer on the Internet. A URL consists of two parts: 1) the protocol identifier and 2) the resource name. The protocol identifier indicates which protocol to use, e.g., "http" for Hypertext Transfer Protocol or "ftp" for File Transfer Protocol. Most browsers default to "http" as the protocol identifier, so it is not necessary to include that part of the URL when entering an Internet address in the browser's toolbar. The resource identifier specifies the IP (Internet Protocol) address or domain name where the Web page is located. An IP address is a string of numbers separated by periods. More familiar and easier to remember than the IP address is a domain name. Domain names are comprised of a name with a top-level domain (TLD) suffix. The IP address and domain name both point to the same place; typing "192.0.32.10" in the address bar of a web browser will bring up the same page as entering the more user-friendly "example.com."

What are some common top-level domain (TLD) suffixes?

Common top-level domain suffixes are:

Top-Level Domain (TLD)	Meaning of the Suffix
.com	Commercial organization, business, or company
.edu	Educational institution
.org	Nonprofit organization (sometimes used by other sites)
.gov	U.S. government agency
.mil	U.S. military agency
.net	Network organizations

How does a search engine work?

Internet search engines are akin to computerized card catalogs at libraries. They provide a hyperlinked listing of locations on the World Wide Web according to the requested keyword or pattern of words submitted by the searcher. A search engine uses computer software called "spiders" or "bots" to automatically search out, inventory, and index Web pages. The spiders scan each Web page's content for words and the frequency of the words, then stores that information in a database. When the user submits words or terms, the search engine returns a list of sites from the database and ranks them according to the relevancy of the search terms.

What is a fuzzy search?

Fuzzy search is an inexact search that allows a user to search for data that are similar to but not exactly the same as what he or she specifies. It can produce results when the exact spelling is unknown, or it can help users obtain information that is loosely related to a topic.

What is the most popular search engine?

As of 2014, the most popular search engine in the United States was Google. The following chart shows the market share of the top five search engines, based on desktop searches. It does not include searches using mobile technology.

Search Engine	Average Market Share
Google	67.6 percent
Bing	18.7 percent
Yahoo	10 percent
Ask	2.4 percent
AOL	1.3 percent

When was Google founded?

In 1996 Stanford University graduate students Larry Page (1973–) and Sergey Brin (1973–) began collaborating on a search engine called BackRub. BackRub operated on the Stanford servers for more than a year until it began to take up too much bandwidth. In 1997, Page and Brin decided they needed a new name for the search engine and decided upon Google as a play on the mathematical term "googol." (A googol is the numeral 1 followed by 100 zeros.) Their goal was to organize the seemingly infinite amount of information on the Web. Google first incorporated as a company in September 1998, and the corporation went public in 2004.

How is information sent over the Internet kept secure?

Public-key cryptography is a means for authenticating information sent over the Internet. The system works by encrypting and decrypting information through the use of a combination of "keys." One key is a published "public key"; the second is a "private key," which is kept secret. An algorithm is used to decipher each of the keys. The method is for the sender to encrypt the information using the public key and the recipient to decrypt the information using the secret, private key.

The strength of the system depends on the size of the key: a 128-bit encryption is about $3 \times 1,026$ times stronger than 40-bit encryption. No matter how complex the en-

Larry Page (right) and Sergey Brin, founders of Google, Inc.

cryption, as with any code, keeping the secret aspects secret is the important part to safeguarding the information.

Users can easily tell whether they are on a secure or non-secure Internet website by the prefix of the Web page address. Addresses that begin with "http://" are not secure while those that begin with "https://" are secure.

How much has Internet usage changed in recent years?

According to industry surveys, Internet usage has surged, increasing by over 700 percent between 2000 and 2014, far exceeding earlier estimates that there would be one billion users by 2005. Africa and the Middle East have shown the greatest rate of growth.

World Internet Usage			
	Internet Users as of December 31, 2000	Internet Users as of June 30, 2014	Growth 2000–2014
Africa	4,514,400	297,885,898	6,498.6%
Asia	114,304,000	1,386,188,112	1,112.7%
Europe	105,096,093	582,441,059	454.2%
Middle East	3,284,800	111,809,510	3,303.8%
North America	108,096,800	310,322,357	187.1%
Latin America/Caribbean	18,068,919	320,312,562	1,672.7%
Oceania/Australia	7,620,480	26,789,942	251.6%
WORLD TOTAL	360,985,492	3,035,749,340	741.0%

*Source: www.internetworldstats.com.

What is email?

Electronic mail, known more commonly as email, uses communication facilities to transmit messages. A user can send a message to a single recipient or to many different recipients at one time. Different systems offer different options for sending, receiving, manipulating text, and addressing. For example, a message can be "registered" so that the sender is notified when the recipient looks at the message (though there is no way to tell if the recipient has actually read the message). Email messages may be forwarded to other recipients. Usually messages are stored in a simulated "mailbox" on the network server or host computer; some systems announce incoming mail if the recipient is logged onto the system. An organization (such as a corporation, university, or professional organization) can provide electronic mail facilities; national and international networks can provide them as well.

Who sent the first email?

In the early 1970s, computer engineer Ray Tomlinson (1941–) noticed that people working at the same mainframe computer could leave one another messages. He imagined great utility of this communication system if messages could be sent to different mainframes. So he wrote a software program over the period of about a week that used file-transfer protocols and send-and-receive features. It enabled people to send messages from one mainframe to another over the Arpanet, the network that became the Internet. To make sure the messages went to the right system, he adopted the @ symbol because it was the least ambiguous keyboard symbol and because it was brief.

What is spam?

Spam, also called junk email, is unsolicited email. Spam is an annoyance to the recipient and may contain computer viruses or spyware. It often advertises products that are usually not of interest to the recipient and are oftentimes vulgar in content. Some estimate that as many as one billion spam messages are sent daily. Many email programs have spam filters or blockers to detect spam messages and either delete them or send them to the "junk" mailbox.

What was the first function of a BlackBerry?

The first BlackBerry devices were released in 1999 as wireless mobile email devices. The BlackBerry was able to sync with corporate email systems. Shortly after the first models were released, they added a "Qwerty" keyboard. As early as 2002, they added voice transmission, five years before Apple introduced the first iPhone. They were a product of Research in Motion, a Canadian company founded by two engineering students, Mike Lazaridis (1961–) and Douglas Fregin.

What is a hacker?

A hacker is a skilled computer user. The term originally denoted a skilled programmer, particularly one skilled in machine code and with a good knowledge of the machine and its operating system. The name arose from the fact that a good programmer could always hack an unsatisfactory system around until it worked.

The term later came to denote a user whose main interest is in defeating secure systems. The term has thus acquired a pejorative sense, with the meaning of one who deliberately and sometimes criminally interferes with data available through telephone lines. The activities of such hackers have led to considerable efforts to tighten security of transmitted data. The "hacker ethic" is that information-sharing is the proper way of human dealing, and, indeed, it is the responsibility of hackers to liberally impart their wisdom to the

The BlackBerry was one of the first successful mobile devices for accessing emails.

software world by distributing information. Many hackers are hired by companies to test their Web security.

More recently, with the growth of the maker movement, hackers are synonymous with "makers"—individuals who are building and making physical objects creatively—not just computer programmers.

What is Web 2.0?

Web 2.0 is not a new version of the World Wide Web but rather a collection of new technologies that changes the way users interact with the Web. When Tim Berners-Lee (1955–) created the World Wide Web, it was a repository of information with static content, and users were generally unable to easily change or add to the content they were viewing. Newer technologies allow users to contribute to the Internet with blogs, wikis, and social networking sites.

A further distinction of Web 2.0 is "cloud computing," where data and applications ("apps") are stored on Web servers, rather than on individual computers, allowing users to access their documents, files, and data from any computer with a Web browser. Apps include many products, such as word processing and spreadsheets, that were traditionally found in software packages. For example, Google Drive allows users to access documents and files from any computer. Users are able to collaborate and edit documents in "real" time without having to send documents back and forth through email.

When was the term Web 2.0 first used?

The term Web 2.0 was first used in 2004 during a brainstorming conference between Tim O'Reilly (1954–) of O'Reilly Media and MediaLive International. Rather than believing, as some had, that the Web had "crashed" following the dot.com collapse, it was a turning point for the Web to become a platform for collaborative effort between Internet users, providers, and enterprises.

What are some examples of user-generated content on the World Wide Web?

Blogs (short for web logs) are akin to modern-day diaries (or logs) of thoughts and activities of the author. In the late 1990s, software became available to create blogs using templates that made them accessible to a wide audience as a publishing tool. Blogs may be created by single individuals or by groups of contributors. Blog entries are organized in reverse chronological order with the most recent entries being seen first. Entries may include text, audio, images, video, and links to other sites. Blog authors may invite reader feedback via comments, which allow for dialogue between blog authors and readers.

Wikis, from the Hawaiian word *wikiwiki,* which means "fast," are Web pages that allow users to add and edit material in a collaborative fashion. The first wikis were developed in the mid-1990s by Ward Cunningham (1949–) as a way for users to quickly add content to Web pages. The advantage of this software was that the users did not need to know complicated languages to add material to the Web. One of the best known wikis is Wikipedia, an online, collaborative encyclopedia. Although entries to Wikipedia need to come from published sources and be based on fact, rather than the writer's opinion, there is no overall editorial authority on the site.

Podcasts are broadcast media that may be created by anyone and are available on demand. Unlike traditional broadcast media (radio and television), podcasts are easily created with a microphone, video camera, computer, and connection to the Web. Podcasting does not require sophisticated recording or transmitting equipment. Most podcasts are broadcast on a weekly, biweekly, or monthly schedule. While traditional broadcast media follow a set schedule, podcasts may be downloaded onto a computer or a portable device such as an MP3 player and listened to whenever it is convenient.

When were chat rooms a popular way to communicate?

Chat rooms were a popular way to communicate in "real-time" via an Internet connection in the 1990s. Users were able to send messages to each other directly. Although CompuServe had launched a CB Simulator service in the 1980s and bulletin board server (BBS) systems were available in the late 1980s, America Online (AOL) chat rooms were not available until 1993. In 1996, AOL introduced a monthly flat rate instead of an hourly rate for chat room usage. The number of users increased dramatically, despite the fact that users relied on dial-up service at much slower speeds than today's Internet connections. AOL introduced Instant Messenger (AIM) in 1997. However, with the introduction of Friendster, MySpace, and Facebook, chat room usage declined. AOL discontinued chat rooms in 2010, and Yahoo Messenger ceased to exist in 2012.

Who coined the term "social networking"?

One of the first individuals to use the term "social network" was John Arundel Barnes (1918–2010), a British sociologist/anthropologist. He developed the concept of a social network in a 1954 paper, *Class and Committees in a Norwegian Island Parish*, to describe the social ties or network of relations in a Norwegian fishing village. The first use of "social networking" and "social media" in the electronic age is hard to pinpoint. Tina Sharkey (1964–), formerly of iVillage, believes she began using the term in 1994, while

What was the first social networking website?

SixDegrees is considered to be the first social networking site. It was founded by Andrew Weinreich and launched in 1997. Although it had millions of registered users with profiles, friend lists, and school affiliations, most did not have easy Internet access, and Internet slowness made it difficult to use. The site was sold in 2000 and ceased to exist in 2001. Other sites soon followed, such as Friendster in 2002, MySpace and LinkedIn in 2003, and Facebook in 2004.

Ted Leonsis (1957–), formerly of AOL, believes it was coined sometime around 1997 by AOL executives. It was popularized in the early twenty-first century.

Who founded Facebook?

Facebook was founded by Mark Zuckerberg (1984–) while a student at Harvard as a social networking site for Harvard students. While membership was initially limited to Harvard students, it soon expanded to include all college students, then a version for high school students, and then to anyone with an email address. There are now over one billion Facebook users, making it the most popular social media website. Facebook users are able to stay connected with family, friends, and business associates.

Facebook founder Mark Zuckerberg

Who developed Twitter?

Twitter was developed by Jack Dorsey (1976–) in 2006. He sent his first tweet on March 21, 2006. The tweet was: "just setting up my twttr." There are now 284 million monthly active Twitter users, and 500 million tweets are sent each day. When the US Airways plane crashed in the Hudson River in 2009, it was shared via Twitter (including a photo of the plane) before the traditional news media reported the story.

How many characters are allowed in a tweet?

Tweets are expressions of ideas or thoughts of a moment. Tweets can contain text, photos, and videos. Tweets are made up of 140-character messages. Users can both follow or read Twitter accounts and send messages.

WEIGHTS, MEASURES, TIME, TOOLS, AND WEAPONS

WEIGHTS, MEASURES, AND MEASUREMENTS

How much does the biblical shekel weigh in modern units?

A shekel is equal to 0.497 ounces (14.1 grams). Below are listed some ancient measurements with some modern equivalents given.

Biblical
Volume
omer = 4.188 quarts (modern) or 0.45 peck (modern) or 3.964 liters (modern)
9.4 omers = 1 bath
10 omers = 1 ephah

Weight
shekel = 0.497 ounces (modern) or 14.1 grams (modern)

Length
cubit = 21.8 inches (modern)

Egyptian
Weight
60 grams = 1 shekel
60 shekels = 1 great mina
60 great minas = 1 talent

Greek
Length
cubit = 18.3 inches (modern)

stadion = 607.2 or 622 feet (modern)

Weight

    obol or obolos = 715.38 milligrams (modern) or 0.04 ounces (modern)

    drachma = 4.2923 grams (modern) or 6 obols

    mina = 0.9463 pounds (modern) or 96 drachmas

    talent = 60 mina

Roman

Length

    cubit = 17.5 inches (modern)

    stadium = 202 yards (modern) or 415.5 cubits

Weight

    denarius = 0.17 ounces (modern)

Volume

    amphora = 6.84 gallons (modern)

What is the SI system of measurement?

French scientists as far back as the seventeenth and eighteenth centuries questioned the hodgepodge of the many illogical and imprecise standards used for measurement, so they began a crusade to make a comprehensive, logical, precise, and universal measurement system called Système Internationale d'Unités, or SI for short. It uses the metric system as its base. Since all the units are in multiples of 10, calculations are simplified. Today, all countries except the United States, Myanmar (formerly Burma), and Liberia use this system. However, some elements within American society do use SI—scientists, exporting/importing industries, and federal agencies.

The SI or metric system has seven fundamental standards: the meter (for length), the kilogram (for mass), the second (for time), the ampere (for electric current), the kelvin (for temperature), the candela (for luminous intensity), and the mole (for amount of substance). In addition, two supplementary units, the radian (plane angle) and steradian (solid angle), and a large number of derived units compose the current system, which is still evolving. Some derived units, which use special names, are the hertz, newton, pascal, joule, watt, coulomb, volt, farad, ohm, siemens, weber, tesla, henry, lumen, lux, becquerel, gray, and sievert. Its unit of volume or capacity is the cubic decimeter, but many still use "liter" in its place. Very large or very small dimensions are expressed through a series of prefixes, which increase or decrease in multiples of ten. For example, a decimeter is 1/10 of a meter, a centimeter is 1/100 of a meter, and a millimeter is 1/1000 of a meter. A dekameter is 10 meters, a hectometer is 100 meters, and a kilometer is 1,000 meters. The use of these prefixes enables the system to express these units in an orderly way and avoid inventing new names and new relationships.

How was the length of a meter originally determined?

It was originally intended that the meter should represent one ten-millionth of the distance along the meridian running from the North Pole to the equator through Dunkirk, France, and Barcelona, Spain. French scientists determined this distance, working nearly six years to complete the task in November 1798. They decided to use a platinum-iridium bar as the physical replica of the meter. Although the surveyors made an error of about two miles, the error was not discovered until much later. Rather than change the length of the meter to conform to the actual distance, scientists in 1889 chose the platinum-iridium bar as the international prototype. It was used until 1960. Numerous copies of it are in other parts of the world, including the U.S. National Bureau of Standards.

How is the length of a meter presently determined?

The meter is equal to 39.37 inches. It is presently defined as the distance traveled by light in a vacuum during 1/299,792,458 of a second. From 1960 to 1983, the length of a meter had been defined as 1,650,763.73 times the wavelength of the orange light emitted when a gas consisting of the pure krypton isotope of mass number 86 is excited in an electrical discharge.

How did the yard as a unit of measurement originate?

In early times the natural way to measure lengths was to use various portions of the body (the foot, the thumb, the forearm, etc.). According to tradition, a yard measured on King Henry I (1068–1135) of England became the standard yard still used today. It was the distance from his nose to the middle fingertip of his extended arm.

Other measures were derived from physical activity, such as a pace, a league (distance that equaled an hour's walking), an acre (amount plowed in a day), a furlong (length of a plowed ditch), etc., but obviously, these units were unreliable. The ell, based on the distance between the elbow and index fingertip, was used to measure out cloth. It ranged from 0.513 to 2.322 meters, depending on the locality where it was used and even on the type of goods measured.

Below are listed some linear measurements that evolved from this old reckoning into U.S. customary measures:

The length of the foot was officially set as the distance from King Henry I's nose to the tip of his extended arm and hand.

U.S. customary linear measures
   1 hand = 4 inches
   1 foot = 12 inches

1 yard = 3 feet
1 rod (pole or perch) = 16.5 feet
1 fathom = 6 feet
1 furlong = 220 yards or 660 feet or 40 rods
1 (statute) mile = 1,760 yards or 5,280 feet or 8 furlongs
1 league = 5,280 yards or 15,840 feet or 3 miles
1 international nautical mile = 6,076.1 feet

Conversion to metric
1 inch = 2.54 centimeters
1 foot = 0.304 meters
1 yard = 0.9144 meters
1 fathom = 1.83 meters
1 rod = 5.029 meters
1 furlong = 201.168 meters
1 league = 4.828 kilometers
1 mile = 1.609 kilometers
1 international nautical mile = 1.852 kilometers

Why is a nautical mile different from a statute mile?

Queen Elizabeth I (1533–1603) established the statute mile as 5,280 feet (1,609 meters). This measure, based on walking distance, originated with the Romans, who designated 1,000 paces as a land mile.

The nautical mile is not based on human locomotion but on the circumference of the Earth. There was wide disagreement on the precise measurement, but by 1954 the United States adopted the international nautical mile of 1,852 meters (6,076 feet). It is the length on the Earth's surface of one minute of arc.

1 nautical mile (int.) = 1.1508 statute miles
1 statute mile = 0.868976 nautical miles

Why is there concern about the international standard measure of a kilogram?

The kilogram is the only base unit of measurement that is still defined by a physical object. The international prototype of the kilogram is a cylinder-shaped metal alloy of 90 percent platinum and 10 percent iridium. It measures 1.5 inches (39 millimeters) in height and diameter. The cylinder was formed in London in 1889 and sent to France to be stored in a vault in the International Bureau of Weights and Measures in Sèvres, near Paris. The General Conference on Weights and Measurements declared the prototype to be the international standard for a kilogram, and it continues to be so. When it was taken out of the vault for the third time in 1989 (the other two times were in 1889 and 1946) and compared to existing copies, it was found to be 50 micrograms (billionths of a kilogram), equal to the mass of a smallish grain of sand, less than the copies. It is un-known why there is a difference in the weight of the prototype and the copies. Although negligible for consumer transactions of weight, as the international standard, it is significant for scientists. Scientists are working to redefine the kilogram in terms of a fundamental constant, such as Planck's constant, instead of a physical object.

How are U.S. customary measures converted to metric measures and vice versa?

Listed below is the conversion process for common units of measure:

To convert from	To	Multiply by
acres	meters, square	4,046.856
centimeters	inches	0.394
centimeters	feet	0.0328
centimeters, cubic	inches, cubic	0.06
centimeters, square	inches, square	0.155
feet	meters	0.305
feet, square	meters, square	0.093
gallons, U.S.	liters	3.785
grams	ounces (avoirdupois)	0.035
hectares	kilometers, square	0.01
hectares	miles, square	0.004
inches	centimeters	2.54
inches	millimeters	25.4
inches, cubic	centimeters, cubic	16.387
inches, cubic	liters	0.016387
inches, cubic	meters, cubic	0.0000164
inches, square	centimeters, square	6.4516
inches, square	meters, square	0.0006452
kilograms	ounces, troy	32.15075
kilograms	pounds (avoirdupois)	2.205
kilograms	tons, metric	0.001
kilometers	feet	3,280.8
kilometers	miles	0.621
kilometers, square	hectares	100
knots	miles per hour	1.151
liters	fluid ounces	33.815
liters	gallons	0.264
liters	pints	2.113
liters	quarts	1.057
meters	feet	3.281
meters	yards	1.094
meters, cubic	yards, cubic	1.308
meters, cubic	feet, cubic	35.315
meters, square	feet, square	10.764
meters, square	yards, square	1.196
miles, nautical	kilometers	1.852
miles, square	hectares	258.999
miles, square	kilometers, square	2.59
miles (statute)	meters	1,609.344

Which countries of the world have not formally begun converting to the metric system?

The United States, Myanmar (formerly Burma), and Liberia are the only countries that have not formally converted to the metric system. As early as 1790, Thomas Jefferson (1743–1826), then secretary of state, proposed the adoption of the metric system. It was not implemented because Great Britain, America's major trading source, had not yet begun to use the system.

To convert from	To	Multiply by
miles (statute)	kilometers	1.609344
ounces (avoirdupois)	grams	28.35
ounces (avoirdupois)	kilograms	0.0283495
ounces, fluid	liters	0.03
pints, liquid	liters	0.473
pounds (avoirdupois)	grams	453.592
pounds (avoirdupois)	kilograms	0.454
quarts	liters	0.946
tons, (short/U.S.)	tonnes	0.907
ton, long	tonnes	1.016
tonnes	tons (short/U.S.)	1.102
tonnes	tons, long	0.984
yards	meters	0.914
yards, square	meters, square	0.836
yards, cubic	meters, cubic	0.765

What are the equivalents for dry and liquid measures?

U.S. customary dry measures

       1 pint = 33.6 cubic inches
       1 quart = 2 pints or 67.2006 cubic inches
       1 peck = 8 quarts or 16 pints or 537.605 cubic inches
       1 bushel = 4 pecks or 2,150.42 cubic inches or 32 quarts
       1 barrel = 105 quarts or 7,056 cubic inches
       1 pint, dry = 0.551 liters
       1 quart, dry = 1.101 liters
       1 bushel = 35.239 liters
       1 tablespoon = 4 fluid drams or 0.5 fluid ounces
       1 cup = 0.5 pints or 8 fluid ounces
       1 gill = 4 fluid ounces
       4 gills = 1 pint or 28.875 cubic inches
       1 pint = 2 cups or 16 fluid ounces

2 pints = 1 quart or 57.75 cubic ounces

1 quart = 2 pints or 4 cups or 32 fluid ounces

4 quarts = 1 gallon or 231 cubic inches or 8 pints or 32 gills or 0.833 British quarts

1 gallon = 16 cups or 231 cubic inches or 128 fluid ounces

1 bushel = 8 gallons or 32 quarts

Conversion to metric

1 fluid ounce = 29.57 milliliters or 0.029 liters

1 gill = 0.118 liters

1 cup = 0.236 liters

1 pint = 0.473 liters

1 U.S. quart = 0.833 British quarts or 0.946 liters

1 U.S. gallon = 0.833 British gallons or 3.785 liters

What is the difference between a short ton, a long ton, and a metric ton?

A short ton or U.S. ton or net ton (sometimes just "ton") equals 2,000 pounds; a long ton or an avoirdupois ton is 2,240 pounds; and a metric ton or tonne is 2,204.62 pounds. Other weights are compared below:

U.S. customary measure

1 ounce = 16 drams or 437.5 grains

1 pound = 16 ounces or 7,000 grains or 256 drams

1 (short) hundredweight = 100 pounds

1 long hundredweight = 112 pounds

1 (short) ton = 20 hundredweights or 2,000 pounds

1 long ton = 20 long hundredweights or 2,240 pounds

Conversion to metric

1 grain = 65 milligrams

1 dram = 1.77 grams

1 ounce = 28.3 grams

1 pound = 453.5 grams

1 metric ton (tonne) = 2,204.6 pounds

What are the U.S. and metric units of measurement for area?

U.S. customary area measures

1 square foot = 144 square inches

1 square yard = 9 square feet or 1,296 square miles

1 square rod or pole or perch = 30.25 square yards or 272.5 square feet

1 rood = 40 square rods

1 acre = 160 square rods or 4,840 square yards or 43,560 square feet

1 section = 1 mile square or 640 acres

1 township = 6 miles square or 36 square miles or 36 sections
1 square mile = 640 acres or 4 roods or 1 section

International area measures

1 square millimeter = 1,000,000 square microns
1 square centimeter = 100 square millimeters
1 square meter = 10,000 square centimeters
1 are = 100 square meters
1 hectare = 100 acres or 10,000 square meters
1 square kilometer = 100 hectares or 1,000,000 square meters

How much does water weigh?

U.S. customary measures

1 gallon = 4 quarts
1 gallon = 231 cubic inches
1 gallon = 8.34 pounds
1 gallon = 0.134 cubic feet
1 cubic foot = 7.48 gallons
1 cubic inch = .0360 pounds
12 cubic inches = .433 pounds

British measures

1 liter = 1 kilogram
1 cubic meter = 1 tonne (metric ton)
1 imperial gallon = 10.022 pounds
1 imperial gallon of salt water = 10.3 pounds

How are avoirdupois measurements converted to troy measurements, and how do the measures differ?

Troy weight is a system of mass units used primarily to measure gold and silver. A troy ounce is 480 grains or 31.1 grams. Avoirdupois weight is a system of units that is used to measure mass, except for precious metals, precious stones, and drugs. It is based on the pound, which is approximately 454 grams. In both systems, the weight of a grain is the same—65 milligrams. The two systems do not contain the same weights for other units, however, even though they use the same name for the unit.

Troy

1 grain = 65 milligrams
1 ounce = 480 grains = 31.1 grams
1 pound = 12 ounces = 5,760 grains = 373 grams

Avoirdupois

1 grain = 65 milligrams

1 ounce = 437.5 grains = 28.3 grams
1 pound = 16 ounces = 7,000 grains = 454 grams

To convert from	To	Multiply by
pounds avdp	ounces troy	14.583
pounds avdp	pounds troy	1.215
pounds troy	ounces avdp	1.097
pounds troy	pounds avdp	0.069
ounces avdp	ounces troy	0.911
ounces troy	ounces avdp	1.097

What are some instruments used for measuring?

Different tools or devices are used for measuring the quantitative values of items in the physical world. Measurement involves comparing the measured quantity with a known standard unit.

Instrument or Device	What It Measures
Anemometer	Wind speed
Barometer	Air pressure
Hygrometer	Specific gravity
Protractor	Angles
Ruler	Length
Scales	Weight
Seismometer	Movements of the Earth
Speedometer	Vehicle speed
Thermometer	Temperature
Voltmeter	Electricity

What units of measurement are named after individuals?

Many units of measurement are named after the individual who discovered or defined the concept.

Which is heavier—a pound of gold or a pound of feathers?

A pound of feathers is heavier than a pound of gold because gold is measured in troy pounds, while feathers are measured in avoirdupois pounds. Troy pounds have 12 ounces; avoirdupois pounds have 16 ounces. A troy pound contains 372 grams in the metric system; an avoirdupois pound contains 454 grams. Each troy ounce is heavier than an avoirdupois ounce.

Unit of Measurement	What It Measures	Named After
Ampere	Electrical current	André-Marie Ampère (1775–1836)
Beaufort scale	Wind speed	Sir Francis Beaufort (1774–1857)
Celsius	Temperature	Anders Celsius (1701–1744)
Curie	Radiation	Marie Curie (1867–1934)
Fahrenheit	Temperature	Gabriel Fahrenheit (1686–1736)
Joule	Energy	James Prescott Joule (1818–1889)
Kelvin	Temperature	William Thomson, Lord Kelvin (1824–1907)
Mercalli scale	Earthquake intensity (describes the severity of an earthquake in terms of its effect on humans and structures)	Giuseppe Mercalli (1850–1914)
Mohs' scale	Hardness (minerals)	Friedrich Mohs (1773–1839)
Ohm/Mho	Electrical resistance	Georg Simon Ohm (1789–1854)
Richter scale	Earthquake magnitude (a mathematical scale to compare the size of earthquakes)	Charles Richter (1900–1985)
Volt	Electromagnetic force	Alessandro Volta (1745–1827)
Watt	Power	James Watt (1736–1819)

How is the distance to the horizon measured?

Distance to the horizon depends on the height of the observer's eyes. To determine that, take the distance (in feet) from sea level to eye level and multiply by three, then divide by two and take the square root of the answer. The result is the number of miles to the horizon. For example, if eye level is at a height of six feet above sea level, the horizon is almost three miles away. If eye level were exactly at sea level, there would be no distance seen at all; the horizon would be directly in front of the viewer.

What is a benchmark?

A benchmark is a permanent, recognizable point that lies at a known elevation. It may be an existing object, such as the top of a fire hydrant, or it may be a brass plate placed on top of a concrete post. Surveyors and engineers use benchmarks to find the elevation of objects by reading, through a level telescope, or from the distance a point lies above some already established benchmark.

What is a theodolite?

This optical surveying instrument used to measure angles and directions is mounted on an adjustable tripod and has a spirit level to show when it is horizontal. Similar to the more commonly used transit, the theodolite gives more precise readings; angles can be read to fractions of a degree. It is comprised of a telescope that sights the main target, a horizontal plate to provide readings around the horizon, and a vertical plate and scale for vertical readings. The surveyor uses the geometry of triangles to calculate the distance from the angles measured by the theodolite. Such triangulation is used in road and tunnel building and other civil engineering work. One of the earliest forms of this surveying instrument was described by Englishman Leonard Digges (d. 1571?) in his 1571 work, *Geometrical Treatise Named Pantometria*.

A theodolite is an instrument commonly used by surveyors.

TIME

How is time measured?

Time is a measurement used to determine the duration of an event or when the event occurred. The passage of time can be measured in three ways. Rotational time is based on the unit of the mean solar day (the average length of time it takes the Earth to complete one rotation on its axis). Dynamic time, the second way to measure time, uses the motion of the moon and planets to determine time and avoids the problem of the Earth's varying rotation. The first dynamic time scale was Ephemeris Time, proposed in 1896 and modified in 1960.

Atomic time is a third way to measure time. This method, using an atomic clock, is based on the extremely regular oscillations that occur within atoms. In 1967, the atomic second (the length of time in which 9,192,631,770 vibrations are emitted by a hot cesium-133 atom) was adopted as the basic unit of time. Atomic clocks are now used as international time standards.

Time has also been measured in other, less scientific terms. Listed below are some other "timely" expressions.

- *Twilight*—The first soft glow of sunlight; the sun is still below the horizon; also the last glow of sunlight.
- *Midnight*—12 A.M.; the point of time when one day becomes the next day and night becomes morning.
- *Daybreak*—The first appearance of the sun.

101

- *Dawn*—A gradual increase of sunlight.
- *Noon*—12 P.M.; the point of time when morning becomes afternoon.
- *Dusk*—The gradual dimming of sunlight.
- *Sunset*—The last diffused glow of sunlight; the sun is below the horizon.
- *Evening*—A term with wide meaning; evening is generally the period between sunset and bedtime.
- *Night*—The period of darkness lasting from sunset to midnight.

What is the basis for modern timekeeping?

Mankind has typically associated the durations of the year, month, week, and day on the earth and moon cycles. The modern clock, however, is based on the number 60. The Sumerians around 3000 B.C.E. employed a base ten counting system and also a base 60 counting system. The timekeeping system inherited this pattern with 60 seconds per minute and 60 minutes per hour. Ten and sixty fit together to form the notion of time: 10 hours is 600 minutes; 10 minutes is 600 seconds; one minute is 60 seconds. One second is not related to any of these factors; instead, scientists based the duration of one second on cesium-133, an isotope of the metal cesium. Officially, one second is the amount of time it takes for a cesium-133 atom to vibrate 9,192,631,770 times.

What are some units of time?

Units of Time

Unit of time	Definition
Millennium or chiliad	1,000 years
Century	100 years
Decade	10 years
Solar year	The amount of time it takes for the Earth to make one complete revolution around the sun.
	365.24219 solar days or 365 days, 5 hours, 48 minutes, 45.51 seconds
Solar day	The amount of time it takes for a place on Earth directly facing the sun to make one complete revolution and return to the same position.
	Approximately 23 hours, 56 minutes
Year	365 1/4 days, 52 weeks, or 12 months
Month	The time elapsed between one full moon and the next full moon.
	4 weeks or anywhere from 28 to 31 days, depending on the month
Fortnight	2 weeks
Week	7 days
Day	The time elapsed between sunrise and the next sunrise (or sunset to the next sunset)
	24 hours

Unit of time	Definition
Hour	60 minutes
Minute	60 seconds
Second	1/60 of a minute
	Officially, 9,192,631,770 oscillations of the atom of cesium-133
Centisecond	0.01 (one-hundredth) of a second
Millisecond	0.001 (one-thousandth) of a second
Microsecond	0.000001 (one-millionth) of a second
Nanosecond	0.000000001 (one-billionth) of a second
Picosecond	0.000000000001 (one-trillionth) of a second
Femtosecond	0.000000000000001 of a second
Chronon	One-billionth of a trillionth of a second

What is the exact length of a calendar year?

The calendar year is defined as the time between two successive crossings of the celestial equator by the sun at the vernal equinox. It is exactly 365 days, 5 hours, 48 minutes, and 46 seconds. The fact that the year is not a whole number of days has affected the development of calendars, which over time generate an accumulative error. The current calendar used, the Gregorian calendar, named after Pope Gregory XIII (1502–1585), attempts to compensate by adding an extra day to the month of February every four years. A "leap year" is one with the extra day added.

When does a century begin?

A century has 100 consecutive calendar years. The first century consisted of years 1 through 100. The twentieth century started with 1901 and ended with 2000. The twenty-first century began on January 1, 2001.

When did January 1 become the first day of the new year?

When Julius Caesar (100–44 B.C.E.) reorganized the Roman calendar and made it solar rather than lunar in the year 45 B.C.E., he moved the beginning of the year to January 1. When the Gregorian calen-

Roman emperor Julius Caesar selected January 1 to be the first day of the new year.

103

dar was introduced in 1582, January 1 continued to be recognized as the first day of the year in most places. In England and the American colonies, however, March 25, intended to represent the spring equinox, was the beginning of the year. Under this system, March 24, 1700, was followed by March 25, 1701. In 1752, the British government changed the beginning date of the year to January 1.

Besides the Gregorian calendar, what other kinds of calendars have been used?

Babylonian calendar—A lunar calendar composed of alternating 29-day and 30-day months to equal roughly 354 lunar days. When the calendar became too misaligned with astronomical events, an extra month was added. In addition, three extra months were added every eight years to coordinate this calendar with the solar year.

Chinese calendar—A lunar month calendar of 12 periods having either 29 or 30 days (to compensate for the 29.5 days from new moon to new moon). The new year begins on the first new moon over China after the sun enters Aquarius (between January 21 and February 19). Each year has both a number and a name (for example, the year 1992, or 4629 in the Chinese era, is the year of the monkey). The calendar is synchronized with the solar year by the addition of extra months at fixed intervals.

Muslim calendar—A lunar 12-month calendar with 30 and 29 days alternating every month for a total of 354 days. The calendar has a 30-year cycle with designated leap years of 355 days (one day added to the last month) in the 30-year period. The Islamic year does not attempt to relate to the solar year (the season). The dating of the beginning of the calendar is 622 C.E. (the date of Muhammad's flight from Mecca to Medina).

Jewish calendar—A blend of the solar and lunar calendar, this calendar adds an extra month (Adar Sheni, or the second Adar) to keep the lunar and solar years in alignment. This occurs seven times during a 19-year cycle. When the extra 29-day month is inserted, the month Adar has 30 days instead of 29. In a usual year, the 12 months alternately have 30, then 29, days.

Egyptian calendar—The ancient Egyptians were the first to use a solar calendar (about 4236 B.C.E. or 4242 B.C.E.), but their year started with the rising of Sirius, the brightest star in the sky. The year, composed of 365 days, was one-quarter of a day short of the true solar year, so eventually, the Egyptian calendar did not coincide with the seasons. It used twelve 30-day months with five-day weeks and five dates of festival.

Coptic calendar—Still used in areas of Egypt and Ethiopia, it has a similar cycle to the Egyptian calendar: 12 months of 30 days followed by five complementary days. When a leap year occurs, usually preceding the Julian calendar leap year, the complementary days increase to six.

Roman calendar—Borrowing from the ancient Greek calendar, which had a four-year cycle based on the Olympic Games, the earliest Roman calendar (about 738 B.C.E.) had 304 days with 10 months. Every second year a short month of 22 or 23 days was added to coincide with the solar year. Eventually, two more months were added at the end of the year (Januarius and Februarius) to increase the year to 354 days. The Roman republican calendar replaced this calendar during the reign of Tarquinius Priscus (616–579 B.C.E.). This new lunar calendar had 355 days, with the month of February having 28 days. The other months had either 29 or 31 days. To keep the calendar aligned with the seasons, an extra month was added every two years. By the time the Julian calendar replaced this one, the calendar was three months ahead of the season schedule.

Julian calendar—In 46 B.C.E., Julius Caesar (100–44 B.C.E.), wishing to have one calendar in use for all the empire, had the astronomer Sosigenes develop a uniform solar calendar with a year of 365 days and one day ("leap day") added every fourth year to compensate for the true solar year of 365.25 days. The year had 12 months with 30 or 31 days except for February, which had 28 days (or 29 days in a leap year). The first of the year was moved from March 1 to January 1.

Gregorian calendar—Pope Gregory XIII (1502–1585) instituted calendrical reform in 1582 to realign the church celebration of Easter with the vernal equinox (the first day of spring). To better align this solar calendar with the seasons, the new calendar would not have a leap year in the century years that were not divisible by 400. Because the solar year is shortening, today, a one-second adjustment is made (usually on December 31 at midnight) when necessary to compensate.

Japanese calendar—It has the same structure as the Gregorian calendar in years, months, and weeks, but the years are enumerated in terms of the reigns of emperors as epochs. The last epoch (for Emperor Akihito [1933–]) is Epoch Heisei, which started January 8, 1989.

Hindu calendars—The principal Indian calendars reckon their epochs from historical events, such as rulers' accessions or death dates, or a religious founder's dates. The Vikrama era (originally from Northern India and still used in western India) dates from February 23, 57 B.C.E., in the Gregorian calendar. The Saka era dates from March 3, 78 C.E., in the Gregorian calendar and is based on the solar year with 12 months of 365 days and 366 days in leap years. The first five months have 31 days; the last seven have 30 days. In leap years, the first six months have 31 days, and the last six have 30. The Saka era is the national calendar of India (since 1957). The Buddhist era starts with 543 B.C.E. (believed to be the date of Buddha's death).

Three other secular calendars of note are the *Julian Day calendar* (a calendar astronomers use, which counts days within a 7,980-year period and must be used with a table), the *perpetual calendar* (which gives the days of the week for the Julian and Gregorian calendars as well), and the *World Calendar*, which is similar to the perpetual calendar, having 12 months of 30 or 31 days, a year-day at the end of each year, and a leap-year-day before July 1 every four years.

There have been attempts to reform and simplify the calendar. One such example is the Thirteen-Month or International Fixed calendar, which would have 13 months of four weeks each. The month *Sol* would come before July; there would be a year-day at the end of each year and a leap-year-day every four years just before July 1. A radical reform was made in France when the French Republican calendar (1793–1806) replaced the Gregorian calendar after the French Revolution. It had 12 months of 30 days and five supplementary days at the end of the year (six in a leap year), and weeks were replaced with 10-day decades.

What is the World Calendar?

Following World War II, the United States led a movement in the United Nations to encourage the international community to adopt a common calendar. Called the World Calendar, it would self-adjust the irregular months, equalize the quarterly divisions of the years, and fix the sequence of weekdays and month dates so that days of the month would always fall on the same day of the week. This calendar would make the normal year 364 days long, divided into four quarters of 91 days each. The quarters could be segmented into three months of 31, 30, and 30 days. Every year would have 52 whole weeks; every quarter would begin with a Sunday and end with a Saturday. An extra day would need to be added to each year. This day, "World's Day," is formally called "W December," observed following the last day of December. Leap Year Day would be a similar insertion in the calendar every fourth year, following June 30, and would be called "W June." Although it was predicted that the calendar would be adopted in 1961, it has repeatedly failed passage in the United Nations.

What is the Julian Day Count?

The Julian Day Count is a system of counting continuing days rather than years. It was developed by Joseph Justus Scaliger (1540–1609) in 1583. The Julian Day Count (named after Scaliger's father, Julius Caesar Scaliger [1484–1558]) is still used today by astronomers, geodesists, financiers, and even some historians who need an unambiguous dating system. Julian Day (JD) 1 was January 1, 4113 B.C.E. On this date the Julian calendar, the ancient Roman tax calendar, and the lunar calendar all coincided. This event would not occur again until 7,980 years later. Each day within this 7,980-year period is numbered. January 1, 2015, at noon was the beginning of JD 2,457,024. The figure reflects the number of days that have passed since its inception. There are online calendar converters to convert Gregorian calendar dates into Julian Day.

How is a modified Julian date calculated?

The modified Julian date (MJD) is an abbreviated form of the Julian Day Count. The MJD begins at midnight (expressed as the

Religious leader and scholar Joseph Justus Scaliger created the Julian Day Count system.

.5 in the equation below), in accordance with more standard conventions of time, rather than noon, and removes the first two digits of the Julian Day. The JD lies between 2400000 and 2500000 for the three centuries beginning November 17, 1758. It is defined as:

$$MJD = JD - 2400000.5$$

Which animal designations have been given to the Chinese years?

There are twelve different names, always used in the same sequence: Rat, Ox, Tiger, Hare (Rabbit), Dragon, Snake, Horse, Sheep (Goat), Monkey, Rooster, Dog, and Pig. The following table shows this sequence.

Chinese Year Cycle

Year	Zodiac Sign
2015	Sheep (Goat)
2016	Monkey
2017	Rooster
2018	Dog
2019	Pig
2020	Rat
2021	Ox
2022	Tiger
2023	Rabbit
2024	Dragon
2025	Snake
2026	Horse
2027	Sheep (Goat)
2028	Monkey
2029	Rooster
2030	Dog
2031	Pig
2032	Rat
2033	Ox
2034	Tiger
2035	Rabbit
2036	Dragon
2037	Snake

The Chinese year of 354 days begins three to seven weeks into the western 365-day year, so the animal designation changes at that time, rather than on January 1.

What were the longest and shortest years on record?

The longest year was 46 B.C.E. when Julius Caesar (100–44 B.C.E.) introduced his calendar, called the Julian calendar, which was used until 1582. He added two extra months and 23 extra days to February to make up for accumulated slippage in the Egyptian calendar. Thus, 46 B.C.E. was 455 days long. The shortest year was 1582, when Pope Gregory XIII (1502–1585) introduced his calendar, the Gregorian calendar. He decreed that October 5 would be October 15, eliminating 10 days to make up for the accumulated error in the Julian calendar. Not everyone changed over to this new calendar at once. Catholic Europe adopted it within two years of its inception. Many Protestant continental countries did so in 1699–1700; England imposed it on its colonies in 1752 and Sweden in 1753. Many non-European countries adopted it in the nineteenth century with China doing so in 1912, Turkey in 1917, and Russia in 1918. To change from the Julian to the Gregorian calendar, add 10 days to dates October 5, 1582, through February 28, 1700; after that date add 11 days through February 28, 1800; 12 days through February 28, 1900; and 13 days through February 28, 2100.

Why were there concerns when the calendar year changed from 1999 to 2000?

The year 2000, the last year of the twentieth century, became known as Y2K or the Millennium bug. There was fear that the practice of data entry into computer software using two digits instead of four digits (e.g., 99 instead of 1999) would reset computers to the year 1900. Thus, the financial and banking industry, utilities, and airlines were concerned that the date would reset to 1900 instead of turning to 2000. The industries spent time and billions of dollars to fix the Y2K bug. However, very few—and only minor—problems occurred as the calendar year changed from 1999 to 2000.

Where do the names of the days of the week come from?

The English days of the week are named for a mixture of figures in Anglo-Saxon and Roman mythology.

Day	Named After
Sunday	The sun
Monday	The moon
Tuesday	Tiu (the Anglo-Saxon god of war, equivalent to the Norse Tyr or the Roman Mars)
Wednesday	Woden (the Anglo-Saxon equivalent of Odin, the chief Norse god)
Thursday	Thor (the Norse god of thunder)
Friday	Frigg (the Norse god of love and fertility, the equivalent of the Roman Venus)
Saturday	Saturn (the Roman god of agriculture)

What is the origin of the week?

The week originated in the Babylonian calendar, where one day out of seven was devoted to rest.

How were the months of the year named?

The English names of the months of the current (Gregorian) calendar come from the Romans, who tended to honor their gods and commemorate specific events by designating them as month names:

- *January* (*Januarius* in Latin) is named after Janus, a Roman two-faced god, one face looking into the past, the other into the future.

- *February* (*Februarium*) is from the Latin word *Februare*, meaning "to cleanse." At the time of year corresponding to our February, the Romans performed religious rites to purge themselves of sin.

- *March* (*Martius*) is named in honor of Mars, the god of war.

Wednesday is named after the Norse god Woden (also known as Odin and Wotan).

- *April* (*Aprilis*), after the Latin word *Aperio*, means "to open" because plants begin to grow in this month.

- *May* (*Maius*) is named after the Roman goddess Maia, as well as from the Latin word *Maiores,* meaning "elders," who were celebrated during this month.

- *June* (*Junius*) is named after the goddess Juno and Latin word *iuniores*, meaning "young people."

- *July* (*Iulius*) was, at first, known as *Quintilis,* from the Latin word meaning five, since it was the fifth month in the early Roman calendar. Its name was changed to July in honor of Julius Caesar (100–44 B.C.E.).

- *August* (*Augustus*) is named in honor of the Emperor Octavian (63 B.C.E.–14 C.E.), the first Roman emperor, known as Augustus Caesar. Originally, the month was known as *Sextilis* (sixth month of early Roman calendar).

- *September* (*September*) was once the seventh month and accordingly took its name from *septem*, meaning "seven."

- *October* (*October*) takes its name from *octo* (eight); at one time it was the eighth month.

- *November* (*November*), from *novem*, means "nine," once the ninth month of the early Roman calendar.

- *December* (*December*), from *decem*, means "ten," once the tenth month of the early Roman calendar.

Why are the lengths of the seasons not equal?

The lengths of the seasons are not exactly equal because the orbit of the Earth around the sun is elliptic rather than circular. When the Earth is closest to the sun in January, gravitational forces cause the planet to move faster than it does in the summer months when it is far away from the sun. As a result, the autumn and winter seasons in the Northern Hemisphere are slightly shorter than spring and summer. The duration of the northern seasons are:

spring	92.76 days
summer	93.65 days
autumn	89.84 days
winter	88.99 days

What are the beginning dates of the seasons for 2015 to 2025?

The four seasons in the Northern Hemisphere coincide with astronomical events. The spring season starts with the vernal equinox (around March 21) and lasts until the summer solstice (June 21 or 22); summer is from the summer solstice to the autumnal equinox (about September 21); autumn from the autumnal equinox to the winter solstice (December 21 or 22); and winter from the winter solstice to the vernal equinox. In the Southern Hemisphere the seasons are reversed; autumn corresponds to spring and winter to summer. The seasons are caused by the tilt of the Earth's axis, which changes the position of the sun in the sky. In winter the sun is at its lowest position (or angle of declination) in the sky; in summer it is at its highest position.

Year	Spring Equinox	Summer Solstice	Fall Equinox	Winter Solstice
2015	Mar. 20	June 21	Sept. 23	Dec. 21
2016	Mar. 20	June 20	Sept. 22	Dec. 21
2017	Mar. 20	June 21	Sept. 22	Dec. 21
2018	Mar. 20	June 21	Sept. 22	Dec. 21
2019	Mar. 20	June 21	Sept. 23	Dec. 21
2020	Mar. 19	June 20	Sept. 22	Dec. 21
2021	Mar. 20	June 20	Sept. 22	Dec. 21
2022	Mar. 20	June 21	Sept. 22	Dec. 21
2023	Mar. 20	June 21	Sept. 23	Dec. 21
2024	Mar. 19	June 20	Sept. 22	Dec. 21
2025	Mar. 20	June 20	Sept. 22	Dec. 21

Occasionally, one sees a date expressed as "B.P. 6500" instead of "B.C.E. 6500." What does this signify?

Archaeologists use B.P. or BP as an abbreviation for years before the present. This date is a rough generalization of the number of years before 1950 and is a date not necessarily based on radiocarbon dating methods.

How is "local noon" calculated?

Local noon, also called "the time of solar meridian passage," is the time of day when the sun is at its highest point in the sky. This is different from noon on clocks. To calculate local noon, first determine the time for sunrise and sunset, add the total amount of sunlight, divide this number by two, and add that to the sunrise. For instance, if sunrise is at 7:30 A.M. and sunset is at 8:40 P.M., the total sunlight is 13 hours and ten minutes, divided by two is 6 hours and 40 minutes (13 hours/2 = 6.5 hours plus 10 minutes), added to 7:30 A.M. equals a local noon of 1:40 P.M.

When is Daylight Savings Time observed in the United States?

Since 2007, Daylight Savings Time in the United States is observed from the second Sunday in March through the first Sunday in November. Historically, following the enactment of the Uniform Time Act of 1966, the dates of observing Daylight Savings Time had been observed from the last Sunday in April through the last Sunday in October, except in times of crisis. An example of a time of crisis was the "energy crisis" of the mid-1970s. In 1974, Daylight Savings Time began on January 6, and in 1975 it began on February 23. This time change was enacted to provide more light in the evening hours. The phrase "fall back, spring forward" indicates the direction in which the clock setting moves during these seasons.

Other countries have adopted DST as well. For instance, countries in the European Union observe summer time from the last Sunday in March to the last Sunday in October. Many countries in the Southern Hemisphere generally maintain DST from October to March; countries near the equator maintain Standard Time.

Which states and territories of the United States are exempt from Daylight Savings Time?

Arizona, Hawaii, Puerto Rico, the U.S. Virgin Islands, Guam, and American Samoa are exempt from Daylight Savings Time. The state of Indiana began observing Daylight Savings Time in 2006.

What are some other names for Daylight Savings Time?

It is also known as fast time or summer time.

How many time zones are there in the world?

There are 24 standard time zones that serially cover the Earth's surface at coincident intervals of 15 degrees longitude and 60 minutes Universal Time (UT), as agreed at the Washington Meridian Conference of 1884, thus accounting, respectively, for each of the 24 hours in a calendar day.

What is unusual about the observance of time zones in Russia and China?

Russia, which spans eleven time zones, is on "advanced time" year round, meaning that it maintains its Standard Time one hour faster than the zone designation. The country also observes Daylight Savings Time from the last Sunday in March until the fourth Sunday in September. China, though it lies across five time zones, keeps one time, which is eight hours faster than Greenwich Time.

When traveling from Tokyo to Seattle and crossing the International Date Line, what day is it?

Traveling from west to east (Tokyo to Seattle), the calendar day is set back (e.g., Sunday becomes Saturday). Traveling east to west, the calendar day is advanced (e.g., Tuesday becomes Wednesday). The International Date Line is a zigzag line at approximately the 180th meridian, where the calendar days are separated.

What is meant by Coordinated Universal Time?

On January 1, 1972, Coordinated Universal Time (UTC) replaced Greenwich Mean Time (GMT) as the time standard to determine local times in worldwide time zones. An international agreement in October 1884 divided the world into twenty-four time zones with the starting point at the Greenwich Meridian (zero degrees longitude, which passes through the Greenwich Observatory). Moving westward from Greenwich, one hour is added for every 15-degree meridian that is crossed. Moving eastward from Greenwich, one hour is subtracted for every 15-degree meridian that is crossed. Time zones are adjusted on land and through inhabited areas to keep countries, states, and cities in the same time zone.

Universal Time is derived from International Atomic Time using a worldwide network of atomic clocks. It is adjusted by adding leap seconds to account for the difference between the definition of the second and the rotation of Earth. This adjustment keeps UTC in conjunction with the apparent position of the sun and the stars. It is the standard used for all general timekeeping applications.

What is a leap second?

Leap seconds are added to keep the difference between Coordinated Universal Time (UTC) and atomic time to less than 0.9 seconds. Leap seconds are necessary for two reasons. First, the atomic second was measured and defined by comparing cesium clocks to the Ephemeris Time scale. The Ephemeris Time scale is an obsolete time scale that

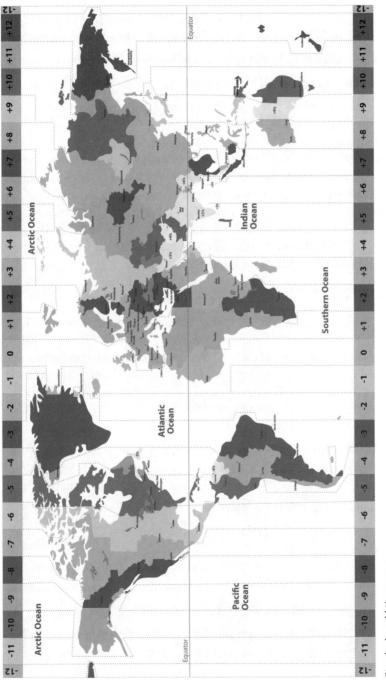

Standard world time zones

113

defined the second as a fraction of the tropical year. The duration of the ephemeris second was slightly shorter than the solar second, and this difference was passed along to the atomic second. Secondly, the speed of the Earth's rotation is not constant. Over long intervals of time, it is slowing down. In order to compensate for this lagging motion, a leap second is added to a specified day. Since 1972, thirty-five leap seconds have been added, most recently on June 30, 2012. The sequence to add a leap second is:

23h 59m 59s followed by 23h 59m 60s followed by 0h 0m 0s.

Atomic time is ahead of UTC by 35 seconds.

Who establishes the correct time in the United States?

The U.S. National Institute of Standards and Technology (NIST) uses a cesium beam clock as its NIST atomic frequency standard to determine atomic time. It is possible to check the time by accessing the NIST website at http://www.nist.gov and selecting "check time" from the options. The clock is accurate to plus or minus one second. The atomic second was officially defined in 1967 by the 13th General Conference of Weights and Measures as 9,192,631,770 oscillations of the atom of cesium-133. This cesium-beam clock of the NIST is referred to as a primary clock because, independently of any other reference, it provides highly precise and accurate time.

What is the United States Time Standard signal?

Universal Time is announced in International Morse Code each five minutes by radio stations WWV (Fort Collins, Colorado) and WWVH (Pu'unene, Maui, Hawaii). These stations are under the direction of the U.S. National Institute of Standards and Technology (formerly called National Bureau of Standards). They transmit 24 hours a day on 2.5, 5, 10, 15, and 20 megahertz. The first radio station to transmit time signals regularly was the Eiffel Tower Radio Station in Paris in 1913.

What is military time?

Military time divides the day into one set of 24 hours, counting from midnight (0000) to midnight of the next day (2400). This is expressed without punctuation.

- Midnight becomes 0000 (or 2400 of the next day)
- 1:00 A.M. becomes 0100 (pronounced oh-one hundred)
- 2:10 A.M. becomes 0210

What do the initials A.M. and P.M. mean?

The initials A.M. stand for *ante meridian*, Latin for "before noon." The initials P.M. stand for *post meridian*, Latin for "after noon."

Sundials are one of the oldest methods of measuring time. They use the position of the sun to indicate time using a gnomon to cast a shadow.

- Noon becomes 1200
- 6:00 P.M. becomes 1800
- 9:45 P.M. becomes 2145

To translate 24-hour time into familiar time, the A.M. times are obvious. For P.M. times, subtract 1200 from numbers larger than 1200; e.g., 1900 − 1200 is 7 P.M.

How is time denoted at sea?

The day is divided into watches and bells. A watch equals four hours, except for the time period between 4 P.M. and 8 P.M., which has two short watches, known as the dog watches. Within each watch, there are eight bells—one stroke for each half hour so that each watch ends on eight bells, except the dog watches, which end at four bells. New Year's Day is marked with 16 bells.

Bell	Time equivalent		
1 bell	12:30	4:30	8:30 a.m. or p.m.
2 bells	1:00	5:00	9:00
3 bells	1:30	5:30	9:30
4 bells	2:00	6:00	10:00
5 bells	2:30	6:30	10:30
6 bells	3:00	7:00	11:00
7 bells	3:30	7:30	11:30
8 bells	4:00	8:00	12:00

115

How does a sundial work?

The sundial, one of the first instruments used in the measurement of time, works by simulating the movement of the sun. A gnomon is affixed to a time (or hour) scale, and the time is read by observing the shadow of the gnomon on the scale. Sundials generally use the altitude of the sun to give the time, so certain modifications and interpretations must be made for seasonal variance.

How do water clocks measure time?

Water clocks, also called clepsydras, meaning "water thief," use the flow of water to measure time. They were used by the ancient Greeks and Romans and also used in China. Water clocks are categorized as either inflow or outflow, depending on whether the time passed is being measured by the water flowing out of a vessel or into a vessel at a constant rate. Lines in the vessel marked the "hours" passed as the vessel filled or emptied. Water clocks were never very accurate due to the difficulty in controlling the rate of the flow of water.

What is a floral clock?

In the Middle Ages it was believed that one could tell the hour of day by observing flowers, which were believed to open and close at specific times. These would be planted in flower dials. The first hour belonged to the budding rose, the fourth to hyacinths, and the twelfth to pansies. The unreliability of this method of timekeeping must have soon become apparent, but floral clocks are still planted in parks and gardens today. The world's largest clock face is that of the floral clock located inside the Rose Building in Hokkaido, Japan. The diameter of the clock is almost 69 feet (21 meters), and the large hand of the clock is almost 28 feet (8.5 meters) in length.

Where did the term grandfather clock originate?

The weight-and-pendulum clock was invented by Dutch scientist Christiaan Huygens (1629–1695) around the year 1656. In the United States, Pennsylvania German settlers considered such clocks, often called long-case clocks, to be a status symbol. In 1876 American songwriter Henry Clay Work (1832–1884) referred to long-case clocks in his song "My Grandfather's Clock," and the nickname stuck.

What is the difference between a quartz watch and a mechanical watch?

Quartz watches and mechanical watches use the same gear mechanism for turning the hour and minute gear wheels. Mechanical watches are powered by a coiled spring known as the mainspring and regulated by a system called a lever escarpment. As the watch runs, the mainspring unwinds. Quartz watches are powered by a battery-powered, electronic, integrated circuit on a tiny piece of silicon chip and regulated by quartz crystals that vibrate and produce electric pulses at a fixed rate.

Who invented the quartz clock?

Warren Marrison (1896–1980) is credited with the invention of the quartz clock. He was a telecommunications engineer at Bell Laboratories searching for quartz frequency standards with his supervisor, Joseph W. Horton (1889–1967). They presented a paper in October 1927 describing a very accurate clock based on the regular vibrations of a quartz crystal in an electrical circuit. The first quartz clock was very large, and the frequency of the quartz crystal could fluctuate due to external conditions. The first commercial quartz clocks were sold by General Radio in the 1930s. A "portable" quartz clock was not available until 1938 when Rohde & Schwarz in Germany produced a quartz clock. It weighed 101 pounds (46 kilograms) and was accurate to within +/− 0.004 seconds per day. Quartz clocks became the standard for accuracy, replacing pendulum clocks. The Seiko Astron, in 1969, was the first commercially available quartz watch. It retailed for $1,250, approximately the same price as a Toyota Corolla.

Who invented the alarm clock?

The earliest alarm clocks were invented by the Greeks in 250 B.C.E. Their system of water clocks used pulleys and wheels for a gear system. At one point, a bell was applied to the gear system. During the fifteenth century, German clockmakers designed a wall clock with a driving weight that would fall onto a bronze ball. Levi Hutchins (1761–1855) of Concord, New Hampshire, is credited with inventing an alarm clock in 1787. His alarm clock, however, rang at only one time—4 A.M.—and would keep ringing until the spring ran out. He invented this device so that he would never sleep past his usual waking time. He never patented or manufactured it. The first modern alarm clock was made by Antoine Redier (1817–1892) in 1847. It was a mechanical device; the electric alarm clock was not invented until around 1890. The earliest mechani-

Although digital alarm clocks are more common, some people still prefer the old-fashioned mechanical alarm clocks like this one.

cal clock was made in 725 C.E. in China by Yi Xing (683–727) and Liang Lingzan (fl. 8th century). The first alarm clock patented in the United States was manufactured by Seth Thomas (1785–1859) in 1876. It was a small, mechanical, wind-up clock that could be set to any hour.

When was the first digital watch produced?

The first watch with a digital display was produced by Hamilton Watch Co. in 1972. When the eighteen-carat-gold Pulsar watch was first introduced, it had a price of $2,100. The time was displayed on a red numeric display caused by a light-emitting diode (LED) by pressing a button on the side. A major obstacle, in addition to the price, was that it required two hands to use the watch; one hand wore the watch and a second hand was required to press the button for the LED display. The LED display required a large amount of the power from the small power cell that was in the watch. During the early 1970s, liquid crystal display technology (LCD) became available commercially. Since LCD required less power than LED, the display was permanent. Texas Instruments produced a digital watch that sold for $20 in 1975 and further reduced the price to $10 by 1976.

Who set a doomsday clock for nuclear annihilation?

The clock first appeared on the cover of the magazine *Bulletin of the Atomic Scientists* in 1947 and was set at 11:53 P.M. The clock, created by the magazine's board of directors, represents the threat of nuclear annihilation, with midnight as the time of destruction. The clock has been reset nineteen times since 1947. In 1953, just after the United States tested the hydrogen bomb, it was set at 11:58 P.M., the closest to midnight ever. In 1991, following the collapse of the Soviet Union, it was moved back to 11:43 P.M., the farthest from midnight the clock has ever been.

However, in 1995, the clock was shifted forward to 11:47, reflecting the instability of the post-Cold War world. The clock was set at seven minutes to midnight (11:53 P.M.) in 2002 because little progress had been made on global nuclear disarmament and because terrorists were seeking to acquire and use nuclear and biological weapons. The clock was set at five minutes to midnight (11:55 P.M.) in 2007 because it was felt the world was at the brink of a second nuclear age. Specifically, both the United States and Russia remained ready to stage a nuclear attack within minutes, North Korea conducted a nuclear test, and there was much concern that Iran would acquire the Bomb. The clock was reset to six minutes to midnight (11:54 P.M.) in 2010 because of an increased spirit of international cooperation.

Talks between Moscow and Washington focused on completing a follow-up agreement to the Strategic Arms Reduction Treaty. President Barack Obama called publicly for a nuclear-weapon-free world. In 2012, the clock was reset to five minutes to midnight (11:55 P.M.) due to the potential for nuclear weapon use in the regional conflicts in the Middle East, Northeast Asia, and South Asia. Furthermore, safer nuclear reactor designers need to be developed with more stringent oversight and training to prevent future nuclear disasters. Finally, climate change due to global warming presents challenges that are not being met quickly enough by technological solutions.

In 2015, the clock was set at three minutes to midnight. There is concern that world leaders are failing to preserve the health and ultimately the existence of humanity. Although there have been some positive developments working towards controlling climate change, it has not been enough. Furthermore, there are new programs to modernize nuclear weapons by both the United States and Russia.

TOOLS AND MACHINES

What are some of the earliest agricultural tools?

The Natufians of Palestine, circa 8000 B.C.E., are believed to have used simple digging and harvesting tools. At this early time in human agricultural history, digging sticks or hoes were used to break the ground. They also had a form of sickle to harvest sown and wild grain.

Later, around 6500 B.C.E., the first rather primitive plow, called the *ard*, was used in the Near East. This important agricultural implement evolved from a simple digging stick—a plow with a handle used by humans to push or pull along the ground. This in turn evolved into the Egyptian plow, with stag antlers or a forked branch tied to a pole to break the ground for planting.

Which tools did Neanderthal man use?

Neanderthal tool kits are termed *Mousterian* from finds at LeMoustier in France (traditionally dated to the early Fourth Glacial Period, around 40,000 B.C.E.). Using fine-grained, glassy, stonelike flint and obsidian, Neanderthals improved on the already old, established *Levallois* technique for striking one or two big flakes of predetermined shape from a prepared core. They made each core yield many small, thin, sharp-edged flakes, which were then trimmed to produce hide scrapers, points, backed knives, stick sharpeners, tiny saws, and borers. These could have served for killing, cutting, and skinning prey and for making wooden tools and clothing.

Who invented the compass?

The first compass has no known inventor. The Chinese, in the first century B.C.E., discovered that lodestone, an iron mineral, always pointed north when placed on a surface. The first Chinese compass was a spoon made of lodestone with markings indicating the four directions. Later, the Chinese enclosed the lodestone in a decorative casing and used a needle to indicate which direction was north. The problem with lodestone, however, was that it easily lost its magnetic properties. After much experimentation with dif-

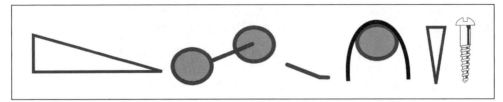

The six simple machines are (left to right): Plane, Wheel and Axle, Lever, Pulley, Wedge, and Screw

ferent metals, the Chinese discovered that adding carbon to iron to make steel gave the needle a strong magnetic charge that lasted for a long time.

What are the six simple machines?

All machines and mechanical devices, no matter how complicated, can be reduced to some combinations of six basic, or simple, machines. The lever, the wheel and axle, the pulley, the inclined plane, the wedge, and the screw were all known to the ancient Greeks, who learned that a machine works because an "effort," which is exerted over an "effort distance," is magnified through "mechanical advantage" to overcome a "resistance" over a "resistance distance." Some consider there to be only five simple machines and regard the wedge as a moving inclined plane.

Who invented the internal combustion engine?

Experimentation and research to develop an internal combustion engine began in the early nineteenth century. French physicist Nicolas Carnot (1796–1832) published *Réflexions* in 1824, in which he introduced the concept of an engine that would use a mixture of "gas" (fuel) and air. He suggested that in an ideal engine, the "Carnot engine," the motive power would be used to produce the temperature difference within the engine. Although his theories were useful in developing the laws of thermodynamics, it was decades before the first practical internal combustion engine was developed. French chemist and inventor Jean-Joseph-Étienne Lenoir (1822–1900) applied Carnot's principles to build the first practical internal combustion engine in 1859. It was a two-cycle, one-cylinder engine with slide valves that used coal gas as the fuel. A battery supplied the electrical charge to ignite the gas after it was drawn into the cylinder. It generated one or two horsepower of energy but was still inefficient and too small to power automobiles efficiently. Nikolaus August Otto (1832–1891) succeeded in overcoming the inefficiencies of Lenoir's engine by designing a carburetor that processed liquid fuel. He received a patent for the "silent Otto," a two-stroke gas engine in 1861. He began to collaborate with Eugen Langen (1833–1895), and they exhibited their engines at the 1867 World's Fair in Paris. He built the "Otto engine" in 1876, the first four-stroke piston cycle internal combustion engine. The "Otto engine" is credited with being the first practical and successful replacement for the steam engine. U.S. Patent #365,701 entitled "Gas Motor Engine," was issued in 1887 to Otto. He was inducted into the National Inventors Hall of Fame in 1981 for his invention of the gas motor engine.

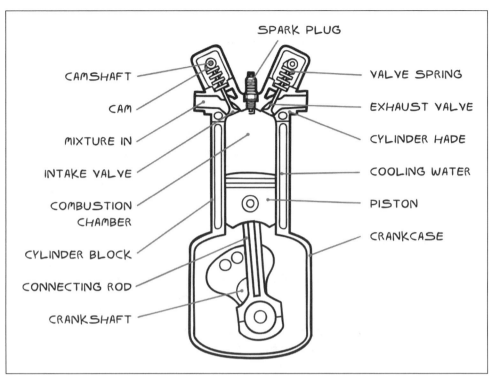

The basic components of the internal combustion engine

How does a four-stroke engine differ from a two-stroke engine?

A four-stroke engine functions by going through four cycles: 1) the intake stroke draws a fuel-air mixture in on a down stroke; 2) the mixture is compressed on an upward stroke; 3) the mixture is ignited, causing a down stroke; and 4) the mixture is exhausted on an up stroke. A two-stroke engine combines the intake and compression strokes (1 and 2) and the power and exhaust strokes (3 and 4) by covering and uncovering ports and valves in the cylinder wall. Two-stroke engines are typically used in small displacement applications, such as chainsaws, some motorcycles, etc. Four-stroke engines are used in cars and trucks and generators.

What does "cc" mean in engine sizes?

Cubic centimeters (cc), as applied to an internal combustion engine, is a measure of the combustion space in the cylinders. The size of an engine is measured by theoretically removing the top of a cylinder, pushing the piston all the way down, and then filling the cylinder with liquid. The cubic centimeters of liquid displaced (spilled out) when the piston is returned to its high position is the measure of the combustion volume of the cylinder. If a motorcycle has four cylinders, each displacing 200 ccs, it has an 800 cc engine. Automobile engines are sized essentially the same way.

What is a donkey engine?

A donkey engine is a small auxiliary engine that is usually portable or semi-portable. It is powered by steam, compressed air, or other means. It is often used to power a windlass or lift cargo on shipboard.

What is a power take-off?

The standard power take-off is a connection that will turn a shaft inserted through the rear wall of a gear case. It is used to power accessories such as a cable control unit, a winch, or a hydraulic pump. Farmers use the mechanism to pump water, grind feed, or saw wood. A power take-off drives the moving parts of mowing machines, hay balers, combines, and potato diggers.

What is holography?

Hungarian-born scientist Dennis Gabor (1900–1979) invented the technique of holography (image in the round) in 1947. Gabor received the Nobel Prize in Physics in 1971 for his pioneering work on the invention and development of holography. In 2012, he was inducted into the National Inventors Hall of Fame for his research in holography. In 1961 Emmett Leith (1927–2005) and Juris Upatnieks (1936–) produced the modern hologram using a laser, which gave the hologram the strong, pure light it needed. Three dimensions are seen around an object because light waves are reflected from all around it, overlapping and interfering with each other. This interaction of these collections of waves, called wave fronts, give an object its light, shade, and depth. A camera cannot capture all the information in these wave fronts, so it produces two-dimensional objects. Holography captures the depth of an object by measuring the distance light has traveled from the object.

A simple hologram is made by splitting a laser light into two beams through a silvered mirror. One beam, called the object beam, lights up the subject of the hologram. These light waves are reflected onto a photographic plate. The other beam, called a reference beam, is reflected directly onto the plate itself. The two beams coincide to create, on the plate, an "interference pattern." After the plate is developed, a laser light is projected through this developed hologram at the same angle as the original reference beam but from the opposite direction. The pattern scatters the light to create a projected, three-dimensional, ghostlike image of the original object in space.

Which industries use robots?

A robot is a device that can execute a wide range of maneuvers under the direction of a computer. Reacting to feedback from sensors or by reprogramming, a robot can alter its maneuvers to fit a changed task or situation. Robots may be identified as industrial robots or service robots. Their work is often classified into three major categories: 1) the assembly and finishing of products, 2) the movement of materials and objects, and 3) the performance of work in an environmentally difficult or hazardous situation. Robots are used to do welding, painting, drilling, sanding, grinding, cutting, polishing, and spraying tasks in manufacturing plants. Much of this work is done by one-armed robots. Welding accounts for nearly one-quarter of robot use in the category of assembly and finishing products. Robots are frequently used for assembly-line work in manufacturing plants. Robots are often used to lift and move large, heavy objects. This greatly reduced the risk of injury to humans. Robots can work in environments that are extremely threatening to humans or in difficult physical environments. They clean up radioactive areas, extinguish fires, disarm bombs, and load and unload explosives and toxic chemicals. Robots can process light-sensitive materials, such as photographic films that require near darkness—a difficult task for human workers. They can perform a variety of tasks in underwater exploration and in mines (a hazardous, low-light environment). In the printing industry, robots perform miscellaneous tasks, such as sorting and tying bundles of output material, delivering paper to the presses, and applying book covers. In research laboratories, small desktop robots prepare samples and mix compounds.

What is the difference between industrial robots and service robots?

Industrial robots are defined by ISO (International Organization for Standardization) 8373. The standard states that an industrial robot is: "An automatically controlled, reprogrammable, multipurpose manipulator programmable in three or more axes, which may be either fixed in place or mobile for use in industrial automation applications."

Reprogrammable indicates that the motions or functions of the robot may be changed without physical alterations to the robot.

Multipurpose indicates that the robot is capable of being adapted to a different application with physical alterations.

Physical alterations is the alteration of the mechanical structure or control system. *Axes* is the direction used to specify the robot motion in either a linear or rotary mode. Industrial robots are used in manufacturing, including the automotive industry, the electrical/electronics industry, the rubber and plastics industry, the metal and machinery industry, and the food and beverage industry.

There is no international standard or definition for service robots. A group of the ISO has been working since 2007 to revise ISO 8373 to include an official definition of service robots. At the present time, the International Federation of Robotics defines a service robot as "a robot which operates semi- or fully autonomously to perform services useful to the well-being of humans and equipment, excluding manufacturing operations."

Service robots are often, but not always, mobile. They may or may not be equipped with an arm structure like an industrial robot. Service robots are further divided into professional service robots and personal and domestic service robots. Professional service robots are used in a wide variety of applications from defense applications, such as unmanned aerial vehicles, to field robots, such as milking robots, to medical robots used to assist in surgery or therapy or for diagnostic purposes. Other professional service robots are used in logistics systems, such as mailing and courier systems, professional cleaning, and underwater systems. Service robots for personal and domestic use are

123

used mainly for household activities, such as vacuuming and floor cleaning, lawn mowing, and entertainment, such as toys and hobbies, education, and training.

What is the world population of robots?

The International Federation of Robotics reported that at the end of 2013, the worldwide stock of operational industrial robots was between 1,332,000 and 1,600,000 units. Asia, including Australia and New Zealand, was the biggest market for new industrial robots in 2013. The average length of service for an industrial robot is twelve years. Worldwide, the automotive industry utilizes the most industrial robots.

It is difficult to estimate the total number of service robots in use since their effective life span in operation ranges from more than twenty years for underwater robots to a short-term use for defense applications. The number of units of professional service robots sold during 2013 was 21,000 units. Personal and domestic service robots have a lower cost and are mass produced. In 2013, approximately four million personal service robots were sold.

What was the first industrial robot?

The first industrial robot, Unimate, was introduced and put to work on the assembly line at a General Motors plant in 1961. Unimate's task was to remove and stack pieces of die-cast metal parts taken hot from the molds. The idea for Unimate was first discussed in 1956 by George Devol (1912–2011) and Joseph Engelberger (1925–). Devol was inducted into the National Inventors Hall of Fame in 2011 for the invention of the industrial robot.

What film was the first to feature a robot?

In 1886, the French movie *L'Ève Future* had a Thomas Edison-like mad scientist building a robot in the likeness of a woman. A British lord falls in love with her in a variation on the Pygmalion theme. True working robots, in contrast to these entertainment devices, are strictly functional in appearance, looking more like machines rather than human beings or animals. However, one early exception to this is the very human-looking "Scribe" built in 1773 by two French inventors, Pierre Jacquet-Droz (1721–1790) and Henri-Louis Jacquet-Droz (1752–1791), a father and son team, who produced "The Automaton" to dip a quill pen into an inkwell and write a text of forty characters maximum.

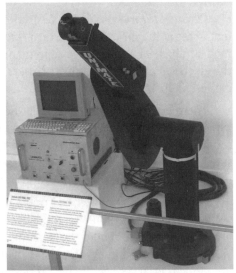

This 1983 Unimate 500 Puma (shown here with control unit and computer terminal for inputting instructions) is on display at the Deutsches Museum in Munich, Germany.

Who was the founder of cybernetics?

Norbert Wiener (1894–1964) is considered the creator of cybernetics. Derived from the Greek word *kubernetes*, meaning steerman or helmsman, cybernetics is concerned with the common factors of control and communication in living organisms, automatic machines, or organizations. These factors are exemplified by the skill used in steering a boat, in which the helmsman uses continual judgment to maintain control. The principles of cybernetics are used today in control theory, automation theory, and computer programs to reduce many time-consuming computations and decision-making processes formerly done by people.

When was the jackhammer invented?

In 1861 a French engineer, German Sommelier, came up with the idea of the pneumatic pick or jackhammer while working on the Mont Cenis tunnel, which would link Italy with France. Engineers at the time said the tunnel project would take thirty years to complete. Sommelier worked with steam-powered drills and compressed air to invent the more efficient pneumatic drill pick that became known as the jackhammer. The tunnel was completed in 1871, some twenty years ahead of schedule.

When was the road roller invented?

A Frenchman, M. Louis Lemoine, invented the steam-driven road roller in 1859. His invention revolutionized road construction, greatly improving the quality of the roadbed. Prior to the road roller, the roadbed was packed manually. Later, a roller harnessed to oxen or horses was introduced.

Is there significance to the color of fire hydrant nozzles?

Color coding of fire hydrants has significance to water and fire departments. The color indicates the water flow available from the hydrant.

Color Coding of Fire Hydrants

Class	Flow (gallons per minute)	Color of Hydrant
AA	1,500 or greater	Light Blue
A	1,000–1,499	Green
B	500–999	Orange
C	Less than 500	Red

Who brought about the standardization of screw threads?

Two Englishmen are usually credited with pursuing standards for nuts, bolts, and screws. Henry Maudslay (1771–1831), working in the machine construction industry, tried to introduce a standard between 1800 and 1810. The number of different-sized nuts and bolts was relatively small at this time. Maudslay influenced his apprentice Joseph Whitworth (1803–1887) to continue to develop standards, and in 1841 the Whitworth thread was adopted in England as a standard.

This type of wrench, on which the jaws are at right angles to the handle, derives its name from its inventor, Charles Moncke (1819–1894), whose last name was eventually corrupted into "monkey."

Why are Phillips screws used?

The recessed head and cross-shaped slots of a Phillips screw are self-centering and allow a closer, tighter fit than conventional screws. Straight-slotted screws can allow the screwdriver to slip out of the groove, ruining the wood.

Screws were used in carpentry as far back as the sixteenth century, but slotted screws with a tapering point were made at the beginning of the nineteenth century. The great advantage of screws over nails is that they are extremely resistant to longitudinal tension. Larger-size screws that require considerable force to insert have square heads that can be tightened with a wrench. Screws provide more holding power than nails and can be withdrawn without damaging the material. Types of screws include wood screws, lag screws (longer and heavier than wood screws), expansion anchors (usually for masonry), and sheet metal screws. The screw sizes vary from 0.25 to six inches (six millimeters to 15 centimeters).

Why is the term penny used in nail sizes?

The term "penny" originated in England and is a measurement relating to the length of nails. One explanation is that it refers to cost, with the cost of 100 nails of a certain size being 10 pence or 10d ("d" being the British symbol for a penny). Another explanation suggests that the term refers to the weight of one thousand nails, with "d" at one time being used as an abbreviation for a pound in weight.

People have been using nails for the last five thousand years or so. Nails are known to have been used in Ur (ancient Iraq) to fasten together sheet metal. Before 1500, nails were made by hand by drawing small pieces of metal through a succession of graded holes in a metal plate. In 1741, sixty thousand people were employed in England making nails.

The first nail-making machine was invented by American Ezekiel Reed. In 1851, Adolphe F. Brown of New York invented a wire-nail-making machine. This enabled nails to be mass produced cheaply.

How does a breathalyzer determine the alcohol level of the breath?

Breathalyzers used by police are usually electronic, using the alcohol blown in through the tube as fuel to produce electric current. The more alcohol the breath contains, the stronger the current. If it lights up a green light, the driver is below the legal limit and has passed the test. An amber light means the alcohol level is near the limit; a red light means it is above the limit. The device has a platinum anode, which causes the alcohol to oxidize into acetic acid with its molecules losing some electrons. This sets up the electric current. Earlier breathalyzers detected alcohol by color change. Orange-yellow crystals of a mixture of sulfuric acid and potassium dichromate in a blowing tube turn to blue-green chromium sulfate and colorless potassium sulfate when the mixture reacts with alcohol, which changes into acetic acid (vinegar). The more crystals that change color, the higher the alcohol level in the body.

How does SCRAM monitor alcohol consumption?

SCRAM (Secure Remote Continuous Alcohol Monitoring) uses transdermal testing to measure the amount of alcohol in a person's body. The system was first introduced

Breathalyzers detect alcohol using a platinum anode. When alcohol molecules come in contact with the anode, they oxidize, emitting electrons that are picked up as an electric current.

in 2003 for general use. When alcohol is consumed, ethanol migrates through the skin and is excreted through perspiration. SCRAM measures the transdermal alcohol content (TAC) by taking a sample of perspiration. There are three components to the SCRAM system: 1) the SCRAM bracelet, which is worn on the ankle, 2) the SCRAM modem, and 3) SCRAMnet. The modem stores alcohol readings, tamper alerts, and body temperature and transmits the information via the Internet to SCRAMnet. The system is able to test for alcohol in a person regardless of their location. It is used in forty-eight states to manage alcohol offenders.

What are Lichtenberg figures?

Lichtenberg figures are patterns appearing on a photographic plate or on a plate coated with fine dust when the plate is placed between electrodes and a high voltage is applied between them. The figures were first produced by Georg Christoph Lichtenberg (1742–1799) in 1777, marking the discovery of electrostatic recording.

WEAPONS

What is an onager?

The onager was the simplest of the early catapults. One type of onager twisted a mass of human hair or animal sinew with one wooden beam inserted into it. Geared winches were used to twist the hair or sinew without letting it unwind. To load it, soldiers manned a windlass, which pulled the beam down until it was horizontal, which added more twist to the fiber. A stone was attached to the end of the beam, and this weapon was fired when a soldier pulled a rope that released the beam from its mooring.

What is the name of the historic weapon consisting of a round, spiked ball attached to a chain?

A flail is a weapon consisting of a stout handle fitted with a short, iron-shod bar or wooden rod with iron spikes. The "morning star" flail or mace featured one or more spiked balls on a chain. The flexibility of the chain made the weapon harder to parry.

Who invented the Bowie knife?

A popular weapon of the American West, the Bowie knife was named after Jim Bowie (1796–1836), who was killed at the Alamo. According to most reliable sources, his brother, Rezin Bowie (1793–1841), might have been the actual inventor. The knife's blade measured up to 2 inches (5 centimeters) wide, and its length varied from 9 to 15 inches (23 to 38 centimeters).

Who invented the Swiss Army knife?

The origins of the Swiss Army knife date back to 1884 when Karl Elsener (1860–1918) opened a small cutlery workshop in Ibach-Schwyz, Switzerland. In 1891, the Swiss Army decided to contract with the company for a pocket knife instead of the German-made pocket knife that the soldiers in the Swiss Army had been issued. The basic components of the first knife delivered to soldiers were a blade, a tin opener, and tools for stripping and reassembling the knife the soldier carried. A second version designed for officers included a corkscrew in addition to the basic tools. The officer's model was also designed with the familiar red handle and Swiss flag cross motif that has continued to be recognized as the Swiss Army knife.

What is napalm and how has it been used?

Napalm was invented in 1943 by Harvard chemist Louis Fieser (1899–1977) in cooperation with the U.S. Army. It is composed of gasoline (33 percent), benzene (21 percent), and polystyrene (46 percent). It was first used in World War II. It burned more slowly and at higher temperatures than pure gasoline and sticks to anything it touches. Napalm depletes the air of oxygen, and, when dropped on bunkers or into caves, it asphyxiates the occupants without burning them. It was used in the Korean War, the Vietnam War, and in Desert Storm, where canisters were dropped on Iraqi fortifications and tank obstacles.

Is chemical warfare a modern phenomenon?

Chemical warfare can be traced back to ancient and medieval times, especially siege warfare. Assaults on castles and walled cities could take months, even years, so innovative means were sought to end these stalemates. Incendiaries and toxic or greasy smokes were common weapons in both attacking and defending besieged castles or cities. Possibly the first recorded use of poison gas occurred in the wars between Athens and Sparta (431–404 B.C.E.). Flaming pitch or tar and sulfur mixtures were used during the Peloponnesian War. These smoke-generating mixtures could be quite toxic when inhaled. The Spartan armies are said to have deployed a toxic metal arsenic in vapor clouds.

What is mustard gas and what happens upon exposure?

Mustard gas is made from a variety of chemicals, including sulfur mustard. It is actually a clear liquid, but when mixed with other chemicals, it looks brown and smells like garlic. Mustard gas was used heavily as a chemical warfare agent in World Wars I and II; it causes skin burns, blisters, and damage to the respiratory tract. Exposure to large amounts can cause death.

Can anthrax be used as a weapon?

A mere two grams of dried anthrax spores, distributed evenly as a powder over a population of 500,000 in an urban setting, could cause death or serious illness for perhaps 200,000 people. However, many comments about the potential devastation of a massive anthrax attack are misleading and unnecessarily alarming. The attackers would have to overcome a series of very challenging technical problems.

First, the attackers would have to have access to a very virulent strain of the bacterium, because spores harvested from the ground or from farm animals could not be spread as a powder. Next, they would have to grind down the supply of spores so that the particles would be small enough to be deeply inhaled but not so small that they wouldn't be immediately exhaled. Lastly, the attackers would have to add an antistatic element to the milled particles, because in the process of milling the spores become electrostatically charged, thus forming large clumps of spores that then would drop harmlessly to the ground.

What other biological substances are similar to anthrax and could be used in bioterrorism?

Several other materials can be introduced and used in biological warfare. Examples of other toxins are:

- *Ricin* is one of the most toxic naturally occurring substances known. It is derived from the seeds of castor bean plants—the same plants that are used to make castor oil.
- *Botulism*, like anthrax, is bacteria found in soil. It occasionally strikes people who eat poorly processed canned food or fish in which the bacteria has grown. The bacteria produce an extremely toxic substance, botulinum, that causes blurred vision, dry mouth, difficulty in swallowing or speaking, weakness, and other symptoms. Paralysis, respiratory failure, and death may follow.

- *Aflatoxin* and *mycotoxin* are both common in crops. Aflatoxin B1 is frequently found in molds that grow on nuts. Iraq and Iran are two of the world's largest producers of pistachio nuts, but the toxin can also be cultivated from molds that grow on corn and other crops. These toxins destroy the immune system in animals and are carcinogenic over a long period of time in humans.

- *Clostridium perfringens* is a common source of food poisoning. It is similar to anthrax since it also forms spores that can live in soil. Though its spores are less nasty in food, the organism causes gas gangrene when it finds its way into open battlefield wounds. Gas gangrene produces pain and swelling as the infected area bloats with gas. It later causes shock, jaundice, and death.

When were tanks first introduced into combat?

The war chariot in ancient times served as an "armored" vehicle carrying weapons and troops. One of the major disadvantages of the war chariot was that the wheels would often break or get stuck in soft or rocky terrain. Another disadvantage was that horse-drawn war chariots became ineffective when the horses were injured or killed. Lt. Col. (later promoted to Major General) E. D. Swinton (1868–1951) was inspired by seeing a Holt tractor towing a gun to propose the development and construction of an armored fighting vehicle in 1914. Concurrently, Captain T. G. Tulloch of the British army explored the development of an armored vehicle. The idea was supported by Sir Winston Churchill (1874–1965), then lord of the admiralty, and the first tanks were ready for deployment in 1916. Forty-nine tanks were first introduced for combat during the Battle

A British tank that was used during World War I.

> ## Which warfare innovations were introduced during the American Civil War?
>
> **B**arbed wire, trench warfare, hand grenades, land mines, armored trains, ironclad ships, aerial reconnaissance, submarine vessels, machine guns, and even a primitive flamethrower were products of the American Civil War, 1861–1865.

of the Somme in 1916. Many of these tanks suffered from mechanical problems. The impact of tanks as a weapon in battle became more significant during World War II.

What was the difference between the British "male" and "female" tanks?

There were two versions of the first British tanks in 1916. The male tanks were armed with a 6-pounder naval gun and four machine guns. The female tanks were armed with only six machine guns. Male tanks weighed 31 tons (28 tonnes) and female tanks weighed 30 tons (27 tonnes). Both were 26 feet (8 meters) long, 13 feet (4 meters) wide, and 8 feet (2.4 meters) high.

Who invented mine barrage?

In 1777, David Bushnell (1742?–1824) conceived the idea of floating kegs containing explosives that would ignite upon contact with ships.

When was the Colt revolver patented?

The celebrated six-shooter of the American West was named after its inventor, Samuel Colt (1814–1862). Although he did not invent the revolver, he perfected the design, which he first patented in 1835 in England and then in the United States the following year. Colt had hoped to mass produce this weapon but failed to get enough backing to acquire the necessary machinery. Consequently, the guns, made by hand, were expensive and attracted only limited orders. In 1847, the Texas Rangers ordered one thousand pistols, enabling Colt to finally set up assembly-line manufacture in a plant in Hartford, Connecticut.

Why is a shot tower used in shot making?

For shot to be accurate and of high velocity, it should be perfectly round. However, early methods for molding lead for shot often resulted in flawed or misshapen balls. In 1782, a British plumber, William Watts, solved this problem when he devised a simple method. This method consisted of pouring the molten lead through a sieve and then allowing the resulting drops to fall from a great height into a pool of water. The air cooled the drops and the water cushioned their fall, preventing them from being deformed. This new technology spread rapidly and shot towers, from 150 to 215 feet (46 to 65 meters) high, appeared across Europe and America. Although steel shot has largely replaced lead due to environmental concerns, about 30 towers still drop lead shot worldwide, including five in the United States. The essentials of the 1782 design remain unchanged.

Who invented the machine gun?

The first successful machine gun, invented by Richard J. Gatling (1818–1903), was patented in 1862 during the Civil War. Its six barrels were revolved by gears operated by a hand crank, and it fired 1,200 rounds per minute. Although there had been several partially successful attempts at building a multifiring weapon, none were able to overcome the many engineering difficulties until Gatling. In his gear-driven machine, cocking and firing were performed by cam action. The U.S. Army officially adopted the gun on August 24, 1866.

The first automatic machine gun was a highly original design by Hiram S. Maxim (1840–1916). In 1884, the clever Maxim designed a portable, single-barreled automatic weapon that made use of the recoil energy of a fired bullet to eject the spent cartridge and load the next.

The original "tommy gun" was the Thompson Model 1928 SMG. This 45-caliber machine gun, designed in 1918 by General John Taliaferro Thompson (1860–1940), was to be used in close-quarters combat. The war ended before it went into production, however, and Thompson's Auto Ordnance Corporation did not do well, until the gun was adopted by American gangsters during Prohibition. The image of a reckless criminal spraying his enemies with bullets from his handheld "tommy gun" became a symbol of the Depression years. The gun was modified several times and was much used during World War II.

Who was known as the Cannon King?

Alfred Krupp (1812–1887), whose father, Friedrich (1787–1826), established the family's cast-steel factory in 1811, began manufacturing guns in 1856. Krupp supplied large weapons to so many nations that he became known as the "Cannon King." Prussia's victory in the Franco–German War of 1870–1871 was largely the result of Krupp's field guns. In 1933 when Adolf Hitler (1889–1945) came to power in Germany, this family business began manufacturing a wide range of artillery. Alfred Krupp (1907–1967), the

How did the bazooka get its name?

It was coined by American comedian Bob Burns (1890–1956). As a prop in his act, he used a unique musical instrument that was long and cylindrical and resembled an oboe. When U.S. Army soldiers in World War II were first issued hollow-tube rocket launchers, they named them bazookas because of their similarity to Burns's instrument.

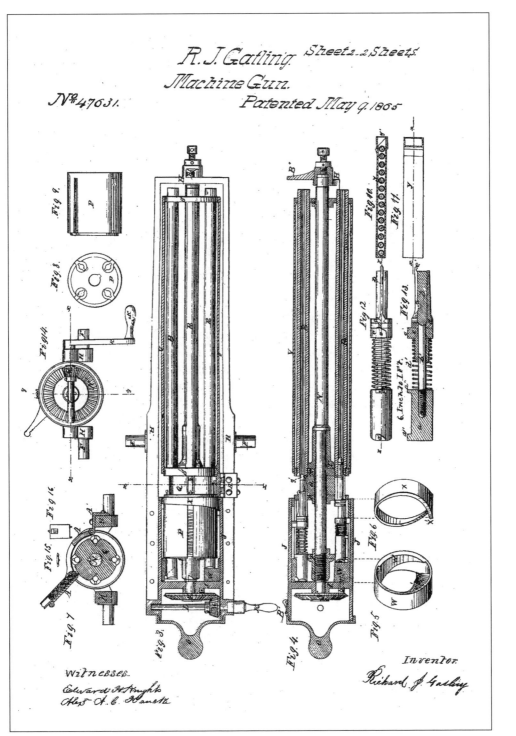

An illustration from the patent for Richard J. Gatling's machine gun.

great-grandson of Alfred, supported the Nazis in power and accrued staggering wealth for the company. The firm seized property in occupied countries and used slave labor in its factories. After the war, Alfred was imprisoned for twelve years and had to forfeit all his property. Granted amnesty in 1951, he restored the business to its former position by the early 1960s. On his death in 1967, however, the firm became a corporation and the Krupp family dynasty ended.

What is considered the most technologically advanced handgun?

Metal Storm's O'Dwyer Variable Lethality Law Enforcement pistol is cited as the most sophisticated handgun. It has no moving parts, and its firing system is totally electronic. The seven-shot single barrel can fire multiple rounds with a single pull of the trigger. It has a built-in electronic security system to limit its use to authorized users. The O'Dwyer is intended for specialist police teams and the military.

What was the Manhattan Project?

The Manhattan Engineer District was the formal code name for the U.S. government project to develop an atomic bomb during World War II. It soon became known as the Manhattan Project—a name taken from the location of the office of Colonel James C. Marshall (1897–1977), who had been selected by the U.S. Army Corps of Engineers to build and run the bomb's production facilities. When the project was activated by the U.S. War Department in June 1942, it came under the direction of Colonel Leslie R. Groves (1896–1970).

The first major accomplishment of the project's scientists was the successful initiation of the first self-sustaining nuclear chain reaction, done at a University of Chicago laboratory on December 2, 1942. The project tested the first experimental detonation of an atomic bomb in a desert area near Alamogordo, New Mexico, on July 16, 1945. The test site was called Trinity, and the bomb generated an explosive power equivalent to between 15,000 and 20,000 tons (15,240 to 20,320 tonnes) of TNT. Two of the project's bombs were dropped on Japan the following month (Hiroshima on August 6 and Nagasaki on August 9, 1945), resulting in the Japanese surrender that ended World War II.

Set off near Alamogordo, New Mexico, on July 16, 1945, the first atomic bomb explosion was the fruition of the Manhattan Project's years of work. The weapon would soon be dropped twice on Japan.

What were "Little Boy" and "Fat Man"?

"Little Boy" was the code name for the first atomic weapon dropped by the United States on Hiroshima. Little Boy was designed at Los Alamos, New Mexico, by a team of scientists headed by J. Robert Oppenheimer (1904–1967). Little Boy measured about 10 feet long (3 meters) and weighed 8,000 pounds (3,629 kilograms) with an explosive yield of a minimum of 12,000 tons (10,866 tonnes) of TNT. It was fueled by uranium-235. "Fat Man" was the code name of the second atomic bomb dropped by the United States on Nagasaki. Fat Man was also designed at Los Alamos by a team of scientists headed by Oppenheimer. Like Little Boy, it measured about 10 feet long, but it weighed 9,000 pounds (4,082 kilograms) and had an explosive yield of 20,000 tons (18,144 tonnes) of TNT. Fat Man was fueled by plutonium. The least powerful nuclear weapons available today are more powerful than either Little Boy or Fat Man.

How does a smart bomb work?

"Smart bomb" is slang for a bomb that can be accurately guided to its target by a laser beam, radar, radio control, or electro-optical system. Although pilots must accurately release the bomb in the general direction of the target, steerable tail fins allow minor adjustments in the bomb's glide path to be made in order to more accurately reach the target. An advantage of smart bombs is that they can be released further away from the target, thus reducing the aircraft's vulnerability to enemy ground fire.

How far can a Pershing missile travel?

The surface-to-surface nuclear missile, which is 34.5 feet (10.5 meters) long and weighs 10,000 pounds (4,536 kilograms), has a range of about 1,120 miles (1,800 kilometers). It was developed by the U.S. Army in 1972. Other surface-to-surface missiles are the Polaris, with a range of 2,860 miles (4,600 kilometers); the Minuteman, 1,120 miles (1,800 kilometers); the Tomahawk, 2,300 miles (3,700 kilometers); the Trident, 4,600 miles (7,400 kilometers); and the Peacemaker, 6,200 miles (10,000 kilometers).

How does a cruise missile work?

Cruise missiles are highly accurate, long-range guided missiles that fly at low altitudes at high subsonic speeds. They are guided by Global Positioning System (GPS), Terrain Contour Matching (TERCOM), or Digital Scene Matching Area Correlation (DSMAC) guidance systems. Cruise missiles are difficult to detect on radar because of their small size and low-altitude flying capability.

What is a nuclear winter?

The term "nuclear winter" was coined by American physicist Richard P. Turco (1943–) in a 1983 article in the journal *Science*, in which he describes a hypothetical post-nuclear war scenario having severe worldwide climatic changes: prolonged periods of darkness, below-freezing temperatures, violent windstorms, and persistent radioactive fallout. This would be caused by billions of tons of dust, soot, and ash being tossed into the atmosphere, accompanied by smoke and poisonous fumes from firestorms. In the case of a severe nuclear war, within a few days, the entire Northern Hemisphere would be under a blanket so thick that as little as 1/10 of 1 percent of available sunlight would reach the Earth. Without sunlight, temperatures would drop well below freezing for a year or longer, causing dire consequences for all plant and animal life on Earth.

Reaction to this doomsday prediction led critics to coin the term "nuclear autumn," which downplayed such climatic effects and casualties. In January 1990, the release of *Climate and Smoke: An Appraisal of Nuclear Winter*, based on five years of laboratory studies and field experiments, reinforced the original 1983 conclusions.

What is the most destructive nonlethal weapon?

Known as the "Blackout Bomb," the BLU-114/B graphite bomb was used by NATO against Serbia in May 1999, disabling 70 percent of its power grid with minimal casualties. The bomb was specifically designed to attack electrical power infrastructures. A similar bomb was used against Iraq in the Gulf War of 1990–1991. It knocked out 85 percent of Iraq's electricity-generating capacity. It works by exploding a cloud of ultrafine carbon-fiber wires over electrical installations, causing temporary short-circuiting and disruption of the electrical supply.

Why are improvised explosive devices (IEDs) called "homemade" devices?

Improvised explosive devices (IEDs) are called homemade devices because they are created with materials that are available to the builder of the IED. They are designed to cause death or injury by using explosives that are hidden and detonated using a variety of trigger mechanisms. IEDs may use commercial, military, or homemade explosives. They may be used with explosives alone or in combination with toxic chemicals, biological toxins, or radiological material. Although IEDs vary in shape and form, they generally contain a common set of components, including an initiation system or fuse, explosive oil, a detonator, a power supply for the detonator, and a container. Triggering methods may be a cell phone, garage door opener, a child's remote-control toy, driving over a rubber hose to produce enough air pressure to activate a switch, or even a manually operated trigger.

What are weapons of mass destruction?

Weapons of mass destruction (WMDs) are defined as any explosive or incendiary device, such as a bomb, grenade, missile, or mine, with a charge of more than four ounces (113 grams). Weapons of mass destruction also include devices that are designed to cause death or bodily injury through the release or impact of toxic chemicals, disease organisms, or radiation/radioactivity at a level dangerous to human life. The goal of WMDs is to severely injure or kill thousands of people with a single use.

These improvised explosive devices (IEDs) found in the Ukraine were controlled by radio signals.

What is a radiological dispersal device?

Better known as a "dirty bomb," this device is designed to release massive amounts of radioactive material as widely as possible. Unlike a nuclear bomb, which is designed to release large amounts of energy such as heat and radiation by splitting or fusing atoms, the dirty bomb is a conventional bomb packed with radioactive materials. Components include materials used in medicine and research as well as low-grade, unenriched uranium. Dirty bombs are not considered true weapons of mass destruction; instead, they are intended to cause mass disruption. The economic and psychological consequences of having an important urban area contaminated with radiation could be severe.

MINING, MATERIALS AND MANUFACTURING

MINING

Why is mining an important technology?

Mining is the process used to extract minerals and ores from the Earth. Minerals and ores are essential building blocks in our modern, technological world. Virtually every industry uses minerals or metals, including the automotive industry, the electronics industry, chemical manufacturing and processing, building and construction, paper, cosmetics, medicine, and agriculture.

Is mining a new technology?

Mining techniques were developed in ancient times to extract precious metals. The ancient Egyptians mined copper and tin. The ancient Greeks and Romans mined silver and gold. Although many of the metals and minerals extracted by the ancient civilizations were easily accessible, they did dig some underground mines using slave labor. The earliest scientific description and study of mining methods was found in the book *De re Metallica*, by German physician Georgius Agricola (1494–1555). *De re Metallica* continued to be the authoritative reference and textbook on mining, mining machinery, and mine ventilation for nearly two hundred years.

What are the basic steps of mining?

The first step is to locate the mineral or ore. Geologists and other engineers locate mineral deposits. The deposits are evaluated for their potential for mining, often considering the economic value of the mineral in correlation with the cost of recovery. The ultimate extraction of minerals from the Earth's surface involves drilling, blasting, hoisting, and hauling away the commodity for further processing.

How does surface mining differ from underground mining?

Surface mining is the preferred method of mining when the desired mineral or ore is close enough to the surface that it can be reached with bulldozers, trucks, and power shovels. Underground mining is used to reach minerals and metals deep beneath the Earth's surface. It requires digging shafts and excavating areas underground. It is usually used only to reach high-value ores, such as gold, which are unreachable using surface mining techniques. Underground mining is a more expensive method to recover minerals and ores. It is also more dangerous than surface mining.

What are the different types of surface mines?

All methods of surface mining remove a layer of rock and earth that lies above the desired mineral or coal. This layer is called the overburden. Different methods of surface mining techniques are used, depending on the location of the mineral or coal in relation to the surface of the land. Surface mining includes area mining, contour mining, mountaintop mining, open-pit mining, and alluvial or placer mining. Area mining is preferred when the coal or mineral lies close to the surface. The overburden is removed using large stripping shovels or draglines. These shovels dig parallel trenches to remove the overburden and expose the coal or mineral. After the coal or other mineral has been extracted, the overburden is replaced back into the exposed area. Area mining is commonly used in the western United States and Midwest.

Contour mining is similar to area mining, but it is used in areas with steep, hilly terrain. Once the coal is extracted, the overburden is used to return the hill to its natural, original slope and terrain.

Mountaintop mining was a common method in Appalachia, especially in eastern Kentucky, western Virginia, southern West Virginia, and parts of eastern Tennessee for coal mining. It is similar to area mining, but instead of the overburden being reused to return the terrain to its natural contour, the mountaintop is leveled, and the overburden is used as fill in the adjacent valleys. There are several environmental concerns with mountaintop mining, including the loss of the natural terrain, loss of vegetation, and violations of the Clean Water Act.

Alluvial or placer mining is used to separate heavy minerals, such as gold, from sand and gravel beds, including stream beds. In the simplest form, the minerals may be recovered by shaking the sand and gravel in simple pans, hence the phrase, "panning for gold." A more sophisticated method includes using a sluice box. A sluice box is a long, shallow box that acts like a sieve. When the sluice box is shaken, the lighter sand and gravel is washed away and the heavier mineral is left behind.

Why are surface mining methods used more frequently than underground mining methods?

Surface mining is both safer and less expensive as a technique than underground mining. Surface mining techniques are used for nearly 90 percent of the rock and mineral resources mined in the United States and more than 60 percent of the coal mined in the United States. Approximately 50 percent of all surface mining in the United States is coal mining. Extraction of sand, gravel, stone, and clay account for another 35 percent of surface mining activities in the United States. Metallic ores account for about 13 percent of the surface mining in the United States.

What are open-pit mines?

Open-pit mines are deep, cone-shaped holes or pits excavated in rock. In order to ensure stability, open-pit mines are continually widened as they are dug deeper. Open-pit mines are preferred when the desired metal or coal is distributed through large volumes of rock or when the coal seams are very thick. As the rock is excavated to expose the desired metal or coal, it is removed from the site. Open-pit mines are not returned to their natural state once the mineral or coal has been extracted.

Where is the largest open-pit mine?

The largest single open-pit mine is the Bingham Canyon Mine in Utah. It is more than three-quarters of a mile (1.2 kilometers) deep and more than two and three-quarters miles wide across at the top—and still growing. It is deep enough to stack two Willis Towers (1,451 feet [442 meters] high) without reaching the top of the mine. It is also wide enough to stack 12 aircraft carriers end to end. More than 19 million tons (17 million metric tons) of copper have been mined from Bingham Canyon Mine since it was first opened in 1903. It produces 300,000 tons (272,155 tonnes) of copper annually. Although it produces mainly copper, it also produces 400,000 ounces (11,340 kilograms) of gold,

The Bingham Canyon Mine in Utah is the biggest open-pit mine. It produces 300,000 tons of copper annually.

4 million ounces (113,398 kilograms) of silver, 30 million pounds (13 million kilograms) of molybdenum, and 1 million tons (907,185 metric tons) of sulfuric acid annually.

How large was the Berkeley Pit open-pit mine?

Mining for gold and silver began in western Montana in the 1860s. By the mid to late 1870s, more and more copper was being discovered in the area, corresponding to the time Alexander Graham Bell (1847–1922) introduced the telephone, which used copper wire to transmit voice communications. As early as the late nineteenth century, the area was known as the "Richest Hill on Earth" for its quantities of silver, gold, and most importantly, copper. In 1955, the Anaconda Company began excavation of the The Berkeley Pit. Berkeley Pit continued to be in operation until 1982. When the mine closed, the pit measured 7,000 feet (2,134 meters) long, 1,600 feet (488 meters) deep, and 5,600 feet (1,707 meters) wide. During its existence, 320 million tons (290 million metric tons) of ore and over 700 million tons (635 million metric tons) of waste rock were mined. At times, 17,000 tons (15,422 metric tons) of ore were extracted from the pit per day. In total, the Berkeley Pit produced enough copper to pave a four-lane highway four inches thick from Butte, Montana, to Salt Lake City, Utah, and 30 miles (48 kilometers) beyond—a distance of over 400 miles (644 kilometers).

What are some of the tools and machinery used in surface mining?

The tools and machinery of surface mining consist of bulldozers, front-end bucket loaders, scrapers, trucks, power shovels, drills, graders, and draglines. Although much of this machinery is used in general construction and road building, draglines are used almost exclusively in surface mining. The primary use of draglines is to remove the overburden. They are some of the largest excavating machines in use in the world. Draglines consist of a boom, a bucket, and a cab mounted on an assembly that has rollers to move along the ground. The newest draglines have AC electric drive systems. The largest draglines have boom lengths of 360–435 feet (110–133 meters) and bucket capacities of 100–152 cubic yards (76–116 cubic meters). They are operated 24 hours per day, seven days a week to remove the overburden.

What are the two basic methods of underground mining?

The two basic methods of underground mining are room and pillar and longwall. In room and pillar mines, the mineral or ore is removed by cutting rooms, or large tunnels, in the solid rock, leaving pillars of the mineral or ore for roof support. The pillars are mined when that part of the mine is closing. Longwall mining takes successive slices over the entire length of a long working face using a machine called a continuous miner.

How are underground mines accessed?

In order to access underground mines, shafts or tunnels must be drilled for entry. The decision to choose one type of mine is most often determined based on how deep the mineral deposit is under the Earth's surface in conjunction with the terrain. There are three types of mines, named for their method of access: 1) drift mines, 2) slope mines, and 3) shaft mines.

- *Drift mines* are entered horizontally into the side of a hill.
- *Slope mines* use an inclined tunnel that slopes from the surface to the mineral deposit.
- *Shaft mines* are usually used only for the deepest mineral deposits. A shaft is a vertical passage excavated from the surface to the mineral deposit.

Where is the deepest underground mine in the world?

The gold mines of South Africa are the deepest mines in the world. The Mponeng Gold Mine, operated by AngloGold Ashanti, is the deepest mine (and also the deepest man-made hole) in the world. The operating depth of the mine ranges from 1.5 miles (2.4 kilometers) to 2.4 miles (3.9 kilometers) below the surface of the Earth. The distance from the surface is comparable to stacking ten Empire State Buildings on top of each other. The mining operation is reached by two elevator cars, called cages. The first travels 1.6 miles (2.6 kilometers) from the surface in four minutes. A second cage travels another mile (1.6 kilometers) below the surface. The final distance to the deepest part of the mine is accessible only by foot or by truck.

What are some of the special challenges of underground mining?

Ventilation, temperature control, lighting, ceiling collapses, earth movements caused by drilling and blasting, and methane gas buildups, which can lead to fires and explosions, are all concerns of underground mining. Throughout the decades and centuries of mining, engineers and inventors have tried to find ways and technologies to overcome these challenges. Devices and inventions important to mine safety include respirators, ventilators, roof support systems with roof bolts and posts, protective head coverings (hard hats), gas detectors, and rock dusting. Rock dusting involves covering the walls of a mine with powdered limestone to aid in the lighting of the mine, minimize explosions, and reduce health hazards. Mine safety continues to be the focus of research and innovation in the

Safety lamps like this one were used by miners so that flames would not accidentally ignite flammable gases found in the tunnels.

twenty-first century. Newer technologies include an inflatable shelter that can provide safe refuge for miners in the event of an underground collapse or explosion.

Who invented the miner's safety lamp?

A source of light was essential for miners working underground. Unfortunately, in the early years of mining, the combination of open flames and flammable gases in the mine led to many injuries and fatalities among miners. Two English scientists, independent of each other, developed a safety lamp for mining in 1815. George Stephenson (1781–1848) designed a lamp encased in a glass cylinder that was capped with a metallic cover with tiny holes. A metal bonnet covered the lamp. There was not enough heat or airflow to ignite any of the flammable gases outside the lamp. Independently, but also in 1815, Sir Humphry Davy (1778–1829) designed a safety lamp that became known as the Davy lamp. The Davy lamp consisted of a metallic gauze screen that surrounded the flame. The metallic screen allowed air to reach the flame so it could burn, but at the same time, it cooled the flame so it could not ignite the gas outside the lamp.

What major technological invention changed the safety lamp?

Although the electric light bulb was invented in 1879, it was not adapted for mining purposes until the early twentieth century. Two mining engineers, John T. Ryan Sr. (1884–1941) and George H. Deike (1879–1963), formed the Mine Safety Appliances Company in 1914 in response to a West Virginia mine disaster in 1912. As was a too common occurrence, methane gas exploded in the mine, killing eighty miners. Ryan and Deike joined forces to develop methods of avoiding methane gas explosions in mines. Their first contribution to mine safety was the electric cap lamp. They enlisted the help of inventor Thomas Edison (1847–1931) to design the Edison Cap Lamp. The lamp consisted of a battery encased in a self-locking steel case worn on the miner's belt. A flexible cord traveled from the battery to the lamp on the miner's cap. The battery could power a six-candlepower lamp for twelve hours. It was recharged at the end of the miner's shift. As an added safety measure, in the event the bulb was broken, the electrical contacts were immediately disconnected so the tungsten filament would cool and would not be able to ignite any flammable gases in the mine. The Mine Safety Appliance Company claimed that in the twenty-five years following the introduction of the electric cap lamp, mine explosions decreased by 75 percent.

What is a miner's canary?

A "miner's canary" refers to the birds used by miners to test the purity of the air in the mines. Canaries are very sensitive to dangerous levels of carbon monoxide, which lead to carbon monoxide poisoning. At least three birds were taken into a mine by exploring parties, and the distress of any one bird was taken as an indication of a dangerous level of carbon monoxide in the mine. Some miners used mice rather than birds. This method of safety was used prior to the more sophisticated equipment used to detect the presence of carbon dioxide today.

COAL

How and when was coal formed?

Coal is a sedimentary rock. It is formed from the remains of plants that have undergone a series of far-reaching changes, turning into a substance called peat, which subsequently was buried. Through millions of years, Earth's crust buckled and folded, subjecting the peat deposits to very high pressure and changing the deposits into coal. The Carboniferous, or coal-bearing, period occurred about 250 million years ago. Geologists in the United States sometimes divide this period into the Mississippian and the Pennsylvanian periods. Most of the high-grade coal deposits are to be found in the strata of the Pennsylvanian period.

What are the types of coal?

The first stage in the formation of coal converts peat into lignite, a dark-brown type of coal. Lignite is then converted into subbituminous coal as pressure from overlying materials increases. Under still greater pressure, a harder coal called bituminous, or soft, coal is produced. Intense pressure changes bituminous coal into anthracite, the hardest of all coals.

Which type of coal is the most abundant in the United States?

Bituminous coal, containing 45 percent to 86 percent carbon, is the most abundant type of coal in the United States. It accounts for nearly half of the total coal production in the United States. The three states that are the largest producers of bituminous coal are West Virginia, Kentucky, and Pennsylvania.

What is cannel coal?

Cannel coal is a type of coal that possesses some of the properties of petroleum. Valued primarily for its quick-firing qualities, it burns with a long, luminous flame. It is made up of coallike material mixed with clay and shale, and it may also look like black shale, being compact and dull black in color.

What are the most common underground mining methods for coal?

In the United States, about two-thirds of the coal recovered by underground mining is by the room and pillar method; the other third is recovered by longwall mining. In other locations, such as Great Britian, the longwall method is used more frequently.

In coal mining what is meant by damp?

Damp is a poisonous or explosive gas in a mine. The most common type of damp is firedamp, also known as methane. Whitedamp is carbon monoxide. Blackdamp (or chokedamp) is a mixture of nitrogen and carbon dioxide formed by mine fires and explosions of firedamp in mines. Blackdamp extinguishes fire and suffocates its victims.

What are some uses for coal?

The primary use of coal is to generate electricity. In our modern, technological society, electricity powers our homes and factories. Different types of coal have different uses. Steam or thermal coal is used to generate electricity. Coal also has an important role in the manufacture of steel. Coking or metallurgical coal is used in the production of steel. In addition, the by-products of coal are used in the chemical and pharmaceutical industries. Some products that contain coal by-products are paints, fertilizers, medicines, plastics, explosives, creosote, building materials, fireproofing materials, synthetic rubber, and even billiard balls. Coal is an important component of carbon fiber, which is used in construction, mountain bikes, and tennis rackets, where an extremely strong but lightweight reinforcement material is required.

How have advances in mining technology impacted the mining of coal?

Improved mining technology and the growth of surface mining of coal has dramatically increased productivity. The amount of coal produced by one miner in one hour has more than tripled since 1978.

What is fly ash?

Fly ash is one of the coal combustion products. It is the very fine portion of ash residue that results from the combustion of coal. The fly ash portion is usually removed electrostatically from the coal combustion gases before they are released to the atmosphere. Air scrubber systems also remove some of the remaining fly ash. Fly ash is used as a raw material in cement and concrete production. It is a component of asphalt and road construction. It is also used as a filler to stabilize soils.

Which country is the largest producer of coal?

China is the world's largest producer of coal, producing an estimated 4,034,459 thousand short tons (3,660,000 thousand metric tons), which was 46 percent of the world total in 2012.

Which countries have the greatest reserves of coal?

Coal is one of the most abundant fossil fuels. Nearly half of all the global reserves of coal are found in the United States (27 percent), Russia (18 percent), and China

A coal mine in China. Much of China's industrial muscle is fueled by coal, which has resulted in huge problems with air pollution.

> ## What is red dog?
>
> **R**ed dog is the residue from burned coal dumps. The dumps are composed of waste products incidental to coal mining. Under pressure in these waste dumps, the waste frequently ignites from spontaneous combustion, producing a red-colored ash, which is used for driveways, parking lots, and roads.

(13 percent). The world's total recoverable reserves of coal were estimated at 949,022 million tons (860,938 million tonnes) at the end of 2012.

MINERALS

What is mineralogy?

Mineralogy is a branch of geology that studies minerals. It focuses on the identification, distribution, and classification and properties—both physical and chemical—of minerals. Not only does mineralogy provide scientists with information on how the earth was formed, but minerals provide the basic components and building blocks of many of our modern materials. Scientists have identified nearly four thousand different minerals.

How does a rock differ from a mineral?

Rocks are sometimes described as an aggregate or combination of one or more minerals. Rocks are categorized into three groups: igneous, sedimentary, or metamorphic. Igneous rocks are formed by the solidification of molten magma that emerges through Earth's crust via volcanic activity. There are thousands of different igneous rock types. Some examples of igneous rocks are granite, obsidian, and basalt. Sedimentary rocks are produced by the accumulation of sediments. The most common sedimentary rock is sandstone, which is composed predominantly of quartz crystals. Metamorphic rocks are formed by the alteration of igneous and sedimentary rocks through heat and/or pressure. Examples of metamorphic rocks are marble, slate, schist, and gneiss. Geologists extend the definition of a rock to include clay, loose sand, and certain limestones.

Mineralogists use the term "mineral" for a substance that has all four of the following features: it must be found in nature; it must be made up of substances that were never alive (organic); it has the same chemical makeup wherever it is found; and its atoms are arranged in a regular pattern to form solid crystals.

What is quartz?

Quartz is the second most abundant mineral in the Earth's continental crust after feldspar. It is one of the most well-known minerals on Earth, is the important constituent of many rocks, and is one of the most varied of all minerals occurring in different forms, habits, and colors. Some forms of quartz, especially the gemstone forms, have their color enhanced by heat treatment and range from yellow-brown to the familiar purple amethyst.

What are the most common mineral-forming elements?

Most minerals are compounds of two or more chemical elements. Although more than one hundred chemical elements have been identified, only ten elements account for nearly 99 percent of the weight of Earth's crust. The ten most common mineral-forming compounds are: oxygen, silicon, aluminum, iron, calcium, sodium, potassium, magnesium, hydrogen, and titanium. Since oxygen and silicon make up nearly three-quarters of the Earth's crust, silicate minerals—compounds of silicon and oxygen—are the most abundant minerals.

What is the difference between silicon, silica, silicate, and silicone?

Silicon, silica, silicone, and silicate sound similar, but they are four different and distinct substances. Silicon is a chemical element. Chemical elements are the simplest form of matter. They are fundamental substances that consist of only one kind of atom. Silicon is considered a metalloid, meaning it shares some properties of metals and has some properties of nonmetals.

Silica is a chemical compound formed from silicon and oxygen. Quartz is the most common form of silica found in nature.

Silicates are compounds of silicon and oxygen with other elements bonded in a paired formation. This pairing is called the silicon-oxygen tetrahedron because it contains four oxygen atoms and one silicon atom. The silicon-oxygen tetrahedron bonds most frequently with sodium, potassium, calcium, magnesium, iron, and aluminum to form silicates.

Silicones are polymers, a type of synthetic compound. Silicones are formed from two or more silicon atoms linked with carbon compounds. Silicones do not take the tetrahedral shape of silicates, but they form chainlike structures called silicon polymers. They were first developed commercially during World War II.

Why is silicon an important element for technology?

Since silicon is a metalloid, it sometimes acts like a metal. At high temperatures, silicon acts like a metal and conducts electricity, but at low temperatures, it acts like an insulator and does not conduct electricity. Silicon is a semiconductor. Silicon was used to move technology into the world of transistors, then into the world of integrated circuits, and ultimately into the world of the computer chip. The silicon used to make computer chips is formed from silica sand. Thin slices of pure silicon are then etched with the intricate electronic circuits needed to run a computer.

Which form of silica is used as a desiccant?

Silica gel is a colloidal form of silicon dioxide. It is used as a desiccant since it absorbs moisture easily. Silica gel packets are used to package many products to protect them from moisture.

What are some uses of silicones?

Silicones come in many forms, ranging from liquids to greases and waxes to resins and solids. Liquid silicones are used as water repellants and defoamers. They are used in water- and heat-resistant lubricants as greases and waxes. They are also used to make heat- and chemical-resistant products such as paints, rubbers, and plastic parts. Silicones are used in the automotive (sealants and coatings), cosmetic (hair care products), health care (medical

One of the many uses of silicone is that it makes for an excellent sealant.

devices), electronics, paper, paints and inks, textile, and household care product industries. Silicone baking products are rated to withstand very high temperatures—some as high as 900° Fahrenheit (482° Celsius).

Do any minerals contain just one chemical element?

Yes, minerals that contain just one chemical element are called native elements. Examples of native elements are diamond and graphite, which contain pure carbon. Gold, silver, and copper are also native elements.

What are the physical traits that are used to identify a mineral?

Minerals are identified by physical traits, including hardness, color, streak, luster, cleavage and fracture, and specific gravity.

- *Hardness* is the ability of a mineral to scratch another mineral.
- *Color* is sometimes a way to identify a mineral, but the color may be the result of impurities.
- *Streak* is related to color but is a more reliable test. It is the color of the streak a mineral leaves on an unglazed porcelain plate.
- *Luster* is the appearance when light reflects off the surface of a mineral. There are different descriptions of luster, including metallic luster, vitreous or glassy luster, resinous luster, pearly luster, and greasy luster.

- *Cleavage and fracture* refer to how a mineral breaks when struck with force. It is called cleavage when the minerals break along smooth planes. Most minerals do not break cleanly but fracture.
- *Specific gravity* is an indirect measure of density. It is the ratio of the weight of the mineral to an equal volume of water.

When was the Mohs scale introduced?

The Mohs scale is a standard of ten minerals by which the hardness of a mineral is rated. It was introduced in 1812 by German mineralogist Friedrich Mohs (1773–1839). The minerals are arranged from softest to hardest. Harder minerals, with higher numbers, can scratch those with a lower number.

The Mohs Scale

Hardness	Mineral	Comment
1	Talc	Hardness 1–2 can be scratched by a fingernail
2	Gypsum	Hardness 2–3 can be scratched by a copper coin
3	Calcite	Hardness 3–6 can be scratched by a steel pocket knife
4	Fluorite	
5	Apatite	
6	Orthoclase	Hardness 6–7 will not scratch glass
7	Quartz	
8	Topaz	Hardness 8–10 will scratch glass
9	Corundum	
10	Diamond	The hardest mineral in the world

Who was the first person to attempt a color standardization scheme for minerals?

German mineralogist Abraham Gottlob Werner (1749–1817) is considered the father of mineralogy. He wrote the first modern textbook of descriptive mineralogy, *On the External Characters of Fossils*, in 1774. His work devised a method of describing minerals by their external characteristics, including color. He worked out an arrangement of colors and color names, illustrated by an actual set of minerals.

Which two gems contain the mineral corundum?

Pure corundum is colorless. However, when there are trace amounts of metals in corundum, they will give color to the mineral. Both rubies and sapphires contain the mineral corundum. Chromium ions replace small amounts of aluminum in rubies, giving them their characteristic red color. In sapphires, iron and titanium ions replace some of the aluminum, producing the characteristic blue color.

In addition to its use as a gemstone, corundum is used as an abrasive to cut and grind metal and stone.

How does the emerald get its color?

Emerald is a variety of green beryl that is colored by a trace of chromium, which replaces the aluminum in the beryl structure. Other green beryls exist, but if no chromium is present, they are, technically speaking, not emeralds.

How is the star in star sapphires produced?

Sapphires are composed of gem-quality corundum. Color appears in sapphires when small amounts of iron and titanium are present. Star sapphires contain needles of the mineral rutile that will display as a six-ray star figure when cut in the unfaceted cabochon (dome or convex) form. The star effect is called asterism. The most highly prized star sapphires are blue, although star sapphires come in almost every color. Black or white star sapphires are less valuable. Since a ruby is simply the red variety of corundum, star rubies also exist.

What is a tiger's eye?

A tiger's eye is a semiprecious quartz gem that has a vertical, luminescent band like that of a cat's eye. To achieve the effect of a cat's eye, veins of parallel blue asbestos fibers are first altered to iron oxides and then replaced by silica. The gem has a rich yellow to yellow-brown or brown color.

What is galena?

Galena is a lead sulphide and the most common ore of lead, containing 86.6 percent lead. Lead-gray in color with a brilliant, metallic luster, galena has a specific gravity of 7.5, a hardness of 2.5 on the Mohs scale, and usually occurs as cubes or a modification of an octahedral form. Galena is one of the minerals that exhibits perfect cleavage in three directions. Mined in Australia, it is also found in Canada, China, Mexico, Peru, and the United States (Missouri, Kansas, Oklahoma, Colorado, Montana, and Idaho).

When was asbestos last mined in the United States?

Asbestos is a generic name for six types of naturally occurring mineral fibers that have been used in commercial products. The six types of asbestos are chrysotile, crocidolite, amosite, anthophyllite asbestos,

A close-up look at asbestos shows its fibrous nature. Inhaling asbestos particles can lead to lung diseases such as cancer and mesothelioma.

tremolite asbestos, and actinolite asbetos. These fibrous minerals possess high tensile strengths, large length-to-width ratios, flexibility, and resistance to chemical and thermal degradation. Chrysotile is the most commonly used form of asbestos. In the United States, it is used for asbestos-cement products, roof coatings, brake pads, shoes and clutches.

Asbestos has not been mined in the United States since 2002. The four countries that supplied 99 percent of all asbestos in the world in 2014 were:

Sources of Asbestos		
Country	Tons	Metric Tons
Russia	1,157,426	1,050,000
China	440,925	400,000
Brazil	320,773	291,000
Kazakhstan	264,555	240,000

What is stibnite?

Stibnite is a lead-gray mineral with a metallic luster. It is the most important ore of antimony, and it is also known as antimony glance. One of the few minerals that fusees easily in a match flame (977°F or 525°C), stibnite has a hardness of two on the Mohs scale and a specific gravity of 4.5 to 4.6. It is commonly found in hydrothermal veins or hot springs deposits. More than 80 percent of all the antimony mined in the world comes from China. The most important use of antimony is for fireproofing and as a flame retardant—especially for children's clothing.

What is cinnabar?

Cinnabar is the main ore of the mineral mercury. Its cinnamon to scarlet-red color makes it a colorful mineral. It is produced primarily in the United States (California, Oregon, Texas, and Arkansas), Spain, Italy, and Mexico. It is often used as a pigment.

What are the characteristics of diamonds?

Diamonds crystallize directly from rock melts rich in magnesium and saturated with carbon dioxide gas that has been subjected to high pressures and temperatures exceeding 2,559°F (1,400°C). These rock melts originally came from deep in Earth's mantle at depths of 93 miles (150 kilometers).

Diamonds are minerals composed entirely of the element carbon, with an isometric crystalline structure. The hardest natural substance, gem diamonds have a density of 3.53, though black diamonds (black carbon cokelike aggregates of microscopic crystals) may have a density as low as 3.15. Diamonds have the highest thermal conductivity of any known substance.

What are some of the industrial uses of diamonds?

Diamonds are well known as gemstones. Diamonds that do not meet the gem-quality standards for clarity, color, shape, or size are used as industrial-grade diamonds. The most common industrial uses of diamonds include cutting and drilling. Their high thermal conductivity enables diamonds to be used in cutting tools because they do not become hot. They are also used in corrosion-resistant coatings, special lenses, computing, and other advanced technologies.

Are there any diamond mines in the United States?

The United States has no commercial diamond mines. The only significant diamond deposit in North America is Crater of Diamonds State Park near Murfreesboro, Arkansas. It is on government-owned land and has never been commercially mined or systematically developed. For a small fee, tourists can dig there and try to find white, brown, or yellow diamonds. The largest crystal found at the park weighed 40.23 carats and was named the "Uncle Sam" diamond.

However, the United States is one of the world's leading producers of synthetic industrial diamonds. Synthetic industrial diamonds are often preferred because they are produced with the exact properties for specific applications.

Which country is the top producer of diamonds?

Russia produced 34,927,650 carats of diamonds, or 27 percent of the world total in 2012.

How can a genuine diamond be identified?

There are several tests that can be performed without the aid of tools. A knowledgeable person can recognize the surface lustre, straightness, and flatness of facets and high light reflectivity. Diamonds become warm in a warm room and cool if the surroundings are cool. A simple test that can be done is exposing the stones to warmth and cold and then touching them to one's lips to determine their appropriate temperature. This is especially effective when the results of this test are compared to the results of the test done on a diamond known to be genuine. Another test is to pick up the stone with a moistened fingertip. If this can be done, then the stone is likely to be a diamond. The majority of other stones cannot be picked up in this way.

The water test is another simple test. A drop of water is placed on a table. A perfectly clean diamond has the ability to almost "magnetize" water and will keep the water from spreading. An instrument called a diamond probe can detect even the most sophisticated fakes. Gemologists always use this as part of their inspection.

How is the value of a gemstone diamond determined?

Demand, beauty, durability, rarity, freedom from defects, and perfection of cutting generally determine the value of a gemstone, but the major factor in establishing the price of gem diamonds is the control over output and price as exercised by the Central Selling Organization's (CSO) Diamond Trading Corporation Ltd. The CSO is a subsidiary of DeBeers Consolidated Mines Ltd.

What are the four "C"s of diamonds?

The four Cs are cut, color, clarity, and carat. Cut refers to the proportions, finish, symmetry, and polish of the diamond. These factors determine the brilliance of a diamond. Color describes the amount of color the diamond contains. Color ranges from colorless to yellow with tints of yellow, gray, or brown. Colors can also range from intense yellow to the more rare blue, green, pink, and red. Clarity describes the cleanness or purity of a diamond as determined by the number and size of imperfections. Carat is the weight of the diamond.

How is a diamond processed from a rough diamond to a polished gemstone?

A professional diamond cutter first assesses a diamond for any imperfections, faults, or inclusions. The goal is to retain the maximum size of the diamond with the least amount of imperfections. A rough diamond is cleaved by carving a sharp groove into the diamond and then striking the groove to break the diamond into two pieces with the goal of minimizing the size of the inclusion on the larger piece of diamond. Sometimes a diamond is sawed by mounting it in a sawing machine. A revolving blade whose edge is coated with diamond powder saws through the diamond. The next step is to cut or "brut" the diamond. The facets are then cut into the stone, giving it its final shape. Contemporary round-cut diamonds have fifty-eight facets. The final step is to polish the diamond, enhancing its brilliance.

Which diamond shape is the most popular for jewelry?

Round diamonds are the most popular shape of diamond used for jewelry.

How are diamonds weighed?

The basic unit is a carat, which is set at 200 milligrams (0.00704 ounces or 1/142 of an avoirdupois ounce). A well-cut, round diamond of one carat measures almost exactly 0.25 inches (6.3 millimeters) in diameter. Another unit commonly used is the point, which is one-hundredth of a carat. A stone of one carat weighs 100 points. "Carat" as a unit of weight should not be confused with the term "karat" used to indicate purity of the gold into which gems are mounted.

Which diamond is the world's largest?

The Cullinan Diamond, weighing 3,106 carats, is the world's largest. It was discov-

Diamonds can be cut into several shapes for jewelry, but the most popular shape is still round.

ered on January 25, 1905, at the Premier Diamond Mine, Transvaal, South Africa. Named for Sir Thomas M. Cullinan (1862–1936), chairman of the Premier Diamond Company, it was cut into nine major stones and ninety-six smaller brilliants. The total weight of the cut stones was 1,063 carats, only 35 percent of the original weight.

Cullinan I, also known as the "Great Star of Africa" or the "First Star of Africa," is a pear-shaped diamond weighing 530.2 carats. It is 2.12 inches (5.4 centimeters) long, 1.75 inches (4.4 centimeters) wide, and 1 inch (2.5 centimeters) thick at its deepest point. It was presented to Britain's King Edward VII (1841–1910) in 1907 and was set in the British monarch's sceptre with the cross. It is still the largest cut diamond in the world.

Cullinan II, also known as the "Second Star of Africa," is an oblong stone that weighs 317.4 carats. It is set in the British Imperial State Crown.

What is cubic zirconia?

Cubic zirconia was discovered in 1937 by two German mineralogists, M. V. Stackelberg and K. Chudoba. It became popular with jewelry designers in the 1970s after Soviet scientists under the direction of V. V. Osika learned how to "grow" the mineral in the Lebedev Physical Institute laboratory. Most of the cubic zirconia on the market is chemically comprised of zirconium dioxide and yttrium oxide. The two compounds are melted together at a very high temperature (almost 5,000°F [2,760°C]) using the skull melt method. This method uses a radio-frequency generator to heat the zirconium oxide. A careful cooling of the mixture produces the flawless crystals that become cubic zirconia gemstones.

What is the difference between cubic zirconium and diamonds?

Cubic zirconium is a gemstone material that is an imitation of diamonds. The word "imitation" is key. The U.S. Federal Trade Commission defines imitation materials as resembling the natural material in appearance only. Cubic zirconia may be cut the same way as diamonds. It is very dense and solid, weighing 1.7 times more than a diamond of the same millimeter size.

Which minerals are used to make cosmetics?

Minerals have been used for cosmetics since ancient times. It is believed that Cleopatra (69 B.C.E.–30 B.C.E.) used pigments ground up from minerals as cosmetics to enhance her beauty. Mineral cosmetics were introduced in the late 1990s and have gained in popularity as a natural product. The ingredients of mineral cosmetics are titanium dioxide, zinc and iron oxides, mica, and ultramarine pigments. These minerals are processed and ground finely to form powders. As an added benefit, natural mineral cosmetics offer some protection from the harmful rays of the sun. The high levels of zinc oxide protect from UVA, and titanium dioxide protects from UVA and UVB. Although traditional cosmetics may also contain many of the same natural mineral ingredients, they also contain other dyes, oils, preservatives, and fragrances, which may be irritating to some users.

Which country is the leading producer of the most mineral commodities?

China is the leading world producer of forty-four mineral commodities, including aluminum, antimony, barite, bismuth, cadmium, coal, fluorspar, germanium, gold, graphite, indium, iron ore, lead, magnesium compounds and metal, mercury, molybdenum, rare earths, strontium, tin, tungsten, and zinc.

METALS

What is metallurgy?

Metallurgy is the study of metals, metallic compounds, and alloys of metals. The focus of metallurgy is the technical process of extracting metals from their ores and refining them for use.

Which elements are metals?

Nearly 75 percent of all elements are classified chemically as metals. The three characteristics common to most metals are: 1) they are good conductors of electricity, 2) they react, physically or chemically, with other elements to form alloys and compounds, and 3) most are malleable. Malleability means they are flexible so they can be hammered or rolled into thin sheets or wires.

What are the alkali metals?

These are the elements on the left-hand side of the periodic table: lithium (Li, element 3), potassium (K, element 19), rubidium (Rb, element 37), cesium (Cs, element 55), francium (Fr, element 87), and sodium (Na, element 11). The alkali metals are sometimes called the sodium family of elements, or Group I elements. Because of their great chemical reactivity (they easily form positive ions), none exist in nature in the elemental state.

What are the alkaline Earth metals?

These are beryllium (Be, element 4), magnesium (Mg, element 12), calcium (Ca, element 20), strontium (Sr, element 38), barium (Ba, element 56), and radium (Ra, element 88). The alkaline Earth metals are also called Group II elements. Like the alkali metals, they are never found as free elements in nature and are moderately reactive metals. Harder and less volatile than the alkali metals, these elements all burn in air.

What are the transition elements?

The transition elements are the 10 subgroups of elements between Group II and Group XIII, starting with period 4. They include gold (Au, element 79), silver (Ag, element 47), platinum (Pt, element 78), iron (Fe, element 26), copper (Cu, element 29), and other metals. All transition elements are metals. Compared to alkali and alkaline Earth metals, they are usually harder and more brittle and have higher melting points. Transition metals are also good conductors of heat and electricity. They have variable valences, and compounds of transition elements are often colored. Transition elements are so named because they comprise a gradual shift from the strongly electropositive elements of Groups I and II to the electronegative elements of Groups VI and VII.

Which elements are the "noble metals"?

The noble metals are gold (Au, element 79), silver (Ag, element 47), mercury (Hg, element 80), and the platinum group, which includes platinum (Pt, element 78), palladium (Pd, element 46), iridium (Ir, element 77), rhodium (Rh, element 45), ruthenium (Ru, element 44), and osmium (Os, element 76). The term refers to those metals highly resistant to chemical reaction or oxidation (resistant to corrosion) and is contrasted to "base" metals, which are not so resistant. The six "base" metals are aluminum, copper, lead, zinc, nickel, and tin. The term "noble metals" has its origins in ancient alchemy, whose goals of transformation and perfection were pursued through the different properties of metals and chemicals. The term is not synonymous with "precious metals," although a metal, like platinum, may be both.

What are the precious metals?

This is a general term for expensive metals that are used for making coins, jewelry, and ornaments. The name is limited to gold, silver, and platinum. Expense or rarity does not make a metal precious, but rather, it is a value set by law that states that the object made of these metals has a certain intrinsic value. The term is not synonymous with "noble metals," although a metal (such as platinum) may be both noble and precious.

What is pitchblende?

Pitchblende is a massive variety of uraninite, or uranium oxide, found in metallic veins. It is a radioactive material and the most important ore of uranium. In 1898, Marie Curie (1867–1934) and Pierre Curie (1859–1906) discovered that pitchblende contained radium, a rare element that has since been used in medicine and the sciences.

Which countries have uranium deposits?

Uranium, a radioactive metallic element, is the only natural material capable of sustaining nuclear fission, but only one isotope, uranium-235, which occurs in one molecule out of 40 of natural uranium, can undergo fission under neutron bombardment. Mined in various parts of the world, it must then be converted during purification to uranium dioxide. The main use of uranium is to produce electricity. Other uses include power for ships, desalination, and military ordnance. Uranium deposits occur throughout the world. The following chart lists the top ten countries with the largest known recoverable resources of uranium.

157

Top 10 Sources of Uranium in 2014

Country	Mine Production (Tons/Metric Tons)	Percent of World Total
Kazakhstan	25,493 (23,127)	41.1%
Canada	10,067 (9,134)	16.2%
Australia	5,513 (5,001)	8.9%
Niger	4,472 (4,057)	7.2%
Namibia	3,588 (3,255)	6.4%
Russia	3,296 (2,990)	5.3%
Uzbekistan	2,646 (2,400)	4.3%
United States	2,115 (1,919)	3.4%
China	1,653 (1,500)	2.9%
Ukraine	1,024 (929)	1.8%

What are some uses of platinum?

The six platinum group elements (PGE) are platinum, palladium, rhodium, ruthenium, iridium, and osmium. They are some of the rarest metals on Earth. The average grade of PGE in ores mined primarily for their PGE concentrations range from 5 to 15 parts per million (ppm). More than 90 percent of the global PGE production comes from South Africa and Russia. The PGE is often as little as 6 to 10 grams/ton of ore. While platinum is a coveted material for jewelry making, more than 95 percent of all platinum group metals are used for industrial purposes. Many different industries use platinum:

Industries That Use Platinum

Industry	Uses of PGE
Automotive	Catalytic converters
Chemical	Manufacture of specialty silicones
	Nitric oxide—the raw material for fertilizers, explosives, and nitric acid
Petrochemical	Platinum-supported catalysts are needed to refine crude oil and produce high-octane gasoline
Electronics	Increase storage capacities in computer hard disk drives
	Multilayer ceramic capacitors
	Hybridized integrated circuits
Glass manufacturing	Fiberglass
	Liquid-crystal and flat-panel displays
Health	Medical implants, such as pacemakers
	Pharmaceuticals, such as cancer-fighting drugs

A secondary supply of platinum, palladium, and rhodium is obtained through recycling jewelry, electronic equipment, and catalytic converters.

Why is aluminum an important metal?

Aluminum is the second most abundant metallic element in the Earth's crust. It is lightweight, weighing about one-third as much as steel or copper, malleable, and ductile, and it has excellent corrosion resistance and durability. It has important uses in many different industries and technologies.

Industries That Use Aluminum

Industry	Use
Aerospace	Fuselage, wing, and supporting structures of aircraft Spacecraft, including the space shuttles and the International Space Station
Automotive	Powertrain, wheel, body The 2015 Ford F-150 has an all-aluminum body.
Building and Construction	Bridge decks, window frames, doors, roofing, flashing, wall panels, insulation The Empire State Building was the first building to contain a large quantity of aluminum components. Aluminum is favored for green building status and LEED certification.
Electrical	Wiring for residential and commercial buildings Utility grid transmission and electrical conduits
Electronics and Appliances	Cookware, washing machines, dryers, refrigerators, HDTVs, laptop computers, capacitors
Foil and Packaging	Beverage cans, aluminum foil, including aluminum foil pans and containers, food packaging, including aseptic packaging, packaging for pharmaceuticals

Aluminum is 100 percent recyclable. Industry studies report that nearly 75 percent of all aluminum produced in the United States since the early 1900s is still in use today.

How is aluminum processed?

The primary source of aluminum is bauxite ore. Bauxite is composed of aluminum oxide compounds, silica, iron oxides, and titanium dioxide. Bauxite ore reserves are plentiful in Africa, Oceania, and South America. Australia has the largest reserve of bauxite. The top five producers of bauxite are:

Bauxite-producing Countries (2014)

Country	Bauxite Production (Tons/Metric Tons)	Percent of World Total
Australia	89,287 (81,000)	34.6%
China	51,809 (47,000)	20.1%
Brazil	35,825 (32,500)	13.9%
Guinea	21,275 (19,300)	8.3%
India	20,945 (19,000)	8.2%
Jamaica	10,803 (9,800)	4.2%
Kazakhstan	6,063 (5,500)	2.4%
Russia	5,842 (5,300)	2.3%

Country	Bauxite Production (Tons/Metric Tons)	Percent of World Total
Suriname	2,976 (2,700)	1.2%
Venezuela	2,425 (2,200)	0.9%

It is often found in topsoil, making it relatively easy and economic to mine and restore the land. Once the bauxite is mined, it is refined in a two-step process to produce aluminum. The first step is known as the Bayer process. The end product of the Bayer process is aluminum oxide, also called alumina. Alumina is a fine, white powder that looks like salt or sugar but is much harder. The second step of the process is the Hall-Héroult process. The alumina that is extracted in the Bayer process is smelted in the Hall-Héroult process to form aluminum.

Is the Bayer process a new technology?

The Bayer process was invented in 1887 by Carl Josef Bayer (1847–1904). There are four steps in the Bayer process: digestion, clarification, precipitation, and calcination. In the digestion step, the bauxite is crushed, washed, dried, and then dissolved with caustic soda at high temperatures. The mixture is then filtered to remove impurities and transferred to tall tanks called precipitators. In the precipitator tanks, the hot solution begins to cool. Aluminum hydroxide seeds are added to the mixture so solid aluminum hy-

Wound sheets of aluminum at a factory await shipment. Aluminum is a useful, versatile, and abundant metal.

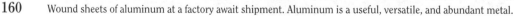

When was aluminum foil first produced?

Aluminum foil was first produced in 1903 in France and in 1913 in the United States. The first commercial use of aluminum foil was to package Lifesavers™ candy Aluminum foil is produced by rolling sheet ingots cast from molten aluminum. The sheets are then re-rolled using heavy rollers to the desired thickness. Alternatively, aluminum foil is made using the continuous casting method and then cold rolling.

droxide crystals form. The aluminum hydroxide settles at the bottom of the precipitator tank. In the final step, calcination, it is removed from the tank and washed to remove any of the remaining caustic soda. It is also heated to remove the excess water. The Bayer process used today is virtually unchanged from when it was invented in 1887. The Hall-Héroult process is named in honor of Charles Martin Hall (1863–1914) and Paul Louis-Toussaint Héroult (1863–1914), a French inventor who independently, but simultaneously, invented the same process as Hall to produce aluminum.

Where was the first large-scale aluminum factory located?

Charles Martin Hall (1863–1914) incorporated the Pittsburgh Reduction Company in 1888 in Pittsburgh, Pennsylvania. Hall was prompted to open the company after implementation of the Hall-Héroult process lowered the cost of producing aluminum from about $100 per pound to $0.18 per pound. In 1891, the company outgrew its Pittsburgh location and opened a facility in New Kensington, Pennsylvania. The name was later changed to the Aluminum Company of America (ALCOA). Hall was inducted into the National Inventors Hall of Fame in 1976 for his technique and process to manufacture aluminum.

Why is titanium important to modern industrial societies?

Titanium is a light, strong, and white metal that is used in the aerospace, paint, paper, plastic, and biomedical industries. Titanium is as strong as steel but 45 percent lighter, and it is twice as strong as aluminum but only 60 percent heavier. Its high strength-to-weight ratio and corrosion resistance makes it very valuable to the aerospace industry. The Mercury, Gemini, and Apollo capsules of the space program were made of titanium alloys. Commercial airliners are built with significant quantities of titanium alloys. Other titanium alloys are used for armored vehicles, eyeglasses, jewelry, bicycles, golf clubs, and other sports equipment. Since titanium is nonreactive in the human body, it is used to make artificial hip and knee joints, pins for setting bones, and other types of biological implants.

Why is titanium dioxide the most widely used white pigment?

Titanium dioxide, also known as titania, titanic anhydride, titanium oxide, or titanium white, has become the predominant white pigment in the world because of its high refractive index, lack of absorption of visible light, ability to be produced in the right size range, and its stability. It is the whitest known pigment, unrivalled for color, opacity, stain resistance, and durability; it is also nontoxic. The main consuming industries are paint, printing inks, plastics, and ceramics. Titanium dioxide is also used in the manufacture of floor coverings, paper, rubber, and welding rods. It is also used as a coloring in toothpaste, skim milk, candy, and sunscreen.

How is titanium processed?

A sample of pure titanium was not produced until 1910 by Matthew A. Hunter (1878–1961). It was not until the 1940s that a commercial process, the Kroll process, was developed by William J. Kroll (1889–1973) to process titanium. The Kroll process begins with titanium oxide. The titanium oxide is reacted with chlorine to get rid of the oxygen. In a second step, liquid magnesium or sodium is reacted with the resulting chloride from the first step to get rid of the chlorine. This produces a porous material, titanium sponge. The titanium sponge is crushed and melted to produce titanium ingots. Kroll was inducted into the National Inventors Hall of Fame in 2000 for his development of the Kroll process to manufacture titanium and titanium alloys.

What is coltan?

Coltan is the shortened name for the metallic ore columbite-tantalite. When refined it becomes a heat-resistant powder, tantalum, which can hold a high electrical charge. These properties make it a vital element in creating capacitors, the electronic elements that control current flow inside miniature circuit boards. Tantalum capacitors are used in almost all cell phones, laptop computers, pagers, and other electrical devices.

Why are the rare earth elements important?

The rare earth elements are the lanthanides in row 6 of the periodic table. Although these elements are quite abundant in the Earth's crust, they are called "rare" because they are difficult to separate from one another in a mineral deposit. The 15 rare earth elements are lanthanum (La), cerium (Ce), praseodymium (Pr), neodymium (Nd), promethium (Pm), samarium (Sm), europium (Eu), gadolinium (Gd), terbium (Tb), dysprosium (Dy), holmium (Ho), erbium (Er), thulium (Tm), lutetium (Lu), and ytterbium (Yb). The rare earth elements are essential for emerging technologies, such as wind turbines, magnets used in electric and hybrid vehicles, and energy-efficient lighting. China has been the dominant producer of the rare earth elements, producing more the 90 percent of the world supply of these minerals. Since 2010, the United States has begun mining these minerals, but it is only a very small proportion of the total demand for these minerals.

Which metal was one of the first extracted and used and continues to be one of the most used metals today?

Copper was one of the first metals ever used by humans. It played an important role in the development of civilization. Copper is found in many different mineral deposits. Copper is usually found in association with sulfur. Porphyry copper deposits account for nearly two-thirds of the world's copper. It is often extracted using open-pit mining methods. Copper has high ductility and malleability (properties that allow it to be stretched and shaped), conducts heat and electricity efficiently, and is resistant to corrosion.

What are some of the uses of copper?

Today, as was common in ancient times, copper is used in coinage in many countries. Copper is used for wiring and plumbing in appliances, heating and cooling systems in buildings, telecommunications links and motors, wiring, radiators, connectors, and brakes and bearings in cars and trucks. Semiconductor manufacturers use copper for circuitry in silicon chips, producing microprocessors that operate faster and use less energy. A newer use of copper is for surfaces that are frequently touched, such as brass doorknobs. Copper has inherent antimicrobial properties that reduces the transfer of germs and disease.

Copper is considered the best material to use in plumbing for many reasons, including its strength, ability to stand up to temperature and pressure changes, ability to keep contaminants out of water, germ resistance, and resistance to the effects of ultraviolet rays from the sun.

Why does copper turn green when exposed to air?

Oxidation is the chemical reaction between oxygen in the air and copper that causes the copper to turn a characteristic blue-green color. In chemical terms, oxidation is a process where an element loses at least one electron when interacting with another element. Oxidation is the process that causes iron to rust. When copper reacts with oxygen, it forms a layer of copper oxide that appears bluish-green in color. Unlike oxidation in other metals, the bluish-green layer is a protective layer called the patina. It does not cause any weakness in the metal.

What is an alloy?

An alloy is a mixture of two or more metals. Alloys may be created by mixing metals while in the molten state or by bonding metal powders. Once two or more metals are mixed to form an alloy, they cannot be separated easily into the individual components of the alloy. The properties and characteristics of an alloy differ from those of the individual components. Alloys are classified as either interstitial or substitutional. In interstitial alloys, smaller elements fill holes that are in the main metallic structure. Steel is an example of an interstitial alloy. In substitutional alloys, some of the atoms of the main metal are substituted with atoms of another metal. Brass is an example of a substitutional alloy.

163

What is the earliest known alloy?

The earliest known alloy is bronze. Bronze dates back to ancient Mesopotamia and was known as early as 3000 B.C.E. The earliest bronze was an alloy of copper and arsenic. Later, tin replaced arsenic to form the alloy bronze. The copper/tin alloy can be traced back to the Sumerians in 2500 B.C.E. Although copper had been used alone for tools and weapons, bronze is much harder than pure copper. Bronze was also used to make coins and for sculpture. The Bronze Age extended until approximately 1000 B.C.E. Bronze is still a popular metal for sculpture.

What is solder?

Solder is an alloy of two or more metals used for joining other metals together. The metal pieces are joined together without melting the entire metal structure. One example of solder is half-and-half, which is composed of equal parts of lead and tin. Other metals used in solder are aluminum, cadmium, zinc, nickel, gold, silver, palladium, bismuth, copper, and antimony. Various melting points to suit the work are obtained by varying the proportions of the metals.

Solder is an ancient joining method. There is evidence of its use in Mesopotamia some 5,000 years ago, and later in Egypt, Greece, and Rome. Use of numerous types of solder is currently wide and varied, and future use looks bright as well. As long as circuitry based on electrical and magnetic impulses and composed of a combination of conductors, semiconductors, and insulators continues to be in use, solder will remain indispensable.

Which element is part of the carbide compounds?

Compounds composed of carbon and one of the metallic elements are called carbides. The carbides are very hard, stable materials that are resistant to chemical attack. They have high melting points and are electri-

Soldering is an ancient method of joining metals together in a very strong bond.

cally conductive. Most react with water and release acetylene gas. Examples of some carbides are calcium carbide, tungsten carbide, boron carbide, and silicon carbide.

- *Calcium carbide* is also known as acetylenogen. It is commonly used in welding and cutting torches.
- *Tungsten carbide* is used for machining and tools, such as drill bits. Since it is so hard, it improves the tool's ability to hold a cutting edge. It is also used as an abrasive.
- *Boron carbide* is extremely hard. It is used in ceramics and nuclear control rods.
- *Silicon carbide* is used as an abrasive. It is used in industry for grinding wheels and saws to cut rock or concrete but is also used to manufacture emery boards and paper.

What is the most common use of zinc?

More than half of all the zinc that is produced is used for galvinization. Zinc has strong anticorrosive properties, and it bonds well with other metals. In galvinizing, thin layers of zinc are added to iron or steel to prevent corrosion. The process of galvanizing is named for Italian physicist Luigi Galvani (1737–1798). The earliest method of galvinization was to dip iron into baths of molten zinc. Other uses of zinc include the production of zinc alloys for the die casting industry, automobile manufacturing, and electrical components. The third significant use of zinc is the production of zinc oxide. Zinc oxide is used to manufacture rubber, batteries, and paints. It is also used as a protective skin ointment.

Which metals are used to make brass?

Brass is an alloy of copper and zinc. Different types of brasses have various proportions of copper and zinc in the alloy. Brass is used for musical instruments, tools, pipes and fittings, and weapons.

What is 24-karat gold?

The term "karat" refers to the percentage of gold versus the percentage of an alloy in a piece of jewelry or a decorative object. Gold is too soft to be usable in its purest form and has to be mixed with other metals. One karat is equal to 1/24th part fine gold. Thus, 24-karat gold is 100 percent pure, and 18-karat gold is 18/24 or 75 percent pure. Nickel, copper, and zinc are commonly added to gold for improved strength and hardness.

Karatage	Percentage of fine gold
24	100
22	91.75
18	75
14	58.5
12	50.25
10	42
9	37.8
8	33.75

Is white gold really gold?

White gold is the name of a class of jeweler's white alloys used as substitutes for platinum. Different grades vary widely in composition, but usual alloys consist of between 20 percent and 50 percent nickel, with the balance gold. A superior class of white gold is made of 90 percent gold and 10 percent palladium. Other elements used include copper and zinc. The main use of these alloys is to give the gold a white color.

How far can a troy ounce of gold, if formed into a thin wire, be stretched before it breaks?

Ductility is the characteristic of a substance to lend itself to shaping and stretching. Gold is one of the most ductile and malleable substances. A troy ounce of gold (31.1035 grams) can be drawn into a fine wire that is 50 miles (80 kilometers) long.

What are the chief gold-producing countries?

China is the leading gold-producing nation in the world, followed by Russia, Australia, the United States, and Peru.

Top 5 Gold-producing Countries

Country	Gold Production (2014)	Gold Reserves (2013)
China	508 tons (461 metric tons)	2,094 tons (1,900 metric tons)
Russia	300 tons (272 metric tons)	5,512 tons (5,000 metric tons)
Australia	298 tons (270 metric tons)	10,913 tons (9,900 metric tons)
United States	220 tons (200 metric tons)	3,307 tons (3,000 metric tons)
Peru	186 tons (169 metric tons)	2,094 tons (1,900 metric tons)

What is fool's gold?

Iron pyrite (FeS_2) is a mineral popularly known as "fool's gold." Because of its metallic luster and pale brass yellow color, it is often mistaken for gold. Real gold is much heavier, softer, not brittle, and not grooved.

What is the composition of U.S. coins currently in circulation?

During colonial times, coins were composed of gold, silver, and copper. The U.S. Mint produced gold coins until 1933, during the Great Depression. In 1965, silver was removed from circulating coins, and the Mint began producing nickels, dimes, quarters, and half dollars composed of copper and nickel. Nickels are composed of 25 percent nickel and 75 percent copper. Quarters, dimes, and half dollars are composed of 8.33 percent nickel, and the balance is copper. Pennies, once copper coins, are now composed of copper-plated zinc.

How does sterling silver differ from pure silver?

Sterling silver is a high-grade alloy that contains a minimum of 925 parts in 1,000 of silver (92.5 percent silver and 7.5 percent of another metal—usually copper).

What is German silver?

Nickel silver, sometimes known as German silver or nickel brass, is a silver-white alloy composed of 52 percent to 80 percent copper, 10 percent to 35 percent zinc, and 5 percent to 35 percent nickel. It may also contain a small percent of lead and tin. There are other forms of nickel silver, but the term "German silver" is the name used in the silverware trade.

Which metal is the main component of pewter?

Tin—at least 90 percent. Antimony, copper, and zinc may be added in place of lead to harden and strengthen pewter. Pewter may still contain lead, but high lead content will both tarnish the piece and dissolve into food and drink to become toxic. The alloy used today in fine quality pieces contains 91 to 95 percent minimum tin, 8 percent maximum antimony, 2.5 percent maximum copper, and 0 to 5 percent maximum bismuth, as determined by the European Standard for pewter.

What is the difference between nonferrous and ferrous metals?

The distinction between ferrous and nonferrous metals is made by mineral economists based on the use of these metals in modern industry. The term "nonferrous" literally means "without iron" (*ferrum* is the Latin term for iron). Based on this definition, it would seem that "nonferrous" metals should include all metallic minerals besides iron. However, nonferrous metals are those that are produced in significant amounts and are important to heavy industry but are not used primarily in the manufacture of steel, jewelry, or coinage. The nonferrous metals are: copper, aluminum, lead, zinc, tin, mercury, and cadmium. The "ferrous" metals are iron or alloys that contain significant amounts of iron.

What is steel?

Steel is an alloy of iron and carbon. The carbon content is less than 2 percent by weight of the metal. Other elements that may be added in smaller quantities are manganese, nickel, cobalt, and chromium.

When did the Iron Age begin?

The earliest iron was probably obtained from meteorites. Iron was known in the ancient Egypt, Babylonia, Assyria, and China civilizations. Each civilization developed a method to smelt and hammer iron. Most of the iron produced during ancient times (circa 1000 B.C.E.) was wrought iron. It was not until the Middle Ages that furnaces were built that could sustain high enough temperatures to melt iron.

Where are the largest iron ore deposits in the United States?

The largest iron ore deposits in the United States are located in Minnesota. The Mesabi Iron Range, discovered in 1866, extends northeasterly for 120 miles (193 kilometers) from west of Grand Rapids to east of Babbit. The Mesabi Iron Range is a narrow belt, approximately 3 miles (4.8 kilometers) of iron-rich sedimentary rocks, known as the Biwabik Iron Formation. Its iron ore operations produce about 40 million tons (36 million metric tons) of high-grade iron ore annually, which is approximately 75 percent of the total U.S. iron ore output. To produce 40 million tons (36 million metric tons) of high-grade iron ore, 240 tons (218 metric tons) of material, including 135 million tons (122 million metric tons) of crude ore and 105 million tons (95 million metric tons) of surface and rock stripping, must be moved. Three companies operate six iron ore mining and processing facilities on the Mesabi Iron Range.

What is slag?

Slag is a nonmetallic by-product of iron production that is drawn from the surface of pig iron in the blast furnace. Slag can also be produced in smelting copper, lead, and other metals. Slag from steel blast furnaces contains lime, iron oxide, and silica. The slag from copper and lead-smelting furnaces contains iron silicate and oxides of other metals in small amounts. Slag is used in cements, concrete, and roofing materials, as well as ballast for roads and railways.

Who was responsible for inventing a process to mass produce steel?

Several individuals were involved in developing and inventing an inexpensive process to mass produce high-quality steel. Two inventors, American William Kelly (1811–1888) and Englishman Henry Bessemer (1813–1898), working simultaneously, but independently, on both sides of the ocean are credited with inventing a process to mass produce steel. Kelly did not publicize his invention. In 1856, Bessemer filed for a patent and described the process in a paper, *Manufacture of Malleable Iron and Steel without Fuel*. Bessemer was inducted into the National Inventors Hall of Fame in 2002 for the development of the Bessemer process.

The basics of the Bessemer process are to force a blast of air through molten pig iron while it is in the crucible or furnace. The oxygen from the air raised the temperature of the molten pig iron, burning out most of the carbon and other impurities. The remaining metal, steel, was more durable and a higher quality than previously produced

William Kelly (left) and Henry Bessemer, working separately, both came up with a way to mass produce steel inexpensively and efficiently.

iron products. Another metallurgist, Robert F. Mushet (1811–1891), discovered that adding spiegeleisen, an iron alloy, would remove the oxygen from the steel, making it less brittle. The invention of the Bessemer converter increased steel production in Britain from about 50,000 tons (45,359 metric tons) during the late 1850s to over 1.3 million tons (1.2 million metric tons) twenty-five years later.

What changes have occurred in the steelmaking process since the Bessemer process was invented?

The first development after the Bessemer process was the open-hearth process. The open-hearth furnace was developed by William Siemens (1823–1883) and Frederick Siemens (1826–1904) in 1861. The Siemens process used heat regeneration by capturing heat that would have been lost in the atmosphere and using it to heat the furnace more efficiently. Three years later, in 1864, Pierre-Émile Martin (1824–1915) modified the open-hearth furnace by changing the location of the chambers that captured the heat. Martin also introduced the use of scrap steel into the steelmaking process. The process became known as the Siemens-Martin process. The Siemens-Martin process was the dominant method of steel manufacturing throughout much of the twentieth century.

Which two methods of steelmaking are currently in use?

Steel is manufactured using either the basic oxygen furnace or the electric arc furnace. The basic oxygen furnace is similar to the Bessemer since it receives materials from the

169

top and pours off the finished steel into ladles. It uses a combination of 70–80 percent raw materials, and the balance is scrap steel. Once the scrap steel is added to the liquid hot metal, oxygen with a purity of greater than 99 percent is "blown" into the molten metal at supersonic speeds. This oxidizes the carbon and burns out the impurities. The molten steel is poured off and sent for further processing. It takes about 40 minutes to produce 250 tons (228 metric tons) of steel.

The electric arc furnace only requires scrap steel. The scrap steel is loaded into the furnace. Two large electrodes are lowered into the furnace. The arcing between the carbon electrodes and scrap steel melts the steel. Electric arc furnaces can make 150–200 tons (136–181 metric tons) of steel in as little as 90 minutes. Electric arc furnaces are used to make high-quality carbon and alloy steels.

How many types of steel are produced?

Since the earliest days of mass producing steel, there are now thousands of types of steel with different physical and chemical properties. Many modern steels are lighter, yet just as strong, if not stronger, than steel that was produced in the last century.

Who invented stainless steel?

Metallurgists in several countries developed stainless steel, a group of iron-based alloys combined with chromium in order to be resistant to rusting and corrosion. Chromium was used in small amounts in 1872 to strengthen the steel of the Eads Bridge over the Mississippi River, but it wasn't until the 1900s that a truly rust-resistant alloy was developed. Metallurgists in several countries developed stainless steel between 1903 and 1912. An American, Elwood Haynes (1857–1925), developed several alloy steels and in 1911 produced stainless steel. Harry Brearley (1871–1948) of Great Britain receives most of the credit for its development. In 1913, he discovered that adding chromium to low carbon steel improved its resistance to corrosion. Frederick Becket (1875–1942), a Canadian American metallurgist, and German scientists Philip Monnartz and W. Borchers were among the early developers.

What is the difference between 18/0, 18/8, and 18/10 stainless steel flatware?

The numbers 18/0, 18/8, and 18/10 refer to the percentage of chromium and nickel in the stainless steel alloy used to make the flatware. The first number, 18, refers to the percentage of chromium while the second number, 0, 8, or 10, refers to the percentage of nickel. Chromium provides durability and has stain- and corrosion-resistant properties. Nickel has some corrosion-resistant properties and adds sheen and luster to the flatware. Flatware made with 18/10 stainless steel is considered the highest quality flatware because it is most durable and is rust-/stain-resistant.

What is continuous casting steel?

Continuous casting was introduced in the 1950s. Prior to the development of continuous casting, molten steel was poured from the furnace into ingots. The ingots were placed in soaking pits for several hours so the proper grain structure could form in the steel. The ingots would be rolled into billets or slabs that were then further shaped or rolled in special finishing mills. Continuous casting removed the need for ingots. Molten steel is poured directly from the furnace, either a basic oxygen furnace or an electric arc furnace, and transformed into billets or slabs. It is then ready for rolling in finishing mills.

What is COR-TEN® steel?

COR-TEN® steel, sometimes written without the hyphen as "Corten steel," is also known as weathering steel. It is a group of steel alloys that was developed to eliminate the need for painting and form a stable, rustlike appearance if exposed to the weather for several years. "Weathering" refers to the chemical composition of these steels, allowing them to exhibit increased resistance to atmospheric corrosion compared to other steels. This is because the steel forms a protective layer on its surface under the influence of the weather.

What are some of the uses of steel?

Steel is one of the most frequently used materials in the modern, technological world.

Industry	Uses
Appliances	Dishwashers, dryers, ranges, refrigerators, washing machines, and the motors to run the appliances are all made of steel
Automotive	Automobile body
Building and Construction	Steel framing for buildings, metal roofs
Containers and Packaging	Steel cans
Defense	Aircraft carriers, helicopters, nuclear submarines, Patriot and Stinger missiles, armor plate for tanks, field artillery pieces
Energy	High-voltage transmission towers, transmission wire, wind turbines
Highways and Bridges	Continuously Reinforced Concrete Pavement (CRCP), highway barriers, bridges
Utility poles	Steel poles are replacing utility poles because they are superior to wood. Steel poles are durable, strong, resistant to insects and rot, free of chemical treatments, and 100 percent recyclable

How much steel is contained in common household appliances?

It is estimated that more than 75 percent of the weight of an appliance is steel.

Appliance	Weight of Steel
Dishwasher	28 pounds (12.7 kilograms)
Microwave	29 pounds (13.1 kilograms)
Washing machine	95 pounds (43.1 kilograms)
Dryer (gas)	100 pounds (45.4 kilograms)
Dryer (electric)	107 pounds (48.5 kilograms)

Appliance	Weight of Steel
Range (electric)	107 pounds (48.5 kilograms)
Range (gas)	149 pounds (67.7 kilograms)
Refrigerator	153 pounds (69.4 kilograms)

Is steel recyclable?

Steel is one of the most easily recyclable materials in the world. Since steel is magnetic, it is relatively easy to recover it from other waste. Electric arc furnaces use only recycled, scrap steel in the production of steel. It is estimated that 25 percent of the weight of a household appliance is from recycled steel. Every industry that uses steel incorporates recycling steel and using recycled steel. There is no limit on the number of times a piece of steel can be melted and formed into a new piece of steel.

Which countries produce the greatest amount of steel?

The world's largest producer of steel in 2014 was China. China produced about 906,871,000 tons (822,700,000 metric tons) of crude steel in 2014. The following table shows the top producers of steel (in thousand tons) for the years 1980–2014:

Steel Production in Millions of Tons/Metric				
Country	1980	1990	2000	2010	2014
---	---	---	---	---	---
China	40.919/37.121	66.35/60.192	128.5/116.573	638.743/579.458	906.871/822.7
Japan	122.792/111.395	110.339/100.098	106.444/96.564	108.599/98.519	121.9/110.6
United States	111.835/101.455	89.726/81.398	101.824/92,373	80.495/73.024	97.195/88.174
Russia*	163.077/147.941	154.436/140.102	59.136/53.647	66.942/60.729	78.772/71.461

*Includes former USSR until 1991

NATURAL MATERIALS

What is fuller's earth?

It is a naturally occurring white or brown clay containing aluminum magnesium silicate. It is a highly adsorbent material. Fuller's earth acts as a catalyst and was named for a process known as fulling—a process used to clean grease from wool fibers before spinning. Currently, it is primarily used for lightening the color of oils and fats and a purifying agent during the refinement of petroleum. It is also a substitute for activated carbon in water-treatment filters. Since it is so adsorbent, it is used to help clean up oil spills.

What are some uses of obsidian?

Obsidian is a volcanic glass that usually forms in the upper parts of lava flows. Embryonic crystal growths, known as crystallites, make the glass an opaque, jet-black color. Red or brown obsidian could result if iron oxide dust is present. There are some well-known formations in existence, including the Obsidian Cliffs in Yellowstone Park and Mount Hekla in Iceland. Obsidian is very sharp and was used as arrowheads or spear points. It is still used today in the era of modern surgery as a scalpel or knife blade because obsidian blades can be sharper and thinner than steel blades. Obsidian is also used as an ornamental stone in jewelry.

Obsidian is a lustrous black mineral that forms from lava flows.

What is amber?

Amber is the fossil resin of trees. The two major deposits of amber are in the Dominican Republic and Baltic states. Amber came from a coniferous tree that is now extinct. Amber is usually yellow or orange in color, semitransparent or opaque with a glossy surface. Many times, amber contains insects that were trapped prior to hardening. It is used by both artisans and scientists.

How is petrified wood formed?

Petrified wood is formed when water containing dissolved minerals such as calcium carbonate and silicate infiltrates wood or other structures. The process takes thousands of years. The foreign material either replaces or encloses the organic matter and often retains all of the structural details of the original plant material. Botanists find these types of fossils to be very important since they allow for the study of the internal structure of extinct plants. After a time, wood seems to have turned to stone because the original form and structure are retained. The wood itself does not turn to stone.

Petrified wood is like the fossilized trunk or branch of a tree. Over time, minerals replace the plant cells, forming a colorful stone that mimics the wood.

173

What is the source of frankincense and myrrh?

Frankincense is an aromatic gum resin obtained by tapping the trunks of trees belonging to the genus *Boswellia*. The milky resin hardens when exposed to the air and forms irregular lumps—the form in which it is usually marketed. Also called olibanum, frankincense is used in pharmaceuticals, as a perfume, as a fixative, and in fumigants and incense. Myrrh comes from a tree of the genus *Commiphora*, a native of Arabia and Northeast Africa. It too is a resin obtained from the tree trunk and is used in pharmaceuticals, perfumes, and toothpaste.

What is ambergris?

Ambergris, a highly odorous, waxy substance found floating in tropical seas, is a secretion from the sperm whale (*Physeter catodon*). The whale secretes ambergris to protect its stomach from the sharp bone of the cuttlefish, a squidlike sea mollusk, which it ingests. Ambergris is used in perfumery as a fixative to extend the life of a perfume and as a flavoring for food and beverages. Natural ambergris has been replaced by synthetic materials.

MANUFACTURING

What are some major manufacturing industries?

Manufacturing is the process that transforms raw materials into products used by individuals and groups in society. Some important manufacturing industries are: glass and ceramics, paper and paperboard, chemicals, plastics, fibers and fabrics, metals, machinery, and electronics.

How has technology influenced manufacturing?

During pre-industrial times, manufacturing was limited by the techniques and tools known to members of a civilization. Technology has been the impetus for the growth of manufacturing as an industry that supplies our finished products. The invention of the steam engine and the availability of power have been credited with being the most influential discovery that developed the modern, technological society. As more tools and techniques were discovered and invented, coupled with an increased knowledge of biological, chemical, and physical processes, manufacturing developed ways for increased productivity. Mechanization, automation, and the introduction of the assembly line all influenced the development of modern manufacturing. During the twenty-first century, technology continues to provide methods to increase productivity with increased efficiency while focusing on the sustainability of raw materials.

Is glass a solid or a liquid?

Even at room temperature, glass appears to be a solid in the ordinary sense of the word. However, it actually is a fluid with an extremely high viscosity, which refers to the internal friction of fluids. Viscosity is a property of fluids by which the flow motion is gradually damped (slowed) and dissipated by heat. Viscosity is a familiar phenomenon in daily life. An opened bottle of wine can be poured: the wine flows easily under the influence of gravity. Maple syrup, on the other hand, cannot be poured so easily; under the action of gravity, it flows sluggishly. The syrup has a higher viscosity than the wine.

Glass is usually composed of mixed oxides based around the silicon dioxide unit. A very good electrical insulator, and generally inert to chemicals, commercial glass is manufactured by the fusing of sand (silica), limestone, and soda (sodium carbonate) at temperatures around 2,552°F to 2,732°F (1,400°C to 1,500°C). On cooling, the melt becomes very viscous, and at about 932°F (500°C, known as glass transition temperature), the melt "solidifies" to form soda glass. Small amounts of metal oxides are used to color glass, and its physical properties can be changed by the addition of substances like lead oxide (to increase softness, density, and refractive ability for cut glass and lead crystal) and borax (to significantly lower thermal expansion for cookware and laboratory equipment). Other materials can be used to form glasses if rapidly cooled from the liquid or gaseous phase to prevent an ordered crystalline structure from forming.

Glass objects might have been made as early as 2500 B.C.E. in Egypt and Mesopotamia, and glass blowing developed around 100 B.C.E. in Phoenicia.

What is crown glass?

In the early 1800s, window glass was called crown glass. It was made by blowing a bubble, then spinning it until flat. This left a sheet of glass with a bump, or crown, in the center. This blowing method of window-pane making required great skill and was very costly. Still, the finished crown glass produced a distortion through which everything looked curiously wavy, and the glass itself was also faulty and uneven. By the end of the nineteenth century, flat glass was mass produced and was a common material. The cylinder method replaced the old method, and it used compressed air to produce glass that could be slit lengthwise, reheated, and allowed to flatten on an iron table under its own weight. New furnaces and better polishing machines made the production of plate glass a real industry. Today, almost all flat glass is produced by a float-glass process, which reheats the newly formed ribbon of glass and allows it to cool without touching a solid surface. This produces inexpensive glass that is flat and free from distortion.

Who invented the float-glass process?

Manufacture of high-quality flat glass, needed for large areas and industrial uses, depends on the float-glass process, invented by Alastair Pilkington (1920–1995) in 1952. The float process departs from all other glass processes where the molten glass flows from the melting chamber into the float chamber, which is a molten tin pool approximately 160 feet (49 meters) long and 12 feet (3.5 meters) wide. During its passage over this molten tin, the hot glass assumes the perfect flatness of the tin surface and develops excellent thickness uniformity. The finished product is as flat and smooth as plate glass without having been ground and polished.

What are the advantages of tempered glass?

Tempered glass is glass that is heat-treated. Glass is first heated and then the surfaces are cooled rapidly. The edges cool first, leaving the center relatively hot compared to the surfaces. As the center cools, it forces the surfaces and edges into com-

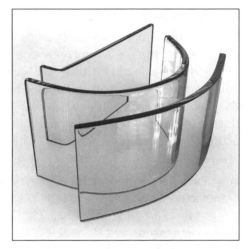

Tempered glass is made by heating and cooling regular glass, a process that makes it four times stronger.

pression. Tempered glass is approximately four times as strong as annealed glass. It is able to resist temperature differences of 200°F to 300°F (90°C to 150°C). Since it resists breakage, it is favored in many applications for its safety characteristics such as automobiles, doors, tub and shower enclosures, and skylights.

How is colored glass produced for stained glass windows?

The production of colored glass usually involves the addition of a metal to the glass. This is accomplished by adding powdered oxide, sulfide, or other compounds of that metal to the glass while it is still molten. The table below lists some of the coloring agents of glass that they produce.

Glass Coloring Agents

Metal	Color
Cadmium Sulfide	Yellow
Gold Chloride	Red
Cobalt Oxide	Blue-violet
Manganese Dioxide	Purple
Nickel Oxide	Violet
Sulfur	Yellow-amber
Chromic Oxide	Emerald green
Uranium Oxide	Fluorescent yellow, green
Iron Oxide	Greens and browns
Selenium Oxide	Reds
Antimony Oxides	White
Copper compounds	Blue, green, red

When were glass blocks invented?

Dating back to 1847, glass blocks were originally used as telegraph insulators. They were much smaller and thicker than structural glass blocks and were used mostly in the southeastern United States until eventually replaced with porcelain and other types of insulating materials. Glass building bricks were invented in Europe in the early 1900s as thin blocks of glass supported by a grid. Structural glass blocks have been manufactured in the United States since Pittsburgh-based Pittsburgh Corning began producing them in 1938. Blocks made at that time measured approximately 8 inches (20 centimeters) square by nearly 5 inches (13 centimeters) in depth and cast a greenish tint as light transmitted through it. Today's glass blocks can be a square foot in size, much more uniformly shaped, and available in many different sizes, textures, and colors.

How is bulletproof glass made?

Bulletproof glass is composed of two sheets of plate glass with a sheet of transparent resin in between, molded together under heat and pressure. When subjected to a severe blow, it will crack without shattering. It was invented by French chemist Édouard Benedictus (1878–1930) in 1903. Benedictus accidentally dropped a beaker that had contained cellulose nitrate, a plastic. The cellulose nitrate had dried in the beaker, creating an adhesive film coating on the inside of the beaker. The beaker cracked, but it retained its shape and did not shatter. Bulletproof glass is basically a multiple lamination of glass and plastic layers.

How were glass windows made to be more efficient?

Glass is durable material and allows a high percentage of sunlight to enter buildings, but it has very little resistance to heat flow. Glazing technology has changed greatly during the past two decades. There are now several types of advanced glazing systems available to help control heat loss or gain. The advanced glazings include double- and triple-pane windows with such coatings as low-emissivity (low-e), spectrally selective, heat-absorbing (tinted), reflective, or a combination of these; windows can also be filled with a gas (typically xenon, argon, or krypton) that helps in insulation.

Who invented thermopane glass?

Thermopane-insulated window glass, which could be produced commercially on a practical basis, was invented by Charles D. Haven in the United States. Haven filed for a U.S. patent on October 12, 1934. Patent #2,030,869 was published on February 18, 1936. Thermopane glass consists of two sheets of glass that are bonded together in such a manner that they enclose a captive air space in between. Often, this space is filled with an inert gas that increases the insulating quality of the window. Glass is also one of the best transparent materials because it allows the short wavelengths of solar radiation to pass through it but prohibits nearly all of the long waves of reflected radiation from passing back through it.

What is low-e glass?

Low-emissivity or low-e glass has a special coating to control heat transfer through insulated windows. Low-e coatings minimize the amount of ultraviolet (UV) and infrared (IR) light (or heat energy) that can pass through the glass without compromising the amount of visible light that is transmitted through the glass. The low-e coating reflects long-wave infrared energy or heat. When the interior heat energy tries to escape through windows during the winter, the low-e coating reflects the heat back to the inside, reducing the radiant heat loss through the glass. During the heat of the summer, the reverse happens and the heat is kept out of the house.

Low-e coatings may be passive coatings or solar control low-e coatings. Passive low-e coatings are manufactured using the pyrolytic process. The coating is applied to the glass while it is being produced on the float line. It fuses to the hot surface, creating a strong bond. Passive low-e coatings are preferred for cold climates because they allow some of the sun's short-wave infrared energy to pass through and help heat the building. Solar control low-e coatings are applied using the MSVD (Magnetron Sputtered Vacuum Deposition) process. The coating is applied to a sheet of finished glass in a vacuum chamber at room temperature. MSVD low-e coatings are preferred in warm climates.

When did heat-resistant glass become available for home use?

Heat-resistant glass became available for home use in 1915 in the United States. Commonly known as Pyrex®, it was developed by Eugene G. Sullivan (1872–1962) and William C. Taylor for Corning Glass Works in 1915. Sullivan was inducted into the National Inventors Hall of Fame in 2011 for the development of Pyrex®. Corning had developed heat-resistant glass for use with incandescent lighting, weatherproof lenses for railroad signal lanterns, and colorants for lens glass since the 1870s. Bessie Littleton asked her husband, Jesse T. Littleton, a physicist at Corning, to bring home some glass to use in place of a broken casserole dish. He brought home the sawed-off bottoms of some battery jars. She baked puddings and pies, and a new product was developed, Pyrex®. Pyrex® is a nonexpanding glass made from borax, alumina, sodium, and soda and fired at over 2,500°F (1,371°C). The reduced sodium content of Pyrex® lessened the chemical interaction with food while the low thermal quality reduced the danger of breakage. By 1919, over 4.5 million pieces of Pyrex® ovenproof bakeware were in use. Other uses of Pyrex® include stronger lenses for reflector telescopes, blanks for mirrors, headlight lenses for automobiles, and in pipelines that transport corrosive substances.

Pyrex® is a form of heat-resistant glass that is used primarily for dishes for home cooking.

What are the major distinguishing characteristics of ceramics?

Ceramics are crystalline compounds of inorganic, nonmetallic elements. They are chemically stable compounds. Ceramics are very hard and among the most rigid of all materials, with an almost total absence of ductility. They have the highest known melting points of materials, with some being as high as 7,000°F (3,870°C) and many that melt at temperatures of 3,500°F (1,927°C). They are good chemical and electrical insulators. Ceramics are oxidation resistant, thus making them the preferred material for use in harsh, corrosive environments.

What are glazes?

Glazes are glassy coatings that help make earthenware containers watertight. Early forms of colored, decorative glazes date back to around 3500 B.C.E.

What is the difference between bone china, fine china, and porcelain?

Bone china is actually made from bone (cow). The bone is finely ground into ash and then mixed with feldspar, ball clay, quartz, and kaolin (a type of clay). The quality of the china is highly dependent on the percent of bone in the mixture; the highest quality bone china contains about 40–45 percent. A similar manufacturing process is used in making fine china but without the bone content. Porcelain is created in the same way, but it is fired at a higher temperature, resulting in much harder china. Porcelain has been found as early as 620 C.E.; these earliest porcelains used kaolin and pegmatite (a type of granite).

Who invented CorningWare®?

Donald Stookey (1915–2014) invented CorningWare®. Originally called Pyroceram, a glass-ceramic material, Stookey discovered the new product by accident. He had placed a plate of Fotoform glass with added nucleating agents in an oven to heat at what he thought was 1,100°F (593°C). The oven malfunctioned and the temperature rose to 1,600°F (871°C). Stookey expected to find a molten mess in the oven, but instead, he found an opaque, milky-white plate. As he removed the plate from the oven, it slipped out of the tongs. Instead of shattering and breaking, it bounced. The age of glass ceramics began. Pyroceram, later called CorningWare®, was first marketed in 1958.

Stookey was inducted into the National Inventors Hall of Fame in 2010 for the invention of glass ceramics. In 1994 Corning was honored with the National Medal of Technology for its innovations in new materials.

What type of high-technology material is used in space vehicles for its heat-resistant properties?

Carbon-carbon composites (CCCs) are baseline materials in high-temperature applications such as re-entry vehicle nose tips, rocket motor nozzles, and the Space Shuttle Orbiter leading edges. Carbon-carbon is a nonmetallic composite material made of carbon fibers in a carbon matrix having operational capabilities ranging from cryogenics to 5,000°F (2,760°C). Numerous fiber types and weave patterns having a wide range of physical properties are available, providing flexibility in the design of carbon-carbon components. Carbon-carbon has a unique combination of light weight, high strength, and temperature resistance.

Who made the first successful synthetic gemstone?

In 1902, Auguste Victor Louis Verneuil (1856–1913) synthesized the first man-made gemstone—a ruby. Verneuil perfected a "flame-fusion" method of producing crystals of ruby and other corundums within a short time period. He displayed the first gem quality, synthetic ruby, in 1902.

Who invented plastic?

In the mid-1850s, Alexander Parkes (1813–1890) experimented with nitrocellulose (also known as guncotton). When he dissolved nitrocellulose in a solvent, such as alcohol or naphtha, it yielded a new material. The material was placed on a heated rolling machine and shaped by dies or pressure. He called the new material "Parkesine." Parkesine was first introduced to the public at the 1862 Great International Exhibition in London. He teamed up with a manufacturer to produce it, but there was no demand for it, and the firm went bankrupt.

What was the first synthetic plastic?

The first synthetic plastic was developed in the 1860s and 1870s. An American, John Wesley Hyatt (1837–1920), experimented with nitrocellulose and camphor with the goal of producing artificial ivory for billiard balls. The collodion material to coat billiard balls became available in 1868 and was patented in 1869. After making improvements to the formula, Hyatt and his

Billiard balls were once made of ivory but are now made of synthetic plastic, much to the relief of elephants everywhere.

brother Isaiah received U.S. Patent #105,338 on July 12, 1870, for the process. In 1872, Isaiah Wyatt coined the term "celluloid" for the product that had become successful for the manufacture of billiard balls and many other novelty and fancy goods, including buttons, letter openers, boxes, hatpins, and combs. The material also became the medium for cinematography: celluloid strips coated with a light-sensitive "film" were ideal for shooting and showing moving pictures. Its popularity began to decline in the middle of the twentieth century when other plastics based on synthetic polymers were introduced. Hyatt was inducted into the National Inventors Hall of Fame in 2006 for his invention of celluloid. Today, ping-pong balls are almost the only product still made with celluloid.

What was Leo Hendrik Baekeland's contribution to the development of plastics?

Celluloid was the only plastic material until Belgian-born scientist Leo Hendrik Baekeland (1863–1944) succeeded in producing a synthetic shellac from formaldehyde and phenol. Called "bakelite," it was the first of the thermosetting plastics. These plastics were synthetic materials that, having once been subjected to heat and pressure, became extremely hard and resistant to high temperatures. They continued to maintain their shape even when heated or subjected to various solvents. Baekeland's first patent in the field of plastics was granted in 1906. He announced his invention to the public on February 8, 1909, in a lecture before the New York section of the American Chemical Society. In 1978, he was inducted into the National Inventors Hall of Fame for the invention of bakelite.

What are the two major types of plastics?

The two major types of plastics are: 1) thermoset or thermosetting plastics and 2) thermoplastics. Once cooled and hardened, thermosetting plastics retain their shapes and cannot return to their original form. They are hard and durable and can be used for automobile parts and aircraft parts. Examples of thermosetting plastics are polyurethanes, polyesters, epoxy resins, and phenolic resins. Thermoplastics are less rigid than thermosetting plastics and can soften upon heating and return to their original form. They are easily molded and formed into films, fibers, and packaging. Examples include polyethylene (PE), polypropylene (PP), and polyvinyl chloride (PVC).

What are some chemical properties of plastics?

Plastics are any of a wide range of synthetic or semi-synthetic organic solids that can be shaped or bent without breaking. Plastics are usually organic polymers of high molecular weight, but they often contain other substances. They are usually synthetic, most commonly derived from petrochemicals, but many are partially natural in origin.

Who invented Teflon®?

In 1938, American engineer Roy J. Plunkett (1910–1994) at DuPont de Nemours discovered polytetrafluoroethylene (PTFE), the polymer of tetrafluoroethylene, by accident. This fluorocarbon is marketed under the name of Fluon in Great Britain and Teflon® in the United States. Plunkett filed for a patent in 1939 and was granted U.S. Patent #2,230,654 in 1941. He was inducted into the National Inventors Hall of Fame in 1985 for the invention of Teflon®. Teflon® was first exploited commercially in 1954. It is resistant to all acids and has exceptional stability and excellent electrical insulating properties. It is used in making piping for corrosive materials, in insulating devices for radio transmitters, in pump gaskets, and in computer microchips. In addition, its nonstick properties make PTFE an ideal material for surface coatings. In 1956, French engineer Marc Grégoire discovered

Styrofoam™ makes for an excellent insulator and has many uses in packaging material, coffee cups, coolers, and more.

a process whereby he could fix a thin layer of Teflon® on an aluminum surface. He then patented the process of applying it to cookware, and the no-stick frying pan was created.

What is Styrofoam™ and what are its uses?

Styrofoam™ is a type of extruded polystyrene foam currently used for thermal insulation and craft applications. It is a trademarked brand of material that is owned and produced by the Dow Chemical Company. In the United States and Canada, the word Styrofoam™ is frequently and incorrectly used to describe expanded (not extruded) polystyrene beads. This type of expanded polystyrene beads is used to produce disposable coffee cups, coolers, and packaging material, which is usually white in color. Styrofoam™ is a good insulator because the material contains millions of trapped gas bubbles that hinder heat conduction. Styrofoam™ materials produced from expanded beads are also excellent insulators. The molecules are so large that they have little movement and hinder heat transfer.

What are composite materials?

Composite materials, or simply composites, consist of two parts: the reinforcing phase and the binder or matrix. Composites may be natural substances, such as wood and bone, or man-made substances. A composite product is different from each of its components and is often superior to each individual component. The binder or matrix of a composite is the material that supports the reinforcing phase. The reinforcing phase is usually in the form of particles, fibers, or flat sheets. Reinforced concrete is an example of a composite material. The steel rods embedded in the concrete (the matrix) are the reinforcing phase, adding strength and flexibility to the concrete. High-performance composites are composites that perform better than traditional structural materials, such as steel. Most high-performance composites have fibers in the reinforcing phase and a polymer matrix. The fibers may be glass, boron, silicon carbide, aluminum oxide, or a type of polymer. The fibers are often interwoven to form bundles. The purpose of the matrix, usually a polymer, in a high-performance composite is to hold the fibers together and protect them.

Who developed fiberglass?

Coarse glass fibers were used for decoration by the ancient Egyptians. Other developments were made in Roman times. Parisian craftsman Ignace Dubus-Bonnel was granted a patent for the spinning and weaving of drawn glass strands in 1836. In 1893, the Libbey Glass Company exhibited lampshades at the World's Columbian Exposition in Chicago that were made of coarse glass thread woven together with silk. However, this was not a true woven glass. Between 1931 and 1939, the Owens Illinois Glass Company and Corning Glass Works developed practical methods of making fiberglass commercially.

Once the technical problem of drawing out the glass threads to a fraction of their original thinness was solved—basically an endless strand of continuous glass filament as thin as 1/5,000 of an inch (0.005 millimeters)—the industry began to produce glass fiber for thermal insulation and air filters, among other uses. When glass fibers were combined with plastics during World War II, a new material was formed. Glass fibers did for plastics what steel did for concrete—gave strength and flexibility. Glass-fiber-reinforced plastics (GFRP) became very important in modern engineering. Fiberglass combined with epoxy resins and thermosetting polyesters are now used extensively in boat and ship construction, sporting goods, automobile bodies, and circuit boards in electronics.

Why is sulfuric acid important?

Sometimes called "oil of vitriol," or vitriolic acid, sulfuric acid (H_2SO_4) has become one of the most important of all chemicals. It was little used until it became essential for the manufacture of soda in the eighteenth century. It is prepared industrially by the reaction of water with sulfur trioxide, which in turn is made by chemical combination of sulfur dioxide and oxygen by one of two processes (the contact process or the chamber process). Many manufactured articles in common use depend in some way on sulfuric acid for their production. Ninety percent of the sulfuric acid manufactured in the United States is used in the production of fertilizers and other inorganic chemicals. It is used extensively in petroleum refining and in the production of fertilizers. It is also used in the production of chemicals, automobile batteries, explosives, pigments, iron and other metals, and paper pulp.

What is aqua regia?

Aqua regia, also known as nitrohydrochloric acid, is a mixture of one part concentrated nitric acid and three parts concentrated hydrochloric acid. The chemical reaction between the acids makes it possible to dissolve all metals except silver. The reaction of metals with nitrohydrochloric acid typically involves oxidation of the metals to a metallic ion and the reduction of the nitric acid to nitric oxide. The term comes from Latin and means "royal water." It was named by the alchemists for its ability to dissolve gold and platinum, which were called the "noble metals."

Who developed the process for making ammonia?

Known since ancient times, ammonia (NH_3) has been commercially important for more than one hundred years. The first breakthrough in the large-scale synthesis of ammonia resulted from the work of Fritz Haber (1863–1934). In 1913, Haber found that ammonia could be produced by combining nitrogen and hydrogen ($N_2 + 3H_2 \rightleftarrows 2NH_3$) with a catalyst (iron oxide with small quantities of cerium and chromium) at 131°F (55°C) under a pressure of about 200 atmospheres. The process was adapted for industrial-quality production by Carl Bosch (1874–1940). Thereafter, many improved ammonia-synthesis systems, based on the Haber-Bosch process, were commercialized using various operating conditions and synthesis loop designs. Haber received the Nobel Prize in Chemistry in 1918 for the synthesis of ammonia, and Bosch received the Nobel Prize in Chemistry in 1931 for the development of chemical high-pressure methods. They were both inducted into the National Inventors Hall of Fame in 2006 for their work toward the production of ammonia.

Ammonia is one of the top five inorganic chemicals produced in the United States. It is used in refrigerants, detergents, and other cleaning preparations, explosives, fabrics, and fertilizers. Most ammonia production in the United States is used for fertilizers. It has been shown to produce cancer of the skin in humans in doses of 1,000 milligrams per kilogram (2.2 pounds) of body weight.

What is the primary use of hydrogen peroxide?

Hydrogen peroxide, a syrupy liquid compound, is used as a strong bleaching, oxidizing, and disinfecting agent. It is usually made either in anthrahydroquinone autoxidation processes or electrolytically. The primary use of hydrogen peroxide is in bleaching wood pulp. A more familiar use is as a 3 percent solution as an antiseptic and germicide. Undiluted, it can cause burns to human skin and mucous membranes, is a fire and explosion risk, and can be highly toxic.

How is dry ice made?

Dry ice is a solid form of carbon dioxide (CO_2) used primarily to refrigerate perishables that are being transported from one location to another. The carbon dioxide, which at normal temperatures is a gas, is stored and shipped as a liquid in tanks that are pressurized at 1,073 pounds per square inch. To make dry ice, the carbon dioxide liquid is withdrawn from the tank and allowed to evaporate at a normal pressure in a porous bag. This rapid evaporation con-

Most useful as a refrigerant, dry ice (frozen carbon dioxide) can also be a lot of fun for making fog effects at Halloween or for theatrical stage effects.

When was dry ice first made commercially?

Dry ice was first made commercially in 1925 by the Prest-Air Devices Company of Long Island City, New York, through the efforts of Thomas Benton Slate (1880–1980). It was used by Schrafft's of New York in July 1925 to keep ice cream from melting. The first large sale of dry ice was made later in that year to Breyer Ice Cream Company of New York.

sumes so much heat that part of the liquid CO_2 freezes to a temperature of $-109°F$ ($-78°C$). The frozen liquid is then compressed by machines into blocks of "dry ice," which will melt into a gas again when set out at room temperature. Although used mostly as a refrigerant or coolant, other uses include medical procedures such as freezing warts, blast cleaning, freeze-branding animals, and creating special effects for live performances and films.

What is creosote?

Creosote is a yellowish, poisonous, oily liquid obtained from the distillation of coal or wood tar (coal tar constitutes the major part of the liquid condensate obtained from the "dry" distillation or carbonization of coal to coke). Crude creosote oil, also called dead oil or pitchoil, is obtained by distilling coal tar and is used as a wood preservative. Railroad ties, poles, fence posts, marine pilings, and lumber for outdoor use are impregnated with creosote in large, cylindrical vessels. This treatment can greatly extend the useful life of wood that is exposed to the weather. Creosote that is distilled from wood tar is used in pharmaceuticals. Other uses of creosote include disinfectants and solvents. In 1986, the U.S. Environmental Protection Agency (EPA) began restricting the use of creosote as a wood preservative because of its poisonous and carcinogenic nature.

What chemical is used in the manufacture of pressure-treated lumber?

The most common chemical used to treat lumber is chromated copper arsenate (CCA). Copper and arsenic are both toxic to many types of organisms, including bacteria, fungi, and insects that attack wood. The chromium helps to bond the copper to the wood to prevent leaching. CCA binds to wood fibers very well and allows the wood to last decades even when it is in contact with the ground.

What is carbon black?

Carbon black is finely divided carbon produced by incomplete combustion of methane or other hydrocarbon gases (by letting the flame impinge on a cool surface). This forms a very fine pigment containing up to 95 percent carbon, which gives a very intense black color that is widely used in paints, inks, and protective coatings and as a colorant for paper and plastics. It is also used in large amounts by the tire industry in the production of vulcanized rubber.

Who invented dynamite?

Dynamite was not an accidental discovery but the result of a methodical search by Swedish technologist Alfred Nobel (1833–1896). Nitroglycerine had been discovered in 1849 by Italian organic chemist Ascanio Sobriero (1812–1888), but it was so sensitive and difficult to control that it was useless. Nobel sought to turn nitroglycerine into a manageable solid by absorbing it into a porous substance. From 1866 to 1867, he tried an unusual mineral, kieselguhr, and created a doughlike explosive that was controllable. He also invented a detonating cap incorporating mercury fulminate with which nitroglycerine could be detonated at will. The use of dynamite for blasting has played a significant role in the development of mining industries and road building.

How are colored fireworks made?

Fireworks existed in ancient China in the ninth century where saltpeter (potassium nitrate), sulfur, and charcoal were mixed to produce the dazzling effects. Magnesium burns with a brilliant white light and is widely used in making flares and fireworks. Various other colors can be produced by adding certain substances to the flame. Strontium compounds color the flame scarlet, barium compounds produce yellowish-green, copper produces a blue-green, lithium creates purple, and sodium results in yellow. Iron and aluminum granules give gold and white sparks, respectively.

Who discovered TNT?

TNT is the abbreviation for 2,4,6-trinitrotoluene, a powerful, highly explosive compound widely used in conventional bombs. Discovered by Joseph Wilbrand (1811–1894) in 1863, it is made by treating toluene with nitric acid and sulfuric acid. This yellow crystalline solid with a low melting point has low shock sensitivity and even burns without exploding. This makes it safe to handle and cast, but once detonated, it explodes violently.

What is rosin?

Rosin is the resin produced after the distillation of turpentine, obtained from several varieties of pine trees, especially the longleaf pine (*Pinus palustris*) and the slash pine (*Pinus caribaea*). Rosin has many industrial uses, including the preparation of inks, adhesives, paints, sealants, and chemicals. Rosin is also used by athletes and musicians to make smooth surfaces less slippery.

Why are essential oils called "essential"?

Called essential oils because of their ease of solubility in alcohol to form essences, essential oils are used in flavorings, perfumes, disinfectants, medicine, and other products. They are naturally occurring, volatile aromatic oils found in uncombined forms within various parts of plants, such as leaves and pods. These oils contain as one of their main ingredients a substance belonging to the terpene group. Examples of essential oils include bergamot, eucalyptus, ginger, pine, spearmint, and wintergreen oils. Extracted by distillation or enfleurage (extraction using fat) and mechanical pressing, these oils can now be made synthetically.

What does one cord of wood yield when processed?

A cord of wood may produce:

- 12 dining room table sets (seating 8 each)
- 250 copies of the Sunday *New York Times*
- 2,700 copies of an average (36-page) daily newspaper
- 1,000 pounds (454 kilograms) to 2,000 pounds (907 kilograms) of paper, depending on the quality and grade of the paper
- 61,370 standard (#10) envelopes

Which woods are used for telephone poles?

The principal woods used for telephone poles are southern pine, Douglas fir, western red cedar, and lodgepole pine. Ponderosa pine, red pine, jack pine, northern white cedar, other cedars, and western larch are also used.

Which woods are used for railroad ties?

Many species of wood are used for ties. The more common are oaks, gums, Douglas fir, mixed hardwoods, hemlock, southern pine, sycamore, and mixed softwoods.

What are the raw materials used to make paper?

Paper is usually made from wood pulp or plant fiber and is chiefly used for written communication. The earliest paper was papyrus, made from reeds by the Egyptians. Paper, whether produced in the modern factory or made by hand, is made up of connected fibers. The fibers can come from a variety of sources, including cloth rags, cellulose fibers from plants, and, most notably, trees. The use of cloth in papermaking has always produced high-quality paper. A large proportion of cotton and linen fibers are responsible for papers that are used for invitations and drawing. Almost half of the fiber used for paper today comes from wood. The

Modern mills like this one are increasingly making paper from recycled waste fibers.

187

remaining material comes from wood fibers from sawmills, recycled newspaper, some vegetable matter, and recycled cloth. Coniferous or "softwood" trees, such as spruce and pine, are sometimes preferred when making paper because the cellulose fibers in the pulp of these trees are longer, resulting in a stronger paper. Deciduous or "hardwood" trees such as poplar and elm are also regularly used in making paper.

When was the first paper mill established in the United States?

The first paper mill in the United States was established in 1690 by William Rittenhouse (1644–1708) near Germantown, Pennsylvania. Rittenhouse left the Netherlands in 1688, where he had been an apprentice papermaker, and settled in Philadelphia, near the print shop of William Bradford (1663–1752). The Rittenhouse mill remained the only paper mill in the United States until 1710 when William Dewees (1677–1745), brother-in-law to William Rittenhouse's son Nicholas (1666–1734), established his own mill.

How is parchment paper made?

Most parchment paper is now vegetable parchment. It is made from a base paper of cotton rags or alpha cellulose, known as waterleaf, which contains no sizing or filling materials. The waterleaf is treated with sulfuric acid, converting a part of the cellulose into a gelatin-like amyloid. When the sulfuric acid is washed off, the amyloid film hardens on the paper. The strength of the paper is increased and will not disintegrate even when fully wet. Parchment paper can withstand heat, and items will not stick to it.

How is sandpaper made?

Sandpaper is a coated abrasive that consists of a flexible-type backing (paper) upon which a film of adhesive holds and supports a coating of abrasive grains. Various types of resins and hide glues are used as adhesives. The first record of a coated abrasive is in thirteenth-century China when crushed seashells were bound to parchment using natural gums. The first known article on coated abrasives was published in 1808 and described how calcined, ground pumice was mixed with varnish and spread on paper with a brush. Most abrasive papers are now made with aluminum oxide or silicon carbide, although the term "sandpapering" is still used. Quartz grains are also used for wood polishing. The paper used is heavy, tough, and flexible, and the grains are bonded with a strong glue.

Why is acid-free paper important?

Acid-free paper is important for the preservation of printed materials on paper. Acidic papers deteriorate quickly. Papers treated with an alkaline reserve, most frequently chalk, neutralize the acids, extending the life span of the paper.

How are paper diapers able to absorb moisture?

In 1948, Johnson & Johnson introduced the first mass-marketed disposable diaper in the United States. More recent technical developments have added sodium polyacrylate, a water-absorbing chemical, in disposable diapers. Sodium polyacrylate forms a gel when

---

When were paper towels first used?

In 1907, Scott Paper Company introduced the Sani-Towels paper towel, the first paper towel in the United States. Later, in 1931, Scott Paper Company introduced the first paper towel for the kitchen. These perforated rolls of towels were 13 inches (33 centimeters) wide and 18 inches (45.7 centimeters) long.

---

mixed with water. This gel has a structure in which water is held the same way that gelatin in jello holds water. Experiments have shown that sodium polyacrylate can absorb as much as several hundred times its weight in tap water.

What gives corrugated cardboard its strength?

Corrugated cardboard (also called pleated) is produced by sticking two paper boards on the sides of a wavy paper called a flute. A flute is like a column in vertical uses and an arch in horizontal uses. This arrangement is what gives the corrugated cardboard its strength. Along with its strength, other features, including the ease of folding and cutting, are what make this product an ideal packaging material. Corrugated paper was first developed as a liner for tall hats and patented in England in 1856. In 1871, Albert Jones of New York City was issued a patent for single-sided corrugated cardboard. This was the first use of this material in shipping, and it was used for wrapping bottles and glass lantern chimneys. In 1874, Oliver Long improved upon the single sided corrugated cardboard by adding a liner sheet on both sides of the wavy interior paper. This is the product as we know it today.

What are natural fibers?

Natural fibers come from plants or animals. Examples of fabrics of animal origin are wool and silk. Wool is made from the fibers of animal coats, including sheep, goats, rabbits, and alpacas. Cotton, linen, hemp, ramie, and jute are all fabrics that have fibers of plant origin.

Fiber	Origin
Wool	Sheep
Mohair	Goat
Angora	Rabbit
Camel hair	Camel
Cashmere	Kashmir goat
Alpaca fleece	Alpaca
Silk	Silkworm cocoons
Cotton	Cotton plant seed pods
Linen	Flax
Hemp	*Cannabis Sativa* stem
Ramie	*Boehmeria nivea* (Chinese grass)
Jute	Glossy fiber

How is silk made?

Silk fiber is a continuous protein filament produced by a silkworm to form its cocoon. The principal species used in commercial silkmaking is the mulberry silkworm (the larva of the silkworm *Bombyx mori*), belonging to the order Lepidoptera. The raw silk fiber has three elements—two filaments excreted from both of the silkworm's glands and a soluble silk gum called sericin, which cements the filaments together. It is from these filaments that the caterpillar constructs a cocoon around itself.

The process of silkmaking starts with raising silkworms on diets of mulberry leaves for five weeks until they spin their cocoons. The cocoons are then treated with heat to kill the silkworms inside (otherwise, when the moths emerged, they would break the long silk filaments). After the cocoons are soaked in hot water, the

Thousands of silkworm cocoons await processing in a Chinese textile factory.

filaments of five to ten cocoons are unwound in the reeling process and twisted into a single thicker filament; still too fine for weaving, these twisted filaments are twisted again into a thread that can be woven.

What is cashmere?

Kashmir goats, which live high in the plateaus of the area from northern China to Mongolia, are covered in a coarse outer hair that helps protect them from the cold, harsh weather. As insulation, these goats have a softer, finer layer of hair or down under the coarse outer hair. This fine hair is shed annually and processed to make cashmere. Each goat produces enough cashmere to make one sweater every four years. The quality of cashmere depends on the thickness of the hair fibers. Cashmere is categorized into three grades: A, B, and C, with A being the best grade.

What are synthetic fibers?

Synthetic fibers are totally made by chemical means or may be fibers of regenerated cellulose. According to U.S. law, the fibers must be labeled in accordance with generic groups as follows:

Synthetic Fiber	Derived From
Acetate	Cellulose acetate
Acrylic	Acrylic resins

Synthetic Fiber	Derived From
Metallic	Any type of fabric made with metallic yarns; made by twisting thin metal foil around cotton, silk, linen, or rayon yarns
Modacrylic	Acrylic resins
Nylon	Synthetic polyamides extracted from coal and petroleum
Rayon	Trees, cotton, and woody plants
Saran	Vinylidene chloride
Spandex	Polyurethane
Triacetate	Regenerated cellulose
Vinyl	Polyvinyl chloride

Who developed polyester?

Wallace H. Carothers (1896–1937) was working for DuPont when he discovered that alcohols and carboxyl acids could be successfully combined to form fibers. This was the beginning of polyester, but it was put on the back burner once Carothers discovered nylon. A group of British scientists, J. R. Whinfield (1901–1966), J. T. Dickson, W. K. Birtwhistle, and C. G. Ritchie, took up Carothers's work in 1939. In 1941 they created the first polyester fiber caller Terylene. In 1946, DuPont bought all legal rights from the British scientists and developed another polyester fiber that they named Dacron®. Polyester was first introduced to the American public in 1951. Carothers was inducted into the National Inventors Hall of Fame in 1984 for his work that led to the invention of nylon, neoprene, and other synthetic fibers.

How are waterproof and water-repellent fabrics manufactured?

Both waterproof and water-repellent fabrics are treated with substances to make them impervious to water. To permanently waterproof a fabric, it is coated with plasticized synthetic resins, and then it is vulcaned or baked. Fabrics treated in this manner become thick and rubberized and may therefore crack. In making a water-repellent fabric, the material is soaked in synthetic resins, metallic compounds, oils, or waxes that allow the fabric to retain its natural characteristics.

What is Kevlar®?

The registered trademark Kevlar® refers to synthetic fiber called liquid crystalline polymers. Discovered by Stephanie Kwolek (1923–2014), Kevlar® is a thin, very strong fiber. It is best known for its use in bulletproof garments.

When were microfibers invented?

Microfibers are very fine fibers with a diameter of less than one denier. "Denier" is the term used to describe the diameter or fineness of a fiber. It is the weight in grams of a 9,000-meter (9,842 yards) length of fiber. Many microfibers are 0.5 to 0.6 denier. As a comparison, microfibers are 100 times finer than human hair and half the diameter of a fine silk fiber. The first microfiber fabric was Ultrasuede™, created by Miyoshi Okamoto in 1970 at Toray Industries in Japan. His colleague, Toyohiko Hikota, perfected the process to create a soft, supple, stain-resistant fabric.

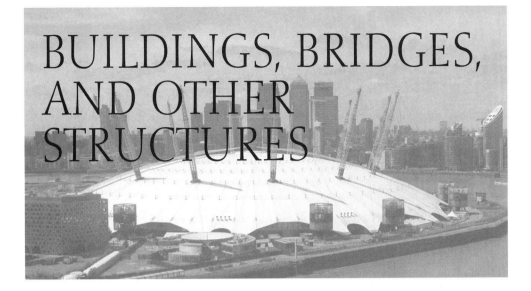

BUILDINGS, BRIDGES, AND OTHER STRUCTURES

BUILDINGS AND BUILDING PARTS

What is "green building"?

Green building, also called sustainable building, is the practice of creating structures and using processes that are responsible environmentally and resource-efficient throughout a building's life cycle from siting to design, construction, operation, maintenance, renovation, and deconstruction. Green buildings are designed to reduce the impact of the built environment on human health and the natural environment by using resources, especially energy and water, efficiently; protecting occupant health; improving indoor environmental quality; and reducing waste, pollution, and environmental degradation. Green building first became a goal of the building industry during the 1990s.

Are there standards for green buildings?

The growth and development of green, sustainable building during the 1990s provided the impetus for the establishment of green building standards and certification programs. The National Association of Home Builders developed a Green Certification program for residential buildings. This program is based on the International Code Council's National Green Building Standard, ICC 700-2012, an American National Standards Institute (ANSI) standard approved in 2009, revised in 2012 and it is being revised in 2015 as of this writing. The certification levels are Bronze, Silver, Gold, and Emerald and also evaluate lot and site development, energy efficiency, resource efficiency, water efficiency, indoor environmental quality, and homeowner education.

The Green Building Initiative (GBI) is another organization dedicated to promoting practical green building approaches for residential and commercial construction. Its assessment and rating program, called Green Globes, was first established in Canada

and brought to the United States in 2004. The rating system evaluates how buildings use
less energy, conserve natural resources, and emit fewer pollutants. Green Globes uses
a system of one to four globes to rate the buildings.

GBI became an ANSI-accredited Standards Developing Organization in 2005. The
ANSI/GBI standard, ANSI/GBI 01-2010, Green Building Assessment Protocol for Commercial Buildings was approved in 2010. A new revision was proposed in 2015 and
should be released in 2016.

What is LEED certification?

LEED (Leadership in Energy and Environmental Design) certification is a voluntary
certification program administered by the U.S. Green Building Council for the construction and operation of high-performance green buildings. The first version of LEED
was launched in 1998 as a pilot program. Both commercial and residential buildings
are eligible to apply for LEED certification. LEED-certified buildings are ranked as certified silver, gold, or platinum (the highest ranking).

Which states are the top-ranked for the number of LEED building projects?

Globally, 4,502 projects received LEED certification in 2014, representing 675.7 million square feet. The top ten states, ranked by number of square feet per person, are:

U.S. States Ranked by Number of LEED Certified Building Projects

	# of projects certified	# of sq. ft. certified (million sq. ft.)	# of sq. ft. certified per person*
Illinois	174	42.46	3.31
Colorado	102	15.82	3.15
Maryland	132	15.58	2.70
Virginia	150	18.62	2.33
Massachusetts	99	14.67	2.20
Hawaii	30	2.66	1.95
California	517	69.76	1.87
Georgia	87	17.75	1.83
Minnesota	39	9.52	1.79

	# of projects certified	# of sq. ft. certified (million sq. ft.)	# of sq. ft. certified per person*
Arizona (tied with New York)	82	11.15	1.74
New York (tied with Arizona)	250	33.69	1.74

*Based on 2010 U.S. Census.

What are green roofs?

Green roofs are roofs constructed with a vegetative layer grown on the rooftop. One benefit of a green roof is reduced energy use since green roofs absorb heat and act as insulators for buildings, reducing the energy needed for heating and cooling. They can also slow stormwater runoff. The overall aesthetic value of a green roof can improve the quality of life.

Green walls are vegetated wall surfaces. The three major types of green walls are green façades, living walls, and retaining living walls. Green façades usually have vines and climbing plants that grow on supporting structures. Living walls have a greater diversity of plant species than green façades. They are either freestanding or attached to

Growing plants on rooftops (green roofs) has a number of benefits: it helps cool the building, reduces energy use, and can provide a restful green space for residents or workers.

195

a structural wall. Green retaining living walls have the additional objective of stabilizing a slope to protect against erosion.

How does a chimney differ from a flue?

A chimney is a brick-and-masonry construction that contains one or more flues. A flue is a passage within a chimney through which smoke, fumes, and gases ascend. A flue is lined with clay or steel to contain the combustion wastes. By channeling the warm, rising gases, a flue creates a draft that pulls the air over the fire and up the flue. Each heat source needs its own flue, but one chimney can house several flues.

How tall is the world's tallest chimney?

The number two stack of the Ekibastuz, Kazakhstan, power plant is 1,378 feet (420 meters) tall.

Which part of a door is a doorjamb?

The doorjamb is not part of the door but is the surrounding case into and out of which the door opens and closes. It consists of two upright pieces, called side jambs, and a horizontal header.

What is crown molding?

Crown molding is a wood, metal, or plaster finishing strip placed on the wall where it intersects the ceiling. If the molding has a concave face, it is called a cove molding. In inside corners, it must be cope-jointed to ensure a tight joint.

What is an R-value?

An R-value, or resistance to the flow of heat, is a special measurement of insulation that represents the difficulty with which heat flows through an insulating material. The higher the R-value, the greater the insulating value of the material. A wall's component R-values can be added up to get its total R-value:

Wall component	R-value
Inside Air Film	0.7
1/2" Gypsum Wallboard	0.5
R-13 Insulation	13.0
1/2" Wood Fiber Sheathing	1.3
Wood Siding	0.8
Outside Air Film	0.2
Total R-value	16.5

Building component	R-value
Standard attic ceiling	19
Standard 4-inch (10-centimeter) thick,	11

Building component	R-value
Typical single-pane glass window	1
Double-glazed window	2
Superwindows*	4

*A superwindow has the inner surface of a pane coated with an infrared radiation reflector material such as tin oxide and argon gas filling in the space between the panes of a double-glazed window.

When was the International Code Council established?

The International Code Council (ICC) was established in 1994. The founders of the ICC are Building Officials and Code Administrators International, Inc. (BOCA), International Conference of Building Officials (ICBO), and Southern Building Code Congress International, Inc. (SBCCI). Historically, each of these organizations had established separate model construction codes that had been used for decades throughout the United States. At the end of the twentieth century, it was determined that a single set of comprehensive and coordinated national model construction codes without regional limitations would be best for the country. The goal of the codes is to construct buildings for residential and commercial use that provide minimum safeguards. Published codes include the International Building Code, International Fire Code, International Plumbing Code, International Mechanical Code, International Energy Conservation Code, and others. The International Codes have been adopted by all fifty states and the District of Columbia.

What is an STC rating and what does it mean?

The STC (Sound Transmission Class) tells how well a wall or floor assembly inhibits airborne sound. The higher the number, the greater the sound barrier. Typical STC ratings are as follows:

STC Number	What It Means*
25	Normal speech can be easily understood
30	Loud speech can be understood
35	Loud speech is audible but not intelligible
42	Loud speech audible as a murmur
45	Must strain to hear loud speech
48	Some loud speech barely audible
50	Loud speech not audible

*The degree to which sound barriers work depends greatly on the elimination of air gaps under doors, through electrical outlets, and around heating ducts.

What is pisé or rammed earth?

Rammed earth is an ancient building technique in which moist earth is compacted into a rough approximation of sedimentary rock. Using forms, rammed earth may be shaped into bricks or entire walls. Dating back as far as 7000 B.C.E., it was used in portions of the 2,000-year-old Great Wall of China as well as temples in Mali and in Morocco. Romans and Phoenicians introduced the technique to Europeans, and it became a popular building technique in France, where it became known as *pisé de terre*. In the United States, it was used for both gracious Victorian homes and low-cost housing.

Today, builders add cement to the earthen mix, which results in stronger, waterproof walls. In traditional rammed-earth construction, hand- or compressor-powered tampers are used to compact a mixture of moistened earth and cement between double-sided wooden forms to about 60 percent of its original volume. In a newer technique, a high-pressure hose sprays the mixture against a single-sided form. Sometimes steel-reinforced bars are also used.

What is a yurt?

Originally a Mongolian hut, the yurt has been adapted in the United States as a low-cost structure that can be used as a dwelling. The foundation is built of wood on a hexagonal or round frame. The wooden, latticework side walls have a tension cable sandwiched between the wall pieces at the top to keep the walls from collapsing. The walls are insulated and covered with boards, log slabs, canvas, or aluminum siding. A shingle roof, electricity, plumbing, and a small heating stove may be installed. The interior can be finished as desired with shelves, room dividers, and interior siding. The yurt is practical and relatively inexpensive to erect.

Originating in the Mongolian steppes, a yurt is a large, round or hexagonal tent with a wooden frame.

When was the first shopping center built?

The first shopping center in the world was built in 1896 at Roland Park, Baltimore, Maryland. The first enclosed shopping mall was Southdale Shopping Center in Edina, Minnesota, which opened in 1956. It originally had seventy-two stores arranged in a two-level design around a center court and free parking for five thousand vehicles.

Where is the largest shopping mall located?

The largest shopping mall in the world is the South China Mall located in Dongguan, China. It encompasses a total area of 9.6 million square feet (892,000 square meters) with gross leasable space of 7.1 million square feet (659,612 square meters). Since opening in 2005, the South China Mall has been mostly vacant and is considered a dead mall. The largest shopping mall in North America is the West Edmonton Mall in Alberta, Canada, which covers 5.3 million square feet (492,000 square meters) on a 121-acre (49-hectare) site. It has over 800 stores and services, with parking for 23,000 vehicles. In addition, there is an amusement park, a water park, ice skating, miniature golf, and other entertainment.

The largest shopping mall in the United States (based on total area) is the Mall of America in Bloomington, Minnesota. The Mall of America covers 4.2 million square feet (390,000 square meters). In addition to the 520 stores on the premises, there is also a theme park, an entertainment area, and an aquarium. It was opened in 1992.

How much does the Leaning Tower of Pisa lean?

The Leaning Tower of Pisa, 184.5 feet (56 meters) tall, is about 17 feet (5 meters) out of perpendicular, increasing by about 0.2 inch (1.25 millimeters) a year. The Romanesque-style tower was started by Bonanno Pisano in 1173 as a campanile or bell tower for the nearby baptistry but was not completed until 1372. Built entirely of white marble, with eight tiers of arched ar-

The Leaning Tower of Pisa in Italy was leaning at an angle of about 5.5 degrees until a restoration project in the 1990s moved it back to slightly less than 4 degrees.

cades, it began to lean during construction. Although the foundation was dug down to 10 feet (3 meters), the builders did not reach bedrock for a firm footing. Ingenious attempts were made to compensate for the tilt by straightening up the subsequent stories and making the pillars higher on the south side than on the north. The most recent attempt at compensation for the tilt, completed in 2001, removed earth from the north side of the foundation—the side opposite the tilt. It is hoped that this "fix" will hold for about three hundred years.

What is the ground area of the Pentagon Building in Arlington, Virginia?

The Pentagon, the headquarters of the U.S. Department of Defense, is the world's largest office building in terms of ground space. Its construction was completed on January 15, 1943—in just over seventeen months. With a gross floor area of over 6.5 million square feet (604,000 square meters), this five-story, five-sided building has three times the floor space of the Empire State Building and is one and a half times larger than the Willis Tower (formerly called the Sears Tower) in Chicago. The World Trade Center complex in New York City, completed in 1973 and destroyed in 2001, was larger with over 9 million square feet (836,000 square meters), but it consisted of two structures (towers). Each of the Pentagon's five sides is 921 feet (281 meters) long with a perimeter of 4,610 feet (1,405 meters). The secretary of defense, the secretaries of the three military departments, and the military heads of the army, navy, and air force are all located in the Pentagon. The National Military Command Center, which is the nation's military communications hub, is located where the Joint Chiefs of Staff convene. It is commonly called the "war room."

What is the largest building in the world?

Identifying the largest building in the world is difficult because it depends on the criteria used to define the largest building. The structure with the largest floor space may not be the same as the structure with the largest usable space. Until recently, the largest commercial building in the world under one roof was the flower auction building of the cooperative VBA in Aalsmeer, the Netherlands. In 1986, the floor plan was extended to 91 acres (37 hectares). It measures 2,546 × 2,070 feet (776 × 639 meters). In 2013, the New Century Global Center opened in Chengdu in southwest China. Chinese officials claim it is the largest freestanding structure in the world with 18 million square feet (1.7 million square meters) of floor space. The Center is a combination of shopping mall, entertainment center, office space, hotel space, and an indoor seaside village with a water park and artificial beach. The largest building by volume is the Boeing assembly plant in Everett, Washington, with a capacity of 472 million cubic feet (13.3 million cubic meters) and covering 98.3 acres (39.8 hectares).

When was the first skyscraper built?

Designed by William Le Baron Jenney (1832–1907), the first skyscraper, the 10-story Home Insurance Company Building in Chicago, Illinois, was completed in 1885. A sky-

What is the "topping out" party that iron workers have?

When the last beam is placed on a new bridge, skyscraper, or building, ironworkers hoist up an evergreen tree, attach a flag or a handkerchief, and brightly paint the final beam and autograph it. This custom of raising an evergreen tree goes back to Scandinavia in 700 C.E., when attaching the tree to the building's ridge pole signaled to all who helped that the celebration of its completion would begin.

scraper—a very tall building supported by an internal frame (skeleton) of iron and steel rather than by load-bearing walls—maximizes floor space on limited land. Three technological developments made skyscrapers feasible: a better understanding of how materials behave under stress and load (from engineering and bridge design); the use of steel or iron framing to create a structure, with the outer skin "hung" on the frame; and the introduction of the first "safety" passenger elevator. Elisha Otis (1811–1861) demonstrated the safety elevator in 1854. He ascended in an elevator "cage" and halfway up he had the hoisting cable cut with an axe. The elevator "cage," did not fall. Otis patented his invention on January 15, 1861 (#31,128). He was inducted into the National Inventors Hall of Fame in 1988 for the invention of the elevator brake.

What criteria are used to determine the tallest building?

The Council on Tall Buildings and Urban Habitat, representing world leaders in the field of the built environment, measures the height (structural top) of a building from the sidewalk level of the main entrance to the structural top of the building. The measurement includes spire but does not include television antennas, radio antennas, or flagpoles. Other measures of height include the height to the highest occupied floor of the building, height to the top of the roof, and height to the tip of a spire, pinnacle antenna, mast, or flagpole.

How many supertall and megatall buildings are in the world?

The Council on Tall Buildings and Urban Habitat defines supertall buildings as buildings over 984 feet (300 meters) in height. Megatall buildings are defined as buildings over 1,968 feet (600 meters) in height. As of 2014, there were 81 supertall and two megatall buildings completed and occupied in the world.

What is the tallest building in the world?

The Burj Khalifa, located in Dubai, is 2,717 feet (828 meters) tall and has 163 floors. Constructed of steel and concrete, it was completed in 2010 as an office/residential/hotel building. Construction began on the Kingdom Tower in Jeddah, Saudi Arabia, in 2013. It is scheduled for completion in 2018 with a minimum height of 3,281 feet (1,000 meters).

What are the tallest buildings in the United States?

One World Trade Center in New York is the tallest building in the United States, completed in 2014. It has 94 floors above ground and is 1,776 feet (541.3 meters) tall. It is the third tallest building in the world.

The other tallest buildings in the United States are Willis Tower (formerly called Sears Tower) in Chicago and Trump International Hotel and Tower in Chicago. Willis Tower, complete in 1974, is 1,451 feet tall (442.1 meters) and has 108 floors. There is an observation deck on the 103rd floor at 1,353 feet (412 meters). Trump International Hotel and Tower was completed in 2009. It has 98 floors and is 1,389 feet (423.2 meters) tall.

The Empire State Building in New York was the tallest building in the United States for many years. It was completed in 1931. It is 1,250 feet (381 meters) tall with 102 floors. There is an observation deck on the 86th floor and on the 102nd floor at 1,224 feet (373 meters).

The Burj Khalifa in Dubai stands an incredible 2,717 feet (828 meters) tall.

How tall was the World Trade Center?

The two towers of the World Trade Center in New York City were 110 stories tall. The North Tower, completed in 1972, stood 1,368 feet (417 meters) high, and the South Tower, completed in 1973, stood 1,362 feet (415 meters) high. Both towers were destroyed in a terrorist attack on September 11, 2001.

What is the tallest self-supporting structure in the world?

The tallest self-supporting structure in the world is the Tokyo Sky Tree in Tokyo, Japan. A telecommunications tower, it was completed in 2012 with a height of 2,080 feet (634 meters). There are observation decks and visitors centers at 1,148 feet (350 meters) and 1,480 feet (451 meters). The tallest structure in North America is the CN Tower in Toronto, Ontario, Canada, at 1,815 feet (553 meters), which was completed in 1976.

What is the tallest structure in the world?

The tallest structure in the world is the KVLY-TV transmitting tower in Blanchard, North Dakota, at 2,063 feet (629 meters) tall.

What material was used to construct the exterior of the Empire State Building?

The exterior of the Empire State Building is made of Indiana limestone and granite with vertical strips of stainless steel.

Who invented the geodesic dome?

A geodesic line is the shortest distance between two points across a surface. If that surface is curved, a geodesic line across it will usually be curved as well. A geodesic line on the surface of a sphere will be part of a great circle. Buckminster Fuller (1895–1983) realized that the surface of a sphere could be divided into triangles by a network of geodesic lines and that structures could be designed so that their main elements either followed those lines or were joined along them.

This is the basis of his very successful geodesic dome: a structure of generally spherical form constructed of many light, straight structural elements in tension and arranged in a framework of triangles to reduce stress and weight. These contiguous tetrahedrons are made from lightweight alloys with high tensile strength.

An early example of a geodesic structure is Britain's Dome of Discovery, built in 1951. It was the first dome ever built with principal framing members intentionally aligned along great circle arcs. The ASM (American Society for Metals) dome, which was built east of Cleveland, Ohio, from 1959 to 1960, is an open latticework geodesic dome. Built in 1965, the Houston Astrodome also forms a giant geodesic dome.

Why was the Mellon Arena an architectural marvel?

When the Civic Arena (later named the Mellon Arena) opened in 1961 in Pittsburgh, Pennsylvania, its retractable roof was a revolutionary architectural design. Built of 2,950 tons (2,676 metric tons) of stainless steel, it was the largest retractable dome roof in the world, measuring 170,000 square feet (15,794 square meters). The roof had no interior supports and was divided radially into eight leaves supported by a cantilever arm that arched 260 feet (79 meters) to support the six movable sections of the roof. The Mellon Arena hosted its final event in 2010 and was demolished in 2012.

When was the first retractable roof stadium opened?

The first retractable roof stadium, Rogers Centre (formerly called the SkyDome) in Toronto, Ontario, Canada, was opened in 1989. Unlike previous stadiums that had removable tops, the SkyDome had a fully retractable roof. The roof consists of four panels, weighing 11,000 tons (9,979 tonnes), and runs on a system of steel tracks and bogies. One panel remains fixed while two of the other panels move forward and backward at a rate of 71 feet (21 meters) per minute to open or close. The fourth panel rotates 180 degrees to completely open or close the roof. It takes 20 minutes for the roof to open or close. Seven other retractable roof stadiums have opened since the SkyDome—Chase Field (formerly called Bank One Ballpark) in Phoenix, Arizona (1998), Safeco Field in Seattle, Washington (1999), Minute Maid Park (formerly called Enron Field) in Houston, Texas (2000), Miller Park in Milwaukee, Wisconsin (2001), Reliant Stadium in Houston (2002), Lucas Oil Stadium in Indianapolis, Indiana (2008), and AT&T Stadium (formerly called Texas Stadium) in Arlington, Texas (2009).

Which building has the largest cable-supported domed roof?

Georgia Dome, which opened in 1992 in Atlanta, is the largest cable-supported domed stadium in the world. The Dome is as tall as a 27-story building. The roof is 290 feet (88 meters) high and consists of 130 Teflon-coated fiberglass panels. The fiberglass panels cover an area equivalent to 8.6 acres (3.5 hectares). The roof's supporting cables are equivalent in length to 11.1 miles (18 kilometers). The Dome has a seating capacity of 71,250. The building covers 8.9 acres (3.6 hectares). The floor of the Georgia Dome has 102,000 square feet (9,476 square meters) of space—enough to hold two military C-5 planes.

When was the first air-supported dome installed?

Air-supported domes consist of a fabric roof supported using a system of fans to boost air pressure inside. A series of diagonally crossing cables help the fabric roof hold its shape. The first air-supported structure was the U.S. Pavilion at Expo 70 in Osaka, Japan.

The O$_2$ Dome or Arena—also known as the North Greenwich Arena—in London, England, is the world's busiest. In 2014 it sold 1.82 million tickets to live events.

It was invented and designed by David Geiger (1935–1989). One of the advantages of air-supported domes is that there are no support columns to obstruct the view inside the facility. Another advantage is their cost-effectiveness. Fabric, air-supported domes are much less costly to construct than fixed or retractable domes.

How many steel masts support the O₂?

The O_2 Dome in London is supported by 12,328-foot (100-meter) steel masts that project through the dome. The overall structure has a diameter of 1,197.5 feet (365 meters) and an internal diameter of 1,050 feet (320 meters). It is 164 feet (50 meters) high at the central point. If the dome was turned upside down, it could hold the same volume as 1,100 Olympic-sized swimming pools.

ROADS, BRIDGES, AND TUNNELS

Who is known as the founder of civil engineering?

Thomas Telford (1757–1834), the first president of the Institute of Civil Engineers, is the founder of the British civil engineering profession. Telford established the professional ethos and tradition of the civil engineer—a tradition followed by all engineers today. He built bridges, roads, harbors, and canals. His greatest works include the Menai Strait Suspension Bridge, the Pontcysyllte Aqueduct, the Gotha Canal in Sweden, the Caledonian Canal, and many Scottish roads. He was the first and one of the greatest masters of the iron bridge.

How were early macadam roads different from modern paved roads?

Macadam roads were developed originally in England and France and are named after Scottish road builder and engineer John Louden MacAdam (1756–1836). The term "macadam" originally designated road surface or base in which clean, broken, or crushed ledge stone was mechanically locked together by rolling with a heavy weight and bonded together by stone dust screenings that were worked into the spaces and then "set" with water. With the beginning of the use of bituminous material (tar or asphalt), the terms "plain macadam," "ordinary macadam," or "waterbound macadam" were used to distinguish the original type from the newer bituminous macadam. Waterbound macadam surfaces are almost never built now in the United States mainly because they are expensive, and the vacuum effect of vehicles loosens them. Many miles of bituminous macadam roads are still in service, but their principal disadvantages are their high crowns and narrowness. Today's roads that carry very heavy traffic are usually surfaced with very durable portland cement.

When was the idea of the Interstate Highway System conceived?

The concept of an interstate highway system was first described to Congress in a 1939 Bureau of Public Roads report to Congress, *Toll Roads and Free Roads*. Five years later, the Federal-Aid Highway Act of 1944 authorized designation of a "National System of Interstate Highways." However, this legislation did not authorize a program to build the highway system. It was not until 1956 that a system of federal aid existed to build and construct the Interstate Highway System. In 1991, the final estimated total cost to build the Interstate Highway System was $128.9 billion. There are 43,500 miles (70,000 kilometers) of road in the Interstate Highway System.

What is the National Highway System?

The National Highway System Designation Act of 1995 established new ways for states to fund transportation projects. The National Highway System (NHS) is comprised of 160,955 miles (260,000 kilometers) of roads that are important to the nation's economy, defense, and mobility. Although this represents only about 4 percent of the nation's roads, these roads carry about 40 percent of all highway traffic, 75 percent of all heavy truck traffic, and 90 percent of all tourist traffic. There are five components to the NHS:

1. The Interstate Highway System, with 43,500 miles (70,000 kilometers), accounts for nearly 30 percent of the NHS.

2. Congress designated 21 high-priority corridors, which account for another 4,500 miles (7,200 kilometers).

3. The Strategic Highway Corridor Network (STRAHNET) identified by the Department of Defense in cooperation with the Department of Transportation. It consists of 15,500 miles (25,000 kilometers) of roads that are critical strategic links to move troops and equipment to airports, ports, and railroad terminals for rapid deployment, if necessary, for the national defense.

4. The Strategic Highway Corridor Network connectors. These connectors consist of 1,865 miles (3,000 kilometers) of roads that link major military installations and other defense-related facilities to the STRAHNET.

5. The nearly 92,000 miles (148,000 kilometers) of roads that serve as important arterial highways for interstate and interregional travel and provide connections to major ports, airports, public transportation facilities, and other intermodal facilities.

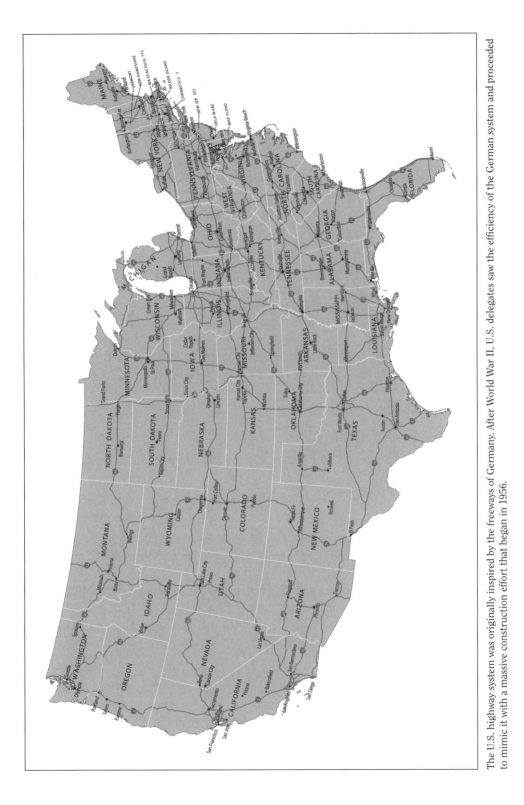

The U.S. highway system was originally inspired by the freeways of Germany. After World War II, U.S. delegates saw the efficiency of the German system and proceeded to mimic it with a massive construction effort that began in 1956.

Which state has the most miles of roads?

Texas lays claim to the most road mileage with 313,210 miles (504,063 kilometers). California is a distant second with 172,499 miles (282,438 kilometers) of roads. On the other end of the scale, only three states have less than 10,000 miles (16,090 kilometers) of roads: Delaware (6,377 miles [10,263 kilometers]), Hawaii (4,416 miles [7,107 kilometers]), and Rhode Island (6,480 miles [10,429 kilometers]). The total mileage count for the United States (including the District of Columbia) is 4,092,730 miles (6,586,610 kilometers).

When was the first U.S. coast-to-coast highway built?

Completed in 1923, after ten years of planning and construction, the Lincoln Highway was the first transcontinental highway connecting the Atlantic coast (New York) to the Pacific (California). Sometimes called the "Main Street of the United States," the highway was proposed by Carl G. Fisher (1874–1939) to a group of automobile manufacturers, who formed the Lincoln Highway Association and promoted the idea. Fisher thought that "Lincoln" would be a good, patriotic name for the road.

Its original length of 3,389 miles (5,453 kilometers) was later shortened by relocations and improvements to 3,143 miles (5,057 kilometers). Crossing twelve states—New York, New Jersey, Pennsylvania, Ohio, Indiana, Illinois, Nebraska, Colorado, Wyoming, Utah, Nevada, and California—in 1925, the Lincoln Highway became, for most of its length, U.S. Route 30.

How are U.S. highways numbered?

The main north-south interstate highways always have odd numbers of one or two digits. The system, beginning with Interstate 5 on the West Coast, increases in number as it moves eastward, and ends with Interstate 95 on the East Coast.

The east-west interstate highways have even numbers. The lowest-numbered highway begins in Florida with Interstate 4, increasing in number as it moves northward, and ends with Interstate 96. Coast-to-coast east-west interstates, such as Routes 10, 40, and 80, end in zero. An interstate with three digits is either a beltway or a spur route.

U.S. routes follow the same numbering system as the interstate system but increase in number from east to west and from north to south. U.S. Route 1, for example, runs along the East Coast; U.S. Route 2 runs along the Canadian border. U.S. route numbers may have anywhere from one to three digits.

Which road has the most lanes?

The urban highway with the most lanes is I-75 in Atlanta, Georgia. It has fifteen lanes. This does not include toll plazas.

Where was the first cloverleaf interchange in the United States located?

Arthur Hale, an engineer from Maryland, was granted U.S. Patent #1,173,505 in 1916 for his design for "Street Crossing." Edward Delano designed the first cloverleaf interchange at the junction of NJ Route 25 (now U.S. Route 1) and NJ Route 4 (now NJ Route 35) in Woodbridge, New Jersey, in 1928. The advantage of the new design consisting of four circular ramps was that motorists were able to merge from one road to another road without braking. By the late twentieth century, the cloverleaf design could no longer handle the volume of traffic at this junction safely. Construction began in 2006 and was completed in 2008 that incorporated a diamond-shaped layout that includes two left-turn lanes on Route 35 (one in each direction), additional lanes and shoulders, ramps, a wider bridge, and a traffic signal.

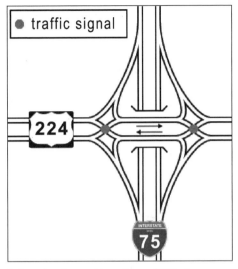

A diverging diamond interchange (DDI) allows drivers to make left turns more easily at busy interchanges.

When was the first diverging diamond interchange (DDI) completed in the United States?

The first diverging diamond interchange (DDI) in the United States was completed in 2009 in Springfield, Missouri. This design allows motorists to make free left turns, meaning they do not cross opposing traffic to complete the left turn. Traffic in a DDI is directed to the opposite side of the roadway after a set of traffic lights. Motorists wanting to turn left then proceed onto the second road to the left without stopping for more lights or crossing opposing traffic. Motorists who want to continue straight proceed to another set of traffic lights. Following the second set of traffic lights, traffic is redirected to the right side of the road. The DDI received the 2010 Francis B. Francois Award for Innovation from the American Association of State Highway and Transportation Officials (AASHTO).

Which city had the first traffic light?

On December 10, 1868, the first traffic light was erected on a 22-foot (6.7-meter) high cast-iron pillar at the corner of Bridge Street and New Palace Yard off Parliament Square in London, England. Invented by J. P. Knight (1828–1886), a railway signaling engineer, the light was a revolving lantern illuminated by gas, with red and green signals. It was turned by hand, using a lever at the base of the pole.

Cleveland, Ohio, installed an electric traffic signal at Euclid Avenue and 105th Street on August 5, 1914. It had red and green lights with a warning buzzer as the color changed.

Around 1913, Detroit, Michigan, used a system of manually operated semaphores. Eventually, the semaphores were fitted with colored lanterns for night traffic. New York City installed the first three-color light signals in 1918; these signals were still operated manually.

What is a Jersey barrier?

Jersey barriers are concrete highway barriers developed by the New Jersey Department of Transportation. Originally only 12 to 18 inches (30 to 46 centimeters) high, they were designed to prevent left turns at certain intersections. Later barriers, reinforced concrete usually poured on the site, were used as temporary traffic safeguards where construction required motorists to cross over into a lane normally used by oncoming traffic. These barriers were 32 inches (81 centimeters) high. Now the barriers line thousands of miles of American highways as permanent fixtures. Barriers are now 54 inches (137 centimeters) high, thus blocking out the glare of oncoming headlights. One of the purposes of a Jersey barrier is to redirect a vehicle that hits it while preventing vehicle rollover.

Why are manhole covers round?

Circular covers are almost universally used on sewer manholes because they cannot drop through the opening from any angle. The circular cover rests on a lip that is smaller than the cover, and the cover and the lip are manufactured together, making for a particularly tight fit. Any other shape—such as a square or rectangle—could slip into the manhole opening. In addition, round manhole covers can be machined more accurately than other shapes, and, once removed, round covers can be rolled rather than having to be lifted.

How do cars, buses, and trucks travel in the English Channel tunnel?

Cars, buses, and trucks are carried on 2,500-foot-long (762-meter-long) trains traveling at 90 miles (145 kilometers) per hour in the two 31-mile (50-kilometer) long tunnels under the English Channel. Automobile operators drive their vehicles onto the trains and remain inside until the passage is complete. The trains carry cars and buses in large wagons. Truck drivers travel separately from their vehicles, which ride in simpler, semi-open wagons. Interspersed with the shuttles are high-speed passenger trains designed to run on the different railroad systems of Britain, Continental Europe, and the tunnel.

What is the world's longest road tunnel?

The world's longest road tunnel runs between Laerdal and Aurland in Norway. Completed in 2000, the tunnel is 15.2 miles (24.5 kilometers) long. There is one lane in each direction. A ventilation and air-purification system ensures a safe level of air quality in the tunnel. Three large caverns or mountain halls divide the tunnel into four sections to help break the monotony of the drive. In the United States, the longest road tunnel is the Anton Anderson Memorial Tunnel near Whittier, Alaska. It is 2.5 miles (4.2 kilometers) long and was completed in June 2000.

Where is the longest bridge-tunnel in the world?

Completed in 1964 after 42 months of work, costing $200 million, and spanning a great distance of open sea, the Chesapeake Bay Bridge-Tunnel is a 17.5-mile (28-kilometer) combination of trestles, bridges, and tunnels that connect Norfolk with Cape Charles in Virginia.

How long is the Afsluitdijk?

The Afsluitdijk, located in the Netherlands, is nearly 20 miles (32 kilometers) long. The dike, part of the Zuider Zee Enclosure Dam project, connects the province of North Holland with the province of Friesland. As early as the seventeenth century, there had been discussion about closing off the Zuider Zee. It was not until after a major storm in 1916 that the Dutch parliament passed legislation authorizing the construction of the Zuider Zee Enclosure Dam project. Construction of the Afsluitdijk began in 1920 and was completed in 1932. As a result of the project, the Zuider Zee was closed off and divided into the saline Waddenzee and the freshwater lake, Ijsselmeer. The dike also serves to protect the Netherlands from the North Sea. Four polders, lands drained and reclaimed from the sea by a series of dikes, were built, enlarging the total land area of the Netherlands by about 626 square miles (1,620 square kilometers) and creating additional land for agricultural, industrial, and recreational purposes. The project is considered a major engineering accomplishment.

What are the various types of bridge structures?

Four basic types of structures can be used to bridge a stream or similar obstacle: rigid beam, cantilever, arch, and suspension systems.

The *rigid beam bridge*, the simplest and most common form of bridging, has straight slabs or girders carrying the roadbed. The span is relatively short and its load rests on its supports or piers.

The *arch bridge* is in compression and thrusts outward on its bearings at each end.

What is a "kissing bridge"?

Covered bridges with roofs and wooden sides are called "kissing bridges," because people inside the bridge could not be seen from outside. Such bridges can be traced back to the early nineteenth century. Contrary to folk wisdom, they were not designed to produce rural "lovers' lanes" but were covered to protect the structures from deterioration.

In the *suspension bridge*, the roadway hangs on steel cables, with the bulk of the load carried on cables anchored to the banks. It can span a great distance without intermediate piers.

Each arm of a *cantilever bridge* is, or could be, freestanding, with the load of the short central truss span pushing down through the piers of the outer arms and pulling up at each end. The outer arms are usually anchored at the abutments and project into the central truss.

Which state in the United States has the most covered bridges?

More than 10,000 covered bridges were built across the United States between 1805 (when the first was erected in Philadelphia) and the early twentieth century. Pennsylvania, with 211 covered bridges, has the most covered bridges in the United States. Ohio, with approximately 133 covered bridges, is the state with the second greatest number of covered bridges. Other covered bridges are scattered throughout the country. Nonauthentic covered bridges—built or covered for visual effect—appear in each state.

What are the world's longest bridge spans?

The world's longest main span in a suspension bridge is the Akashi-Kaikyo bridge in Japan at 6,532 feet (1,991 meters), linking the Kobe and Awajishima Islands. The overall length of the bridge is 12,828 feet (3,910 meters). Construction began in 1988, and the bridge was opened in 1998.

Currently the planet's longest suspension bridge, the Akashi-Kaikyo Bridge in Japan is 6,532 feet (1,991 meters) long.

Major Suspension Bridges

Bridge Name	Location	Year opened	Length of Main Span (Feet/Meters)
Akashi-Kaikyo	Japan	1998	6,532/1,991
Xihoumen	China	2009*	5,413/1,650
Great Belt Link	Denmark	1997	5,328/1,624
Yi Sun-sin	Korea	2013	5,069/1,545
Runyang	China	2005	4,888/1,490

*Main span completed in 2007 but not opened to traffic until 2009.

- The world's longest *cable-stayed bridge* span is Russky Bridge, opened in 2012. It connects Vladivostok, Russia, to Russky Island. It is 3,888 feet (1,185 meters) long with a central span of 3,622 feet (1,104 meters).

- The world's longest *cantilever bridge* is the Quebec Bridge over the St. Lawrence River in Quebec, Canada, opened in 1917, which has a span of 1,800 feet (549 meters) between piers, with an overall length of 3,239 feet (987 meters).

- The world's longest *steel-arch bridge* is the Chaotianmen Bridge in China, completed in 2008, with a span of 1,811 feet (552 meters).

- The world's longest *concrete-arch bridge* is the Wanxian Bridge in China, completed in 2008. The bridge measures 1,378 feet (420 meters).

- The world's longest *stone-arch bridge* is the 3,810-foot (1,161-meter) Rockville Bridge, completed in 1901, north of Harrisburg, Pennsylvania, with 48 spans containing 216 tons (219 tonnes) of stone.

Who built the Brooklyn Bridge?

John A. Roebling (1806–1869), a German-born American engineer, constructed the first truly modern suspension bridge in 1855. Towers supporting massive cables, tension anchorage for stays, a roadway suspended from the main cables, and a stiffening deck below or beside the road deck to prevent oscillation are all characteristics of Roebling's suspension bridge. Roebling was inducted into the National Inventors Hall of Fame in 2004 for his engineering feat of building a suspension bridge.

In 1867, Roebling was given the ambitious task of constructing the Brooklyn Bridge. In his design he proposed the revolutionary idea of using steel wire for cables rather than the less-resilient iron. Just as construction began, Roebling died of tetanus when his foot was crushed in an accident, and his son, Washington A. Roebling (1837–1926), assumed responsibility for the bridge's construction. Fourteen years later, in 1883, the bridge was completed. At that time, it was the longest suspension bridge in the world, spanning the East River and connecting New York's Manhattan with Brooklyn. The bridge has a central span of 1,595 feet (486 meters), with its masonry towers rising 276 feet (841 meters) above high water. Today, the Brooklyn Bridge is among the best known of all American civil engineering accomplishments.

Where is the longest suspension bridge in the United States?

Spanning New York (City) Harbor, the Verrazano-Narrows Bridge is the longest suspension bridge in the United States. With a span of 4,260 feet (1,298 meters) from tower to tower, its total length is 7,200 feet (2,194 meters). Named after Giovanni da Verrazzano (1485–1528), the Italian explorer who discovered New York Harbor in April 1524, it was erected under the direction of Othmar H. Ammann (1879–1965) and completed in 1964. To avoid impeding navigation in and out of the harbor, it provides a clearance of 216 feet (66 meters) between the water level and the bottom of the bridge deck. Like other suspension bridges, the bulk of the load of the Verrazano-Narrows Bridge is carried on cables anchored to the banks.

Are there any floating bridges in the United States?

The four floating pontoon bridges in the United States are all located in the state of Washington. They are: the Governor Albert D. Rosellini Bridge–Evergreen Point (SR 520 bridge), the Lacey V. Murrow–Lake Washington Bridge, the Homer M. Hadley Memorial Bridge, and the Hood Canal Bridge (SR 104).

The SR 520 bridge, opened in 1963, is the longest floating bridge in the world. The floating section is 7,578 feet (2,310 meters) long. Construction to refurbish and enlarge the bridge began in 2012 and is expected to be completed 2016.

Originally known at the Lake Washington Floating Bridge, the Lacey V. Murrow–Lake Washington Bridge was opened in 1940. In 1990, during construction to refurbish the bridge, a large section sank to the bottom of the lake. The bridge was completed and reopened in 1993 with a length of 6,620 feet (2,020 meters). It carries the eastbound lanes of Interstate 90.

The Homer M. Hadley Memorial Bridge carries the westbound lanes of Interstate 90 across Lake Washington. It was opened in 1989 with a total length of 5,811 feet (1,771 meters).

The Hood Canal Bridge (SR 104), connecting the Olympic and Kitsap peninsulas, has a floating section of 6,521 feet (1,988 meters). It was opened in 1961. During a storm in 1979, the west end of the bridge sank. It was reopened in 1982. The east half was refurbished between 2003 and 2009. The Hood Canal Bridge is a saltwater floating bridge.

How do floating pontoon bridges support the weight of heavy traffic?

Pontoon bridges date back to ancient times when it was necessary to transport men and equipment for military purposes across bodies of water. These were usually temporary bridges that were dismantled when they were no longer needed. Permanent floating bridges are used in situations where there are environmental challenges to construct traditional bridges. These challenges include crossing a very wide and very deep body of water that has a soft lake- or ocean-bottom soil. The soft bottom soil makes it difficult to erect traditional bridge piers. The depth of the water would require extremely tall bridge piers. Corrosion from saltwater is another challenge for bridge construction.

Modern floating bridges are constructed with steel and concrete—materials that are not generally thought of as ones that float. Pontoon bridges float because of the scientific principle of buoyancy. Watertight, concrete pontoons are connected rigidly end-to-end, with the roadway built on top of the pontoons. The hollow pontoons are very buoyant, allowing them to support a weight (roadway and traffic) equivalent to the amount of water they displace—even when the pontoons are constructed from a heavy material such as concrete. The pontoons are held in place by steel cables and anchors made of reinforced concrete. The anchors sink into the loose material at the bottom of the lake. At each end of the bridge, there are additional anchors attached directly into the ground.

What is the longest causeway?

The Lake Pontchartrain Causeway in Louisiana is the longest overwater highway bridge in the world. The road over the lake, called a causeway, consists of two twin spans 80 feet

The Lake Pontchartrain Causeway in Louisiana is about twenty-four miles long.

(24 meters) apart. The northbound span is 23.87 miles (38.19 kilometers) long, and the southbound span is 23.86 miles (38.2 kilometers) long. The first span was opened in 1956, and the second span was opened in 1969.

Who designed the Golden Gate Bridge?

Joseph B. Strauss (1870–1938), formally named chief engineer for the project in 1929, was assisted by Charles Ellis (1876–1949) and Leon Moisseiff (1872–1943) in the design. An engineering masterpiece that opened to traffic in May 1937, this suspension bridge spans San Francisco Bay, linking San Francisco with Marin County, California. It has a central span of 4,200 feet (1,280 meters) with towers rising 746 feet (227 meters).

MISCELLANEOUS STRUCTURES

How large is the Great Pyramid at Giza?

The Great Pyramid at Giza, possibly the most impressive of the seven wonders of the ancient world—and the only one still standing—has sides of 754 feet (230 meters) and a height of 479 feet (146 meters). Composed of 2.3 million stone blocks ranging in weight from 2.5 to 15 tons (2.3 to 13.6 tonnes), the Great Pyramid covers an area of thirteen acres (52,609 square meters), approximately the size of 200 tennis courts. The Great Pyramid is an engineering accomplishment since the corners of the pyramid are perfect right angles, the sides are aligned to the cardinal points of the compass, and the foundation is perfectly level to support the weight of the structure.

What is the tallest national monument?

The Gateway Arch in St. Louis, Missouri, is the tallest national monument. At 630 feet (192 meters), it is 75 feet (23 meters) taller than the Washington Monument. It was designed by Eero Saarinen (1910–1961) in the shape of an inverted catenary curve using stainless steel. Some interesting facts and figures about the Gateway Arch are:

General Facts

- Outer width of Outside North Leg to Outer South Leg: 630 feet (192 meters)
- Maximum height: 630 feet (192 meters)
- Shape of arch section: equilateral triangle
- Dimension of arch at base: 54 feet (16.46 meters)
- Deflection of arch: 18 inches in 150 mph wind (0.46 meters in 240 kmh wind)
- Number of sections in arch: 142
- Thickness of plates for outer skin: 0.25 inches (6.3 millimeters)
- Type of material used in arch: exterior stainless steel; #3 finish type 304

Weight of Steel in Arch

- Stainless steel plate exterior skin: 886 tons (804 metric tonnes)
- Carbon steel plate interior skin, 3/8 inch (9.5 millimeters) thick: 2,157 tons (1,957 metric tons)
- Steel stiffeners: 1,408 tons (1,277 metric tons)
- Interior steel members, stairs, trains, etc.: 300 tons (272 metric tons)
- Total steel weight: 5,199 tons (4,644 metric tons)

Weight of Concrete in Arch

- Between skins to 300 feet (91 meters): 12,127 tons (11,011 metric tons)
- In foundation below ground: 25,980 tons (23,569 metric tons)
- Total concrete weight: 38,107 tons (34,570 metric tons)
- External protection: Six 0.50 inch × 20 inch (13 × 510 millimeter) lightning rods and one aircraft obstruction light

How much damage did the earthquake of 2011 cause to the Washington Monument?

The 5.8 magnitude earthquake centered approximately 90 miles (145 kilometers) southwest of Washington, D.C., in August 2011, caused cracks in the exterior marble and underlying masonry, spalls, and the loss of some joint mortar and small pieces

The Gateway Arch in St. Louis, Missouri, is a symbol of how the city was once the last major stop before American pioneers made the hazardous journey to settle the West.

of masonry. The damage was evaluated by engineers rappelling down the exterior of the monument. Closed for repairs for nearly three years, the Washington Monument re-opened to visitors in May 2014.

The Washington Monument was built in two phases in 1848–1854 and 1876–1884. When the Washington Monument was dedicated on February 21, 1885, it was the tallest building in the world at 555 feet, 5.125 inches (169 meters, 13 centimeters). It was constructed with over 36,000 marble stones from two different quarries during the two phases of construction. The color of the different marble is noticeably different on the bottom third than the top two-thirds of the monument. The monument is 55 feet, 1.5 inches (16.8 meters, 3.8 centimeters) wide at the base and tapers to 34 feet, 5.5 inches (10.4 meters, 14 centimeters) at the top. The walls are 15 feet (4.6 meters) thick at the base and only 18 inches (45.7 centimeters) thick at the top. The total weight of the monument is 90,854 tons (82,421 metric tons). There are 897 steps from the base to the top. When construction of the Washington Monument was completed in 1884, an aluminum cap in the shape of a pyramid was placed on the top of the monument to act as a lightning rod.

What is a Texas tower?

A Texas tower is an off-shore platform used as a radar station. Built on pilings sunk into the ocean floor, it resembles the offshore oil derricks or rigs first developed in the Gulf of Mexico off the coast of Texas. In addition to radar, the components of a typical tower may include crew quarters, a helicopter landing pad, fog signals, and oceanographic equipment.

When was the first offshore oil rig erected?

In 1869, Thomas Fitch Rowland (1831–1907) received a patent (U.S. Patent #89,794) for a fixed, working platform for drilling offshore to a depth of nearly 50 feet (15 meters). The earliest working offshore oil wells were piers that were constructed from the shore extending out over the water in 1896 in California. Although there were some oil wells drilled from platforms in Grand Lake St. Marys in Ohio as early as 1891, the first oil wells over water without a pier connection to the shore were Caddo Lake wells in Louisiana. They began producing oil in 1911. In 1938 Pure Oil Co. and Superior Oil Co. built a drilling platform in the Gulf of Mexico. It was destroyed by a hurricane in 1940. In 1947, the first permanent oil well platform was erected in the Gulf of Mexico. The Kermac #16 well was 10 miles (16 kilometers) out in the open sea and in nearly 20 feet (6 meters) of water.

Will building a seawall protect a beach?

It may for a while, but during a storm the sand cannot follow its natural pattern of allowing waves to draw the sand across the lower beach, making the beach flatter. With a seawall, the waves carry off more sand, dropping it into deeper water. A better alternative to the seawall is the revetment. This is a wall of boulders, rubble, or concrete block, tilted back away from the waves. It imitates the way a natural beach flattens out under wave attack.

What are the dimensions of the Eiffel Tower?

The Eiffel Tower was designed by Alexandre Gustave Eiffel (1832–1923) and architect Stephen Sauvestre (1847–1919). The main engineers for the structure were Maurice Koechlin (1856–1946) and Èmile Nouguier (1840–1898).

One of the most famous landmarks in the world, the Eiffel Tower in Paris, France, was built for the 1889 Paris Exposition.

- Number of iron structural components in tower: 15,000
- Number of rivets: 2,500,000
- Weight of foundations: 306 tons (277,602 kilograms)
- Weight of iron: 8,092 tons (7,341,214 kilograms)
- Weight of elevator systems: 1,042 tons (946,000 kilograms)
- Total weight: 9,441 tons (8,564,816 kilograms)
- Pressure on foundations: 58–64 pounds per square inch (4–4.5 kilograms per square centimeter), depending on pier
- Height of first platform: 189 feet (58 meters)
- Height of second platform: 379 feet, 8 inches (116 meters)
- Height of third platform: 905 feet, 11 inches (276 meters)
- Total height in 1889: 985 feet, 11 inches (300.5 meters)
- Total height with television antenna: 1,052 feet, 4 inches (320.75 meters)
- Number of steps to the top: 1,671
- Maximum sway at top caused by wind: 4.75 inches (12 centimeters)
- Maximum sway at top caused by metal dilation: 7 inches (18 centimeters)
- Size of base area: 2.54 acres (10,282 square meters)
- Dates of construction: January 26, 1887, to March 31, 1889
- Cost of construction: 7,799,401.31 francs ($1,505,675.90)

How many locks are on the Panama Canal?

The 40-mile-long (64-kilometers-long) Panama Canal was completed in 1914, connecting the Atlantic and Pacific Oceans. There are three locks between Gatun Lake and Limon Bay at Colon on the Atlantic side and three locks between Gaillard Cut and Balboa on the Pacific side. Ships are raised 85 feet (26 meters) in their passage from one ocean to the other.

Where is the highest dam in the United States and what is its capacity?

Oroville, the highest dam in the United States, is an earth-fill dam that rises 754 feet (230 meters) and extends more than a mile across the Feather River, near Oroville, California. Built in 1968, it forms a reservoir containing about 3.5 million acre-feet (4.3 million cubic meters) of water. The next highest dam in the United States is the Hoover Dam on the Colorado River, on the Nevada–Arizona border. It is 726 feet (221 meters) high and was, for twenty-two years, the world's highest. Presently, nine dams are higher than the Oroville—the highest currently is the 1,098-foot (335-meter) Rogun(skaya) earth-fill dam that crosses the Vakhsh River in Tajikistan. Built between 1981 and 1987, this dam has a volume of 92.9 million cubic yards (71 million cubic meters).

How big is the Hoover Dam?

Formerly called Boulder Dam, the Hoover Dam is located between Nevada and Arizona on the Colorado River. The highest concrete-arch dam in the United States, the dam is 1,244 feet (379 meters) long and 726 feet (221 meters) high. It has a base thickness of 660 feet (201 meters) and a crest thickness of 45 feet (13.7 meters). It stores 21.25 million acre-feet of water in the 115-mile-long (185-kilometer-long) Lake Mead reservoir.

The dam was built because the Southwest was faced with constantly recurring cycles of flood and drought. Uncontrolled, the Colorado River had limited value, but once regulated, the flow would assure a stabilized, year-round water supply, and the low-lying valleys would be protected against floods. On December 21, 1928, the Boulder Canyon Project Act became law, and the project was completed on September 30, 1935—two years ahead of schedule. For twenty-two years, Hoover Dam was the highest dam in the world.

How tall is the figure on the Statue of Liberty, and how much does it weigh?

The Statue of Liberty, conceived and designed by French sculptor Frédéric-Auguste Bartholdi (1834–1904), was given to the United States to commemorate its first centennial. Called (in his patent, U.S. Design Patent #11,023, issued February 18, 1879) "Liberty Enlightening the World," it is 151 feet (46 meters) high, weighs 125 tons (113 metric tons), and stands on a pedestal and base that are 154 feet (47 me-

The iconic Statue of Liberty was a gift to the United States from the French government. Designed by Frédéric-Auguste Bartholdi, it sits on Liberty Island in Upper New York Bay.

ters) tall. Her flowing robes are made from more than 300 sheets of hand-hammered copper over a steel frame. The copper sheets are 3/32 of an inch (2.38 millimeters) thick, or the equivalent of two pennies placed together. Copper was chosen since it resists corrosion from the surrounding seawater. Constructed and finished in France in 1884, the statue's exterior and interior were taken apart piece by piece, packed into 200 mammoth wooden crates, and shipped to the United States in May 1885. The statue was placed by Bartholdi on Bedloe's Island, at the mouth of New York City Harbor. On October 28, 1886, ten years after the centennial had passed, the inauguration celebration was held.

It was not until 1903 that the inscription "Give me your tired, your poor, / Your huddled masses yearning to breathe free …" was added. The verse was taken from *The New Colossus*, composed by New York City poet Emma Lazarus (1849–1887) in 1883. The statue, the tallest in the United States and the tallest copper statue in the world, was refurbished for its own centennial, at a cost of $698 million, reopening on July 4, 1986. The original cast-iron frame was replaced with stainless steel. One visible difference is that the flame of her torch is now 24-karat gold leaf, just as in the original design. In 1916 the flame was redone into a lantern of amber glass. Concealed within the rim of her crown is that the observation deck that can be reached by climbing 354 steps or by taking the more recently installed hydraulic elevator.

From what type of stone was Mount Rushmore National Memorial carved?

Granite. The monument, in the Black Hills of southwestern South Dakota, depicts the 60-foot-high (18-meter-high) faces of four U.S. presidents: George Washington (1732–1799), Thomas Jefferson (1743–1826), Abraham Lincoln (1809–1865), and Theodore Roosevelt (1858–1919). Sculptor Gutzon Borglum (1867–1941) designed the monument, but he died before the completion of the project; his son, Lincoln Borglum (1912–1986), finished it. From 1927 to 1941, 360 people, mostly construction workers, drillers, and miners, "carved" the figures using dynamite.

Who invented the Ferris wheel and when?

Originally called "pleasure wheels," the first such rides were described by English traveler Peter Mundy in 1620. In Turkey he saw a ride for children consisting of two vertical wheels, 20 feet (6 meters) across, supported by a large post on each side. Such rides were called "ups-and-downs" at the St. Bartholomew Fair of 1728 in England, and in 1860 a French pleasure wheel was turned by hand and carried sixteen passengers. They were also in use in the United States by then, with a larger, wooden wheel operating at Walton Spring, Georgia.

Wanting a spectacular attraction to rival that of the 1889 Paris Centennial celebration—the Eiffel Tower—the directors of the 1893 Columbian Exposition had a design competition. The prize was won by American bridge builder George Washington Gale Ferris (1859–1896). In 1893 he designed and erected a gigantic revolving steel wheel whose top reached 264 feet (80.5 meters) above ground. The wheel—825 feet (251.5

meters) in circumference, 250 feet (76 meters) in diameter, and 30 feet (9 meters) wide—was supported by two 140-foot (43-meter) towers. Attached to the wheel were 36 cars, each able to carry 60 passengers. Opening on June 21, 1893, at the exposition in Chicago, Illinois, it was extremely successful. Thousands lined up to pay fifty cents for a 20-minute ride—a large sum in those days considering that a merry-go-round ride cost only four cents. In 1904 it was moved to St. Louis, Missouri, for the Louisiana Purchase Exposition. It was eventually sold for scrap.

How does a giant observation wheel differ from a Ferris wheel?

Giant observation wheels are generally much larger than traditional Ferris wheels. In order to accommodate their large size, observation wheels are supported by a single A-frame or slim, two-leg support system instead of two towers on each side of the axles. The rim of the large observation wheels may be constructed of hollow tubes with steel tubes connected to a frame of triangular shapes or a slim ladder truss design. Traditional Ferris wheels consisted of gondolas or carriages suspended from the rim for passengers. They are kept level by gravity. Newer observation wheels have enclosed passenger capsules that are kept level by mechanical means. This allows passengers the opportunity to stand and walk around in the capsule.

Where is the largest observation wheel in the world?

The largest observation wheel in operation is the Las Vegas High Roller. Designed by the engineering and design firm Arup, the High Roller has a circumference of 550 feet (168 meters) and a diameter of 529 feet (161 meters). It is supported by five steel legs on three 8-foot (2.4-meter) concrete mats. Two of the legs are to the north of the wheel, two are to the south, and the brace leg is to the east of the wheel. Each passenger cabin is supported by a 6-foot, 1-inch (1.8 meters, 2.54 centimeters) wide, solid tubular rim and has a diameter of 20 feet (6 meters). The cabins extend 6 feet (1.8 meters) from the rim. A single cabin can carry up to 40 passengers. There are 28 spherical cabins on the High Roller with a total passenger capacity of 1,120 individuals. The High Roller moves at a continuous rate of approximately 1 foot per second (.3 meters per second) and completes a full rotation every 30 minutes. The slow rate of movement allows passengers to enter a cabin on one side and exit on the other side without having to stop the wheel.

The Las Vegas High Roller ferris wheel has a diameter of 529 feet.

The New York Wheel is planned to be 630 feet (192 meters) tall and will surpass the Las Vegas High Roller. A ground breaking ceremony was held for the New York Wheel in 2015, which has a projected completion date of 2017.

How many roller coasters are there in the world?

There was a total of 3,558 roller coasters (including both wooden and steel) in the world in 2014.

World Roller Coasters

Continent	No. of Roller Coasters
Africa	59
Asia	1663
Australia	26
Europe	861
North America	787
South America	162

Roller coasters have had a long history of thrill-giving. During the fifteenth and sixteenth centuries, the first known gravity rides were built in St. Petersburg and were called "Russian Mountains." A wheeled roller coaster, called the "Switchback," was used as early as 1784 in Russia. By 1817, the first roller coaster with cars locked to the tracks operated in France. The first U.S. roller coaster patent was granted to J. G. Taylor in 1872, and LaMarcus Thompson (1848–1919) built the first known roller coaster in the United States at Coney Island in Brooklyn, New York, in 1884. The present record holder for the longest steel-tracked roller coaster in the world is the "Steel Dragon 2000" at Nagashima Spa Land in Mie, Japan. It is 8,133 feet (2,479 meters) in length. The longest wood-tracked roller coaster in the world is "The Beast" at Kings Island in Cincinnati, Ohio. It is 7,400 feet (2,253 meters) in length. The fastest steel roller coaster is "Formula Rossa" at Ferrari World in Abu Dhabi, United Arab Emirates. It reaches a speed of 149.1 miles per hour (239.9 kilometers per hour). The fastest wooden roller coaster is "Goliath" at Six Flags Great America in Gurnee, Illinois. It reaches a speed of 72 miles per hour (115.9 kilometers per hour).

CARS, BOATS, PLANES AND TRAINS

CARS AND MOTOR VEHICLES

Who invented the automobile?

Although the idea of self-propelled road transportation originated long before, Karl Benz (1844–1929) and Gottlieb Daimler (1834–1900) are both credited with the invention of the gasoline-powered automobile because they were the first to make their automotive machines commercially practicable. Benz and Daimler worked independently, unaware of each other's endeavors. Both built compact, internal-combustion engines to power their vehicles. Benz built his three-wheeler in 1885; it was steered by a tiller. Daimler's four-wheeled vehicle was produced in 1887. Daimler was inducted into the National Inventors Hall of Fame in 2006 for his design of automobile and motorcycle engines.

Earlier self-propelled road vehicles include a steam-driven contraption invented by Nicolas-Joseph Cugnot (1725–1804), who rode it on the Paris streets at 2.5 miles (4 kilometers) per hour in 1769. Richard Trevithick (1771–1833) also produced a steam-driven vehicle that could carry eight passengers. It first ran on December 24, 1801, in Camborne, England. Londoner Samuel Brown built the first practical 4-horsepower, gasoline-powered vehicle in 1826. Belgian engineer J. J. Étienne Lenoir (1822–1900) built a vehicle with an internal combustion engine that ran on liquid hydrocarbon fuel in 1862, but he did not test it on the road until September 1863, when it traveled a distance of 12 miles (19.3 kilometers) in three hours. Austrian inventor Siegfried Marcus (1831–1898) invented a four-wheeled, gasoline-powered handcart in 1864 and a full-sized car in 1875; the Viennese police objected to the noise that the car made, and Marcus did not continue its development. Édouard Delamare-Deboutteville invented an 8-horsepower vehicle in 1883, which was not durable enough for road conditions.

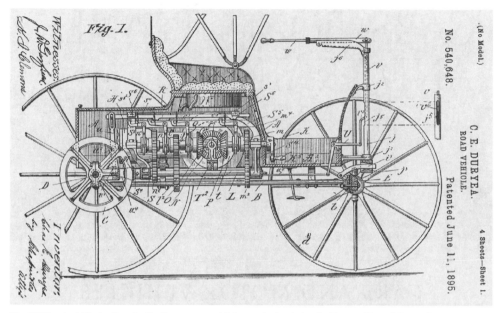

The 1895 patent illustration for the Duryea car, which was the basis for the Duryea Motor Wagon Company.

Who started the first American automobile company?

Charles Duryea (1861–1938), a cycle manufacturer from Peoria, Illinois, and his brother, Frank Duryea (1869–1967), founded America's first auto-manufacturing firm and became the first to build cars for sale in the United States. The Duryea Motor Wagon Company, set up in Springfield, Massachusetts, in 1895, built gasoline-powered horseless carriages similar to those built by Benz in Germany.

However, the Duryea brothers did not build the first automobile factory in the United States. Ransom Eli Olds (1864–1950) built it in 1899 in Detroit, Michigan, to manufacture his Oldsmobile. More than 10 vehicles a week were produced there by April 1901 for a total of 433 cars produced in 1901. In 1902 Olds introduced the assembly-line method of production and made over 2,500 vehicles in 1902 and 5,508 in 1904. In 1906, 125 companies made automobiles in the United States. In 1908, American engineer Henry Ford (1863–1947) improved the automobile assembly-line techniques by adding the conveyor belt system that brought the parts to the workers on the production line; this made automotive manufacture quick and cheap, cutting production time to 93 minutes. His company sold 10,660 vehicles that year. Ford was inducted into the National Inventors Hall of Fame in 1982 for his many contributions to the development of the automobile industry.

How did the term horsepower originate?

Horsepower is the unit of energy needed to lift 550 pounds (247.5 kilograms) the distance of 1 foot (30.48 centimeters) in one second. Near the end of the eighteenth century, Scottish engineer James Watt (1736–1819) made improvements in the steam engine and wished to determine how its rate of pumping water out of coal mines compared with that of horses, which had previously been used to operate the pumps. In order to define a horsepower, he tested horses and concluded that a strong horse could lift 150 pounds (67.5 kilograms) 220 feet (66.7 meters) in one minute. Therefore, one horsepower was equal to $150 \times 220/1$ or 33,000 foot-pounds per minute (also expressed as 745.2 joules per second, 7,452 million ergs per second, or 745.2 watts).

The term "horsepower" was frequently used in the early days of the automobile because the "horseless carriage" was generally compared to the horse-drawn carriage. Today this unit is still used routinely to express the power of motors and engines, particularly of automobiles and aircraft. A typical automobile requires about 20 horsepower to propel it at 50 miles (80.5 kilometers) per hour.

How does a gasoline engine differ from one using ethanol or gasohol?

A gasoline engine uses only gasoline as its fuel. The ethanol or gasohol engine uses either ethanol (a fuel derived from plant sources) or a combination of gasoline and alcohol as its fuel source. The fuel system must have gaskets and components compatible with the fuel that will be used.

What was the first mass-produced alternative fuel vehicle in the United States?

The Honda Civic GX Natural Gas Vehicle began production in 1998. This vehicle is still rated as having the cleanest-burning internal combustion engine of its time.

How does an electric car work?

An electric car uses an electric motor to convert electric energy stored in batteries into mechanical work. Various combinations of generating mechanisms (solar panels, generative braking, internal combustion engines driving a generator, fuel cells) and storage mechanisms are used in electric vehicles.

Is the electric automobile a recent idea?

During the last decade of the nineteenth century, the electric vehicle became especially popular in cities. People had grown familiar with electric trolleys and railways, and technology had produced motors and batteries in a wide variety of sizes. The Edison Cell, a

---

Has there ever been a nuclear-powered automobile?

In the 1950s, Ford automotive designers envisioned the Ford Nucleon, which was to be propelled by a small atomic reactor core located under a circular cover at the rear of the car. It was to be recharged with nuclear fuel. The car was never built.

nickel-iron battery, became the leader in electric vehicle use. By 1900, electric vehicles nearly dominated the pleasure car field. In that year, 4,200 automobiles were sold in the United States. Of these, 38 percent were powered by electricity, 22 percent by gasoline, and 40 percent by steam. By 1911, the automobile starter motor did away with hand-cranking gasoline cars, and Henry Ford (1863–1947) had just begun to mass produce his Model Ts. By 1924, not a single electric vehicle was exhibited at the National Automobile Show, and the Stanley Steamer, a steam-powered vehicle, was scrapped the same year.

The energy crises of the 1970s and 1990s and concern for the environment (as well as "Clean Air" legislation) led to renewed interest and research towards developing an all-electric and/or hybrid vehicle. The first hybrid vehicle available on the mass market in the United States was the two-door Honda Insight in 1999. Toyota began marketing the four-door Prius in 2000. Ford released the Escape Hybrid, the first American hybrid and first SUV hybrid, in September 2004.

How popular are electric vehicles in the United States?

In 2014, 570,475 electric-drive vehicles, including hybrids, plug-in hybrids, and battery vehicles, were sold in the United States, accounting for 3.47 percent of the market share of vehicles sold for the year.

How many different types of electric vehicles are available in the United States?

There are five types of electric vehicles available in the United States:

- Hybrid electric vehicles (HEV) use both an electric battery and an internal combustion engine.
- Plug-in hybrid electric vehicles (PHEV) "plug in" to the electric grid to obtain electricity to power the vehicle. They also have an internal combustion engine.
- Battery electric vehicles (BEV) are powered exclusively from electricity from the onboard battery. The battery is charged by being plugged into the grid.
- Extended-range electric vehicles (EREV) will operate on a battery similar to BEVs. Once the charge in the battery is depleted, they switch to an internal combustion engine.
- Fuel-cell electric vehicles (FCEV) convert chemical energy from a fuel, such as hydrogen, into electricity.

How is the fuel efficiency of a plug-in hybrid vehicle measured?

The U.S. Environmental Protection Agency (EPA) calculates two values of fuel economy for plug-in hybrid vehicles. One is a standard miles per gallon (MPG) calculated when the battery is depleted and the vehicle is using gasoline only. When the vehicle is operating on electricity, the value is calculated as miles per gallon of gasoline-equivalent (MPGe). It represents the number of miles the vehicle can go using a quantity of fuel with the same energy content as a gallon of gasoline.

How large is a Smart car?

The Smart Fortwo is only 106.1 inches (2.69 meters) long, making it the smallest car currently sold in North America. (It is more than 3 feet [1 meter] shorter than a Mini Cooper.) The Smart Company began in 1993 as a joint venture between Daimler-Benz and the Swiss watchmaker Swatch. The goal was to combine the engineering expertise of Mercedes-Benz with the design philosophy of Swatch to build a vehicle for an urban environment with an emphasis on being easy to park and fuel economy. The first model, the City-Coupe, debuted at the Frankfurt Auto Show in

The Smart car is big enough for two passengers, and that's about it! It is a good choice for driving in congested cities with limited parking space.

1997 and was available for sale in Europe in 1998. The safety concerns of such a small car were solved by creating the "tridon" safety cell. The tridon safety cell consists of three layers of steel reinforced at strategic points to absorb and redistribute crash energy away from the occupants of the vehicle. Steel door beams and reinforced axles shield side impacts. The Smart Fortwo model was the first model to become available in the United States in 2008. The EPA fuel economy guide lists the Smart Fortwo model as having a fuel economy of 33 miles per gallon in the city and 41 miles per gallon on the highway, making it the most fuel-efficient non-hybrid vehicle available.

Smart Fortwo Dimensions

Length:	106.1 inches (269.5 centimeters)
Width:	61.4 inches (156 centimeters)
Height:	60.7 inches (154 centimeters)
Wheel Base:	73.5 inches (187 centimeters)
Front head room:	39.7 inches (101 centimeters)
Front shoulder room:	48 inches (122 centimeters)
Front hip room:	45.4 inches (115 centimeters)
Front leg room:	41.2 inches (105 centimeters)

When was the Michelin tire introduced?

The first pneumatic (air-filled) tire for automobiles was produced in France by André Michelin (1853–1931) and Édouard Michelin (1859–1940) in 1885. The first radial-ply tire, the Michelin X, was made and sold in 1948. In radial construction, layers of cord materials called plies are laid across the circumference of the tire from bead to bead (perpendicular to the direction of the tread centerline). The plies can be made of steel wires or belts that circle the tire. Radial tires are said to give longer tread life, better handling, and a softer ride at medium and high speeds than bias or belted bias tires (both of which have plies laid diagonally). Radials give a firm, almost hard, ride at low speeds.

When were tubeless automobile tires first manufactured?

The B. F. Goodrich Company, located in Akron, Ohio, announced the manufacture of tubeless tires on May 11, 1947. Dunlop was the first British firm to make tubeless tires in 1953.

What do the numbers mean on automobile tires?

The numbers and letters associated with tire sizes and types are complicated and confusing. The "Metric P" system of numbering is probably the most useful method of indicating tire sizes. For example, if the tire was numbered P185/75R-14, then "P" means the tire is for a passenger car; the number 185 is the width of the tire in millimeters; 75 indicates the aspect ratio, i.e., that the height of the tire from the rim to the road is 75 percent of the width; R indicates that it is a radial tire; and 14 is the wheel diameter in inches (13 and 15 inches are also common sizes). Speed ratings are indicated by a letter located in the size markings:

Symbol	Maximum Speed (mph/kph)
S	112/180
T	118/190
U	124/200
H	130/210
V	149/240
W	168/270
Y	186/300
Z	186+/300+

What is the difference between all-weather tires and snow tires? Are snow tires necessary?

There are three main differences between all-weather tires and snow tires: 1) tread design, 2) tread depth, and 3) the rubber compound. The goal of the tread design and depth of winter tires is to keep snow and ice away from the point of contact between the tire and the road surface. The depth of the treads helps to increase flexing, which channels snow and slush across the tire's surface and away from the contact point with the road surface. Another big difference is in the rubber compound itself. The rubber compound for all-weather tires hardens and stiffens in cold temperatures. Consequently, these tires do not perform well in cold weather because they have reduced grip and do not conform to the road surface well in lower temperatures. Winter tires use rubber compounds, which are softer, gripping the road surface better in colder temperatures.

Snow tires, better known as winter tires, are recommended in areas where temperatures dip below 45°F (7°C). Studies have shown that it takes longer to brake in snow

A rumble seat is a nice feature of this 1928 REO Flying Cloud automobile on display at the Charlotte AutoFair classic car show.

and ice using all-weather tires instead of winter tires. However, the softer rubber compound used for snow tires is not recommended during the warm-weather seasons. It is important to change back to all-weather tires when the temperatures are warmer.

What is a rumble seat?

A rumble seat is a folding, external seat situated in the rear deck of some older, two-door coupes, convertibles, and roadsters.

What was the first car manufactured with an automatic transmission?

The first of the modern generation of automatic transmissions was General Motors' Hydramatic, first offered as an option on the Oldsmobile during the 1940 season. Between 1934 and 1936, a handful of 18-horsepower Austins were fitted with the American-designed Hayes infinitely variable gear. The direct ancestor of the modern automatic gearbox was patented in 1898.

Where was the first automobile license plate issued?

Leon Serpollet of Paris, France, obtained the first license plate in 1889. They were first required in the United States by New York State in 1901. Registration was required within thirty days. Owners had to provide their names and addresses as well as a description of their vehicles. The fee was one dollar. The plates bore the owner's initials and

> ### Which vehicle had the first modern automobile air conditioner?
>
> The first air-conditioned automobile was manufactured by the Packard Motor Car Company in Detroit, Michigan, and was exhibited publicly November 4–12, 1939, at the 40th Automobile Show in Chicago, Illinois. Air in the car was cooled to the temperature desired, dehumidified, filtered, and circulated. The first fully automatic air conditioning system was Cadillac's "Climate Control," introduced in 1964.

were required to be over 3 inches (7.5 centimeters) high. Permanent plates made of aluminum were first issued in Connecticut in 1937.

What information is available from the vehicle identification number (VIN), body number plate, and engine on a car?

These coded numbers reveal the model and make, model year, type of transmission, plant of manufacture, and sometimes even the date and day of the week a car was made. The form and content of these codes is not standardized (with one exception) and often changes from one year to the next for the same manufacturer. The exception (starting in 1981) is the tenth digit, which indicates the year. For example: W = 1998, X = 1999, 2 = 2002, etc. Various components of a car may be made in different plants, so a location listed on a VIN may differ from one on the engine number. The official shop manual lists the codes for a particular make of car.

How many motor vehicles are registered in the United States?

According to the Federal Highway Administration, there were 249,478,143 motor vehicles registered in the United States in 2011. Of the total, 125,656,528 were automobiles, 118,455,587 were trucks, 666,064 were buses, and 8,437,502 were motorcycles.

How much does it cost to operate an automobile in the United States?

The average cost per mile, calculated as cents per mile, to operate an automobile in 2015 in the United States is listed in the following chart. Figures are based on 15,000 miles of driving per year:

Cost of Operating a Vehicle in the United States (2015)

Operating Costs (cents/mile)	Small Sedan	Medium Sedan	Large Sedan	SUV	Minivan
Gas/oil	9.18	10.87	13.58	14.60	13.65
Maintenance	4.68	5.20	5.11	1.65	5.19
Tires	0.68	1.11	1.158	1.38	0.84
Total Operating Cost cents/mile	14.54	17.18	20.19	21.60	19.70

Ownership Costs (annual amount/year in $)

Insurance	$1,071	$1,106	$1,167	$1,058	$999
License/Registration/Taxes	$489	$671	$836	$827	$688
Depreciation	$2,515	$3,687	$4,759	$4,646	$4,039
Finance charge	$473	$675	$858	$848	$694
Total ownership cost/year	$4,548	$6,139	$7,620	$7,379	6,420
Total costs/mile (cents) (operating costs and ownership costs)	44.9	58.1	71.0	70.8	62.5

How do antilock braking systems (ABS) work?

The antilock brake system (ABS) was first developed by Bosch. Bosch was awarded a German patent for an "Apparatus for preventing lock-braking of wheels in a motor vehicle" in 1936. In 1978, the Mercedes-Benz S-class became the first vehicle equipped with ABS. It was not until 1985 that the Bosch ABS braking systems were installed in U.S. vehicles. The term antilock brake is derived from the German term "antiblockier-system." The ABS system places wheel-speed sensors at each wheel. A control unit automatically modulates brake pressure at each wheel when the sensors detect a lock-up on a wheel, preventing the wheels from locking. Locked wheels create vehicle instability and can cause skidding.

What newer innovations have been introduced to improve automobile safety?

Traction control systems and electronic stability systems, both originally developed by Bosch, help drivers maintain control of their vehicles. Introduced in 1986, the traction control system prevents the wheels from slipping during accelerations by reducing the drive torque at each wheel. Electronic stability control (ESC) systems, introduced in 1995, automatically apply braking power when they detect a situation of the driver "oversteering" and possibly "spinning out" or "understeering" and possibly "plowing out." In both situations, ESC allows the driver to maintain control of the vehicle and stay on the road.

What is the braking distance for an automobile at different speeds?

Average stopping distance is directly related to vehicle speed. On a dry, level, con-

The antilock brake system (ABS) in cars is a sophisticated system that controls brake pressure and prevents dangerous skidding.

233

crete surface, the minimum stopping distances are as follows (including driver reaction time to apply brakes):

Speed mph/kph	Reaction time distance Feet/Meters	Braking distance Feet/Meters	Total distance Feet/Meters
10/16	11/3.4	9/2.7	20/6.1
20/32	22/6.7	23/7.0	45/13.7
30/48	33/10.1	45/13.7	78/23.8
40/64	44/13.4	81/24.7	125/38.1
50/80	55/16.8	133/40.5	188/57.3
60/97	66/20.1	206/62.8	272/82.9
70/113	77/23.5	304/92.7	381/116.1

How far will a car skid on various road surfaces at different speeds?

	Skidding Distances (Feet/Meters)			
Speed (mph)	Asphalt	Concrete	Snow	Gravel
30	40 ft/12 m	33 ft/ 10 m	100 ft/ 30 m	60 ft/ 18 m
40	71 ft/ 22 m	59 ft/ 18 m	178 ft/54 m	107 ft/33 m
50	111 ft/34 m	93 ft/28 m	278 ft/85 m	167 ft/51 m
60	160 ft/49 m	133 ft/41 m	400 ft/122 m	240 ft/73 m

How can one find out about safety recalls on automobiles?

The National Highway Traffic Safety Administration (NHTSA) keeps records of recalls and takes reports of safety problems experienced by consumers. You can call the department's 24-hour hotline at 1-800-424-9393 or 1-888-327-4236, check its website at www.safecar.gov, write to NHTSA, Department of Transportation, Washington, DC 20590, or contact the manufacturer or your dealer. Be sure to include the make, model, year, and vehicle identification number of the vehicle, and a description of the problem or part in question. You will receive any recall information the NHTSA has, either by phone or by a mailed printout.

What day of the week do most fatal automotive accidents occur?

Ongoing studies have shown that Saturday continues to be the day of the week with the most fatal accidents. The National Highway Traffic Safety Administration collected these data for 2011:

Hour of Day	Sun	Mon	Tues	Wed	Thurs	Fri	Sat	Total
Midnight–2:59 A.M.	955	350	257	302	353	534	995	3,746
3–5:59 A.M.	608	234	221	217	251	325	550	2.406
6–8:59 A.M.	308	446	440	452	462	457	400	2,965
9–11:59 A.M.	342	418	429	409	375	427	450	2,850

Hour of Day	Sun	Mon	Tues	Wed	Thurs	Fri	Sat	Total
Noon–2:59 P.M.	589	541	530	506	581	576	654	3,977
3–5:59 P.M.	648	661	678	694	650	703	714	4,748
6–8:59 P.M.	682	609	608	581	657	712	884	4,733
9–11:59 P.M.	485	473	450	463	561	801	889	4,122
Unknown	38	27	24	24	20	32	45	210
Total	4,655	3,759	3,637	3,648	3,910	4,567	5,581	29,757

When did seat belts become mandatory equipment on U.S. motor vehicles?

The U.S. National Highway Safety Bureau first required the installation of lap belts for all seats and shoulder belts in the front seats of cars in 1968. However, most Americans did not regularly use safety belts until after 1984, when the first state laws were introduced that penalized drivers and passengers who did not use the device. Nationally, seat belt use reached 87 percent in 2013, and continued to remain at 87 percent in 2014, according to the National Occupant Protection Use Survey. According to the National Highway Traffic Safety Administration, seat belts saved nearly 63,000 lives during the five-year period from 2008–2012.

Who invented the three-point safety belt?

The three-point safety belt, named for its three attachment points—two hips and one shoulder—was invented in 1958 by Nils Bohlin (1920–2002), who was working as a safety engineer at Volvo. In 1959, the three-point safety belt became standard in all Volvos in Sweden. U.S. Patent #3,043,652 for the safety belt, also known as a combination safety belt because it combines a strap across the chest and hips, was filed in August 1959 and published in July 1962. The three-point safety belt became standard equipment in Volvos in the United States in 1963. Seat belt use is one of the most effective means of preventing death and injury, saving over 13,000 lives per year. Bohlin was inducted into the National Inventors Hall of Fame in 2002 for his invention of the safety belt.

When was the air bag invented?

Patented ideas on air bag safety devices began appearing in the early 1950s. U.S. Patent #2,649,311 was granted on August 18, 1953, to John W. Hetrick for an inflated safety cushion to be used in automotive vehicles. The Ford Motor Company studied the use of air bags around 1957, and other undocumented work was carried out by Assen Jordanoff (1896–1967) before 1956. There are other earlier uses of an air bag concept, including a rumored method of some World War II pilots inflating their life vests before a crash.

In the mid-1970s, General Motors geared up to sell 100,000 air bag-equipped cars a year in a pilot program to offer them as a discounted option on luxury models. GM dropped the option after only 8,000 buyers ordered air bags in three years. As of September 1, 1989, all new passenger cars produced for sale in the United States are required to be equipped with passive restraints (either automatic seat belts or air bags).

Federal law required driver and passenger frontal air bags in all cars beginning with the 1998 model year and on all light trucks beginning with the 1999 model year. According to the National Highway Traffic Safety Administration, frontal air bags saved 2,213 lives, age 13 and older, between 2008 and 2012.

How does an air bag work to prevent injury in an automobile crash?

When a frontal collision occurs, sensors trigger the release and reaction of sodium azide with iron, which produces large quantities of nitrogen gas. This gas fully inflates the bag in about two-tenths of a second after impact to create a protective cushion. The air bag deflates immediately thereafter, and the harmless nitrogen gas escapes through holes in the back.

An air bag will deploy only after the car has an impact speed of 11 to 14 miles (17 to 22 kilometers) per hour or greater. It will not be set off by a minor fender bender, by hitting a cement stop in a parking space, or if someone kicks the bumpers. The National Highway Traffic Safety Administration (NHTSA) recommends that children under 13 years of age always be in the rear seat in an age-appropriate child restraint seat to avoid injury from an air bag.

How have air bags changed since the earliest models?

All passenger cars and light trucks produced after September 1, 2006, are required to have "advanced" frontal air bags. While the earliest models of frontal air bags saved thousands of lives, they were also responsible for injuries and even deaths of individuals who were either unrestrained (not using a seat belt properly) or too close to the air bags as they were inflating. Newer, advanced frontal air bags have sophisticated sensory devices that inflate with less power or energy or volume, thus reducing the number of serious injuries or deaths caused by air bags, especially in children and small-stature adults.

When were side-impact air bags first introduced?

Side-impact air bags (SABs) were first introduced by Volvo in the 1995 model year. As of 2015, there are no federal regulations requiring side-impact air bags in either passenger cars or light trucks. There are three types of side-impact air bags: 1) chest (torso) SABs, 2) head SABs, and 3) head/chest combination (combo) SABs. As the name implies, each type is designed to protect the adult occupants of a vehicle in the event of a serious side-impact crash. Chest

It is estimated by the National Highway Traffic Safety Administration that air bags save about two thousand lives annually in America.

(torso) SABs are mounted in the side of the seat back in the door and are designed to protect the chest. Head SABs may be either curtain SABs or tubular SABs and are mounted in the roof rail above the side windows. They are designed to protect the head. Some of these provide protection from ejection if the vehicle rolls over. Some SABs are designed to stay inflated longer than frontal air bags to provide protection in a rollover crash. Certain new models of convertibles have head SABs mounted in the windowsill. Also mounted in the side of the seat, combo SABs are larger than chest SABs and are designed to protect both the head and chest.

Which colors of cars are the safest?

There are very few scientific studies that address car color and safety. Background color, e.g., trees, snow, sand in the desert, weather conditions, and daylight, all have an impact on visibility. Another consideration is the personalities of the drivers of different colored vehicles. For example, drivers of red cars may be more aggressive than drivers of white or beige vehicles. More important than the color of the vehicle is the driver's attention, ability to concentrate and focus on the driving, and alertness.

What car colors are the most popular?

The annual PPG Industries color popularity report for 2014 lists white as the most popular car color worldwide. The most popular car colors worldwide are:

Most Popular Car Colors Worldwide

Color	Popularity (Percent of Total Cars)
White/White pearl	28
Black/Black effect	18
Silver	13
Gray	13
Natural	10
Red	9
Blue	7
Green	1
Other	1

In the North American market, the most popular car color is white. The most popular car colors in North America are:

North American Popular Car Colors

Color	Popularity (Percent of Total Cars)
White/White pearl	23
Black	18
Gray	16
Silver	15
Red	10

Color	Popularity (Percent of Total Cars)
Blue	9
Natural	7
Green	2

How does VASCAR work?

Invented in 1965, VASCAR (Visual Average Speed Computer and Recorder) is a calculator that determines a car's speed from two simple measurements of time and distance. No radar is involved. VASCAR can be used at rest or while moving to clock traffic in both directions. The patrol car can be behind, ahead of, or even perpendicular to the target vehicle. The device measures the length of a speed trap and then determines how long it takes the target car to cover that distance. An internal calculator does the math and displays the average speed on an LED readout. Most police departments now use several forms of moving radar, which are less detectable and more accurate.

How does police radar work?

Austrian physicist Christian Doppler (1803–1853) discovered that the reflected waves bouncing off a moving object are returned at a different frequency (shorter or longer waves, cycles, or vibrations). This phenomenon, called the Doppler effect, is the basis of police radar. Directional radio waves are transmitted from the radar device. The waves bounce off the targeted vehicle and are received by a recorder. The recorder compares the difference between the sent and received waves, translates the information into miles per hour, and displays the speed on a dial.

How do laser speed guns differ from radar?

Laser speed guns rely on the reflection time of light rather than the Doppler effect. They measure the round-trip time for light to reach a vehicle and reflect back. Laser speed guns shoot a very short burst of infrared laser light at the target (a moving vehicle) and wait for it to reflect off the vehicle. The advantages of laser speed guns are that the cone of light the guns emit is very small so they can target a specific vehicle and they are very accurate. The disadvantage is that they require better aim.

When was a speed trap first employed to apprehend speeding automobile drivers?

In 1905, William McAdoo (1853–1930), police commissioner of New York City, was stopped for traveling at 12 miles (19 kilometers) per hour in an 8-miles-per-hour (13-kilometers-per-hour) zone in rural New England. The speed-detection device consisted of two lookout posts, camouflaged as dead tree trunks, spaced 1 mile (1.6 kilometers) apart. A deputy with a stopwatch and a telephone kept watch for speeders. When a car appeared to be traveling too fast, the deputy pressed his stopwatch and telephoned ahead to his confederate, who immediately consulted a speed-mileage chart and phoned ahead to another constable manning a road block to apprehend the speeder. McAdoo invited the New England constable to set up a similar device in New York City.

One of the most famous speed traps was in the Alabama town of Fruithurst on the Alabama–Georgia border. In one year, this town of 250 people collected over $200,000 in fines and forfeitures from unwary "speeders."

What is the difference between a medium truck and a heavy truck?

Medium trucks weigh 14,001 to 33,000 pounds (6,351 to 14,969 kilograms). They span a wide range of sizes and have a variety of uses, from step-van route trucks to truck tractors. Common examples include beverage trucks, city cargo vans, and garbage trucks. Heavy trucks weigh 33,001 pounds (14,969 kilograms) or greater. Heavy trucks include over-the-road 18-wheelers, dump trucks, concrete mixers, and fire trucks. These trucks have come a long way from the first carrying truck, built in 1870 by John Yule, which moved at a rate of 0.75 mile (1.2 kilometers) per hour.

When was the parking meter introduced?

Carlton C. Magee (1872–1946), editor of the *Oklahoma City Daily News* and a member of the Chamber of Commerce traffic committee, became concerned about the parking problem in larger cities. He proposed a device to charge people for parking spaces. He entered into a partnership with Gerald A. Hale, a professor at Oklahoma Agricultural and Mechanical College, to perfect the mechanism. In 1932, Magee applied for a patent on a parking meter. In July 1935, meters were installed on some streets in Oklahoma City. Parking meters continue to help solve traffic and parking problems in major cities throughout the world.

What are some recent changes to parking meters?

Technological advances of the late twentieth and early twenty-first centuries have altered dramatically the design and function of the traditional coin-operated parking meter. No longer is it necessary to have one meter for each parking space. Instead, many cities now use master meters, also called multispaced meters, which as the name implies is a single meter that controls and manages an unlimited number of parking spaces. Many of these newer meters are also coinless, accepting debit and credit cards for payment, thus eliminating the need for drivers to always have coins. Washington, D.C., became one of the first U.S. cities to introduce pay-by-phone in 2011. Customers are able to pay for parking with credit cards by using their smartphones. Some cities have introduced dynamic pricing where the cost of parking changes based on availability and time of day. Other innovations are infrared sensors embedded in the street to let customers know where there are vacant spaces. Customers are less frustrated because they can find parking more easily.

What is the origin of the term "taxicab"?

The term "taxicab" is derived from two words—*taximeter* and *cabriolet*. The taximeter, an instrument invented by Friedrich Wilhelm Bruhn (1853–1927) in 1891, automatically

recorded the distance traveled and/or the time consumed. This enabled the fare to be accurately measured. The cabriolet was a two-wheeled, one-horse carriage, which was often rented.

The first taxicabs for hire were two Benz-Kraftdroschkes operated by "Droschkenbesitzer" Dütz in the spring of 1896 in Stuttgart, Germany. In May 1897 Friedrich Greiner started a rival service. In a literal sense, Greiner's cabs were the first "true" taxis because they were the first motor cabs fitted with taximeters.

Who invented the first motorcycle?

The earliest motorcycles were bicycles outfitted with steam engines. In 1867, Sylvester Howard Roper (1823–1896) created a "motocycle" or steam velocipede by attaching a two-cylinder steam engine powered by a charcoal burner to a bicycle. Large, bulky, and noisy, steam-powered bicycles never achieved widespread popularity. Concurrently, in Europe, Pierre Michaux (1813–1883) and Louis-Guillaume Perreaux (1816–1889) combined their efforts to attach a single-cylinder steam engine to a bicycle. Gottlieb Daimler (1834–1900) was the first to install an internal combustion engine powered by gasoline on a bicycle, creating the first true motorcycle in 1885. His first motorcycle had a two-speed transmission.

When did Harley-Davidson begin producing motorcycles?

The first Harley-Davidson motorcycle was available to the public in 1903.

BOATS AND SHIPS

What is dead reckoning?

Dead reckoning is the determination of a craft's current latitude and longitude by advancing its previous position to the new one on the basis of assumed distance and direction traveled. The influences of current and wind as well as compass errors are taken into account in this calculation, all done without the aid of any celestial or physical observation. This is a real test of a navigator's skill.

Nuclear-powered submarines, which must retain secrecy of movement and cannot ascend to the surface, use the SINS system (Ship's Inertial Navigation System), developed by the U.S. Navy. It is a fully self-contained system, which requires no receiving or transmitting apparatus and thus involves no detectable signals. It consists of accelerometers, gyroscopes, and a computer. Together they produce inertial navigation, which is a sophisticated form of dead reckoning.

What is the nautical meaning of the phrase "by and large"?

On a sailing ship or sailboat, new sailors at the helm are usually ordered to sail "by and large," meaning to sail into the wind at a slightly larger angle than those with more experience might choose. Sailing almost directly into the wind is most efficient, but doing so may cause the sail to flap back against the mast, resulting in loss of speed and control. Sailing "by and large" thus meant they were on the right, if not the perfect, course. Eventually, the phrase became generally used as a synonym for "approximately."

Why is the right side of a ship called "starboard"?

In the time of the Vikings, ships were steered by long paddles or boards placed over the right side. They were known in Old English as steorbords, evolving into the word starboard. The left side of a ship, looking forward, is called port. Formerly, the left side was called larboard, originating perhaps from the fact that early merchant ships were always loaded from the left side. Its etymology is Scandinavian, being lade (load) and bord (side). The British Admiralty ordered port to be used in place of larboard to prevent confusion with starboard.

Which type of wood was used to build Noah's ark?

According to the Bible, Noah's ark was made of gopher wood. This is identified as *Cupressus sempervirens*, one of the most durable woods in the world. Also called the

---

What is the name of the carved wooden figure of a woman on a sailing ship?

A carved wooden figure at the top of the stem of a sailing ship, usually in the shape of a woman, is called a figurehead.

Mediterranean cypress, the tree is native throughout southern Europe and western Asia. It grows up to 80 feet (24 meters) tall. Similar to this tree is the Monterey cypress (*Cupressus macrocarpa*), which is restricted to a very small area along the coast of central California. It can become as tall as 90 feet (27 meters) with horizontal branches that support a broad, spreading crown. When old, this tree looks very much like the aged cedars of Lebanon.

Where does the term "mark twain" originate?

Mark twain is a riverboat term meaning two fathoms (a depth of 12 feet or 3.6 meters). A hand lead is used for determining the depth of water where there are less than 20 fathoms. The lead consists of a lead weight of 7 to 14 pounds (3 to 6 kilograms) and a line of hemp or braided cotton, 25 fathoms (150 feet or 46 meters) in length. The line is marked at 2, 3, 5, 7, 10, 15, 17, and 20 fathoms. The soundings are taken by a leadsman who calls out the depths while standing on a platform projecting from the side of the ship, called "the chains." The number of fathoms always forms the last part of the call. When the depth corresponds to any mark on the lead line, it is reported as "By the mark 7," "By the mark 10," etc. When the depth corresponds to a fathom between the marks on the line, it is reported as "By the deep 6," etc. When the line is a fraction greater than a mark, it is reported as "And a half 7," "And a quarter 5"; a fraction less than a mark is "Half less 7," "Quarter less 10," etc. If the bottom is not reached, the call is "No bottom at 20 fathoms."

"Mark Twain" was also the pseudonym chosen by American humorist Samuel L. Clemens (1835–1910). Supposedly, he chose the name because of its suggestive meaning since it was a riverman's term for water that was just barely safe for navigation. One implication of this "barely safe water" meaning was, as his character Huck Finn would later remark, "Mr. Mark Twain … he told the truth, mostly." Another implication was that "barely safe water" usually made people nervous or at least uncomfortable.

When were steam-powered ships first introduced?

John Fitch (1743–1798) built one of the first steamboats in 1788. It made regularly scheduled trips on the Delaware River. The *Charlotte Dundas* was the first commercial steamboat. She was a small, wooden, single-paddle steamer. Her initial voyage in 1802 was on the Forth and Clyde Canal in Scotland. The 20-mile (32-kilometer) trip took six hours as she towed two 70-ton lighters. The *Charlotte Dundas* was withdrawn from service the following year because the canal's governors feared there would be damage to the canal.

The North River Steamboat of Clermont, often referred to simply as "Clermont," is acknowledged as the first commercially successful steamship. The *Clermont* was designed by American engineer Robert Fulton (1765–1815). Fulton and Robert Livingston (1746–1813) demonstrated a 65-foot (20-meter) paddle steamer on the River Seine in Paris in 1803. The steamer had an 8-horsepower engine that was designed in France. The 133-foot (45-meter) *Clermont* had a single-cylinder engine that drove a pair of 15-foot (4.6-meter) paddles, one on each side of the hull. The initial trip from New York to Albany and back, a distance of 240 miles (384 kilometers), took 62 hours, with an average

The *Charlotte Dundas* was the first steamboat put to practical use, making its first trial trip in 1802.

speed of 3.8 miles per hour (6 kilometers per hour). Fulton was inducted into the National Inventors Hall of Fame in 2006 for his invention of the steam engine.

How is a ship's tonnage calculated?

Tonnage of a ship is not necessarily the number of tons that the ship weighs. There are at least six different methods of rating ships; the most common are:

Displacement tonnage—used especially for warships and U.S. merchant ships—is the weight of the water displaced by a ship. Since a ton of sea water occupies 35 cubic feet (1 cubic meter), the weight of water displaced by a ship can be determined by dividing the cubic footage of the submerged area of the ship by 35. The result is converted to long tons (2,240 pounds or 1,017 kilograms). Loaded displacement tonnage is the weight of the water displaced when a ship is carrying its normal load of fuel, cargo, and crew. Light displacement tonnage is the weight of water displaced by the unloaded ship.

Gross tonnage (GRST) or *gross registered tonnage* (GRT)—used to rate merchant shipping and passenger ships—is a measure of the enclosed capacity of a vessel. It is the sum in cubic feet of the vessel's enclosed space divided by 100 (100 such cubic feet is considered one ton). The result is gross (registered) tonnage. For example, the old *Queen Elizabeth* did not weigh 83,673 tons but had a capacity of 8,367,300 cubic feet (236,935 cubic meters).

Deadweight tonnage (DWT)—used for freighters and tankers—is the total weight in long tons (2,240 pounds or 1,017 kilograms) of everything a ship can carry when fully loaded. It represents the amount of cargo, stores, bunkers, and passengers that are required to bring a ship down to her loadline, i.e., the carrying capacity of a ship. 243

Net registered tonnage (NRT)—used in merchant shipping—is the gross registered tonnage minus the space that cannot be utilized for paying passengers or cargo (crew space, ballast, engine room, etc.).

What does the term "loadline" mean in shipping?

A loadline or load waterline is an immersion mark on the hull of a merchant ship. This indicates her safe load limit. The lines vary in height for different seasons of the year and areas of the world. Also called the "Plimsoll line" or "Plimsoll mark," it was accepted as law by the British Parliament in the Merchant Shipping Act of 1875, primarily at the instigation of Samuel Plimsoll (1824–1898). This law prevented unscrupulous owners from sending out unseaworthy and overloaded, but heavily insured, vessels (so-called "coffin ships"), which risked the crew's lives.

What is the role of tugboats in transportation?

Tugboats are working vessels that move larger ships through narrow channels or move ships that are not capable of moving themselves. Tugboats may be used in harbors to guide ships through narrow shipping channels or to docks. They are often seen on rivers towing barges. Oceangoing tugs tow large ships, floating oil rigs, dredgers, or mining equipment. These often have towlines of 1,000 to 2,000 feet (300 to 600 meters). Specially designed tugboats are used as icebreakers to break through thick sheets of ice.

Who created the Liberty ship?

The Liberty ship of World War II was the brainchild of Henry J. Kaiser (1882–1967), an American industrialist who had never run a shipyard before 1941. The huge loss of merchant tonnage during the war created an urgent need to protect merchant vessels transporting weapons and supplies, and the Liberty ship was born. It was a standard merchant ship with a deadweight tonnage of 10,500 long tons (10,668 tonnes) and a service speed of 11 knots. Liberty ships were built to spartan standards, and production was on a massive scale. Simplicity of construction and operations, rapidity of building, and large cargo-carrying capacity were assets. To these, Kaiser added prefabrication and welding instead of riveting. The ships were a deciding factor on the side of the Allies. In four years, 2,770 ships with a deadweight tonnage of 29,292,000 long tons were produced.

When was the first hospital ship built?

It is believed that the Spanish Armada fleet of 1587–1588 included hospital ships. England's first recorded hospital ship was the *Goodwill* in 1608, but it was not until after 1660 that the Royal Navy made it a regular practice to set aside ships for hospital use. The U.S. government outfitted six hospital ships, some of which were permanently attached to the fleet, during the Spanish–American War of 1898. Congress authorized the construction of the USS *Relief* in August 1916. It was launched in 1919 and delivered to the navy in December 1920.

Why did the *Titanic* sink?

On its maiden voyage from Southampton, England, to New York, the British luxury liner *Titanic* sideswiped an iceberg at 11:40 P.M. on Sunday, April 14, 1912, and was badly damaged. The 882-foot (269-meter) long liner, whose eight decks rose to the height of an 11-story building, sank two hours and forty minutes later. Of the 2,227 passengers and crew, 705 escaped in 20 lifeboats and rafts; 1,522 drowned.

Famous as the greatest disaster in transatlantic shipping history, circumstances made the loss of life in the sinking of the *Titanic* exceptionally high. Although Capt. Edward John Smith (1850–1912) was warned of icebergs in shipping lanes, he maintained his speed of 22 knots and did not post additional lookouts. Later inquiries revealed that the liner *Californian* was only 20 miles (32 kilometers) away and could have helped had its radio operator been on duty. The *Titanic* had an insufficient number of lifeboats, and those available for use were badly managed, with some leaving the boat only half full. The only ship responding to distress signals was the ancient *Carpathia*, which saved 705 people.

Contrary to a long-held belief, the *Titanic* had not been sliced open by the iceberg. When Dr. Robert Ballard (1942–) from Woods Hole Oceanographic Institution descended to the site of the sunken vessel in the research vessel *Alvin* in July 1986, he found that the ship's starboard bow plates had buckled under the impact of the collision. This caused the ship to be opened up to the sea.

The "unsinkable" *Titanic* sank on its maiden voyage in 1912, drowning over 1,500 passengers and crew.

Ballard found the bow and the stern more than 600 yards (548 meters) apart on the ocean's floor and speculated on what happened after the collision with the iceberg: "Water entered six forward compartments after the ship struck the iceberg. As the liner nosed down, water flooded compartments one after another, and the ship's stern rose even higher out of the water until the stress amidships was more than she could bear. She broke apart ..." and the stern soon sank by itself.

More recent investigations have shown that defective rivets resulted in a structural weakness in the *Titanic*. Rivets from the hull were recently analyzed by a corrosion laboratory and were found to contain unusually high concentrations of slag, making them brittle and prone to failure. The weakened rivets popped and the plates separated.

What is the world's largest passenger ship?

The largest passenger ship ever built is the cruise ship *Oasis of the Seas*. The 225,000-ton ship is 1,187 feet (362 meters) long and has 16 decks with a capacity for 5,400 guests (double occupancy) in 2,700 staterooms. There are an additional 28 lofts. There are seven distinct areas, called neighborhoods, including areas for entertainment, spas and fitness, rock-climbing walls, pools, basketball courts, an area for children and youth, and a Central Park neighborhood complete with live plants. *Oasis of the Seas* was launched in 2009. Her sister ship, *Allure of the Seas*, is the same size. She was launched in 2010.

Was the luxury liner *Queen Mary* ever in the same port at the same time as the *QM2*?

The *Queen Mary* was one of the most magnificent and elegant luxury ocean liners that cruised the Atlantic Ocean. Her maiden voyage from Southampton, England, to New York was May 27, 1936. Her final voyage departed from Southampton on October 31, 1967, and arrived in Long Beach, California, on December 9, 1967, after traveling around Cape Horn. She has since been docked in Long Beach as a tourist attraction and hotel. The *QM2* departed Southampton on her maiden voyage to Fort Lauderdale, Florida, on January 12, 2004. The two ships met once in a royal rendezvous on February 23, 2006. The *QM2* arrived at Long Beach at noon. Each ship sounded a whistle to greet and salute each other at 12:30 P.M.

When were the first nuclear-powered vessels launched?

A controlled nuclear reaction generates tremendous heat, which turns water into steam for running turbine engines. The USS *Nautilus* was the first submarine to be propelled by nuclear power, making her first sea run on January 17, 1955. It has been called the first true submarine since it can remain underwater for an indefinite period of time. The *Nautilus*, 324 feet (99 meters) long, has a range of 2,500 miles (4,023 kilometers) submerged, a diving depth of 700 feet (213 meters), and can travel submerged at 20 knots.

The first nuclear warship was the 14,000-ton cruiser USS *Long Beach*, launched on July 14, 1959. The USS *Enterprise* was the first nuclear-powered aircraft carrier.

Launched on September 24, 1960, the *Enterprise* was 1,101.5 feet (336 meters) long and was designed to carry one hundred aircraft.

The first nuclear-powered merchant ship was the *Savannah*, a 20,000-ton (18,144-tonne) vessel, launched in 1962. The United States built it largely as an experiment, and it was never operated commercially. In 1969, Germany built the *Otto Hahn*, a nuclear-powered ore carrier. The most successful use of nuclear propulsion in non-naval ships has been as icebreakers. The first nuclear-powered icebreaker was the Soviet Union's *Lenin*, commissioned in 1959.

PLANES AND OTHER AIRCRAFT

What was the name of the Wright brothers' airplane?

The name of the Wright brothers' plane was the *Wright Flyer*. A wood and fabric bi-plane, the *Flyer* was originally used by the brothers as a glider and measured 40 feet, 4 inches (12 meters) from wing tip to wing tip. For their historic flight, Wilbur (1867–1912) and Orville (1871–1948) Wright outfitted it with a four-cylinder, 12-horsepower gasoline engine and two propellers, all of their own design. On December 17, 1903, at Kitty Hawk, North Carolina, Orville Wright made the first engine-powered, heavier-than-air craft flight lying in the middle of the lower wing to pilot the craft, which flew 120 feet (37 meters) in 12 seconds. The brothers made three more flights that day, with Wilbur Wright completing the longest one—852 feet (260 meters) in 59 seconds. The Wright brothers were inducted into the National Inventors Hall of Fame in 1975 for their invention of the airplane and their pioneering work in aviation.

How does the wing on an aircraft generate lift?

Bernoulli's principle, named for Swiss mathematician and physicist Daniel Bernoulli (1700–1782), states that an increase in a fluid medium's velocity results in a decrease in pressure. An aircraft's wing is shaped so that the air (a fluid medium) flows faster past the upper surface than the lower surface, thus generating a difference in pressures, causing lift. Centuries later, Bernoulli's principle is still the basis for our understanding of flight and other everyday applications.

Why did the dirigible *Hindenburg* explode?

Despite the official U.S. and German investigations into the explosion, the cause of the explosion still remains a mystery today. The most plausible explanations are structural failure, St. Elmo's Fire, static electricity, or sabotage. The *Hindenburg*, built following the great initial success of the *Graf Zeppelin*, was intended to exceed all other airships in size, speed, safety, comfort, and economy. At 803 feet (245 meters) long, it was 80 percent as long as the ocean liner *Queen Mary*, 135 feet (41 meters) in diameter, and could carry 72 passengers in its spacious quarters.

The *Hindenburg* disaster of 1935 put an end to the idea of commercial travel via dirigible.

In 1935, the German Air Ministry virtually took over the Zeppelin Company to use it to spread Nazi propaganda. After its first flight in 1936, the airship was very popular with the flying public. No other form of transport could carry passengers so swiftly, reliably, and comfortably between continents. During 1936, 1,006 passengers flew over the North Atlantic Ocean in the *Hindenburg*. On May 6, 1937, while landing in Lakehurst, New Jersey, its hydrogen, a highly flammable gas, burst into flames, and the airship was completely destroyed. Of the 97 people aboard, 62 survived.

What is a hovercraft?

A hovercraft is a vehicle supported by a cushion of air, thus capable of operating on land or water. In 1968, regular commercial hovercraft service was established across the English Channel.

Who made the first nonstop transatlantic flight?

The first nonstop flight across the Atlantic Ocean, from Newfoundland, Canada, to Ireland, was made by two British aviators, Capt. John W. Alcock (1892–1919) and Lt. Arthur W. Brown (1886–1948), on June 14–15, 1919. The aircraft, a converted twin-engined Vickers Vimy bomber, took 16 hours, 27 minutes to fly 1,890 miles (3,032 kilometers).

Later, Charles A. Lindbergh (1902–1974) made the first solo crossing flight on May 20–21, 1927, in the single-engined Ryan monoplane *Spirit of St. Louis*, with a wing spread of 46 feet (15 meters) and a nose-to-tail length of 27.6 feet (8.4 meters). His flight from New York to Paris covered a distance of 3,609 miles (5,089 kilometers) and lasted 33.5 hours. The first woman to fly solo across the Atlantic was Amelia Earhart (1897–1937), who flew from Newfoundland to Ireland May 20–21, 1932.

When was the first nonstop, unrefueled, round-the-world airplane flight?

Dick Rutan (1938–) and Jeana Yeager (1952–) flew the *Voyager*, a trimaran monoplane, in a closed-circuit loop westbound and back to Edwards Air Force Base, California, December 14–23, 1986. The flight lasted nine days, three minutes, and forty-four seconds, and covered 24,986.7 miles (40,203.6 kilometers). The first successful round-the-world flight was made by two Douglas World Cruisers between April 6 and September 28, 1924. Four aircraft originally left Seattle, Washington, and two went down. The two successful planes completed 27,553 miles (44,333 kilometers) in 175 days—with 371 hours, 11 minutes being their actual flying time. Between June 23 and July 1, 1931, Wiley Post (1898–1935) and Harold Gatty (1903–1957) flew around the world, starting from New York, in their Lockheed Vega *Winnie Mae*.

What is a Mach number?

A Mach number is the equivalent of the speed of sound. For example, Mach 2 is twice the speed of sound and Mach 0.5 is half the speed of sound.

Who made the first supersonic flight?

Supersonic flight is flight at or above the speed of sound. The speed of sound is 760 miles (1,223 kilometers) per hour in warm air at sea level. At a height of about 37,000 feet (11,278 kilometers), its speed is only 660 miles (1,062 kilometers) per hour. The first person credited with reaching the speed of sound (Mach 1) was Major Charles E. (Chuck) Yeager (1923–) of the U.S. Air Force. In 1947, he attained Mach 1.45 at 60,000 feet (18,288 meters) while flying the Bell X-1 rocket-engine research plane designed by John Stack and Lawrence Bell (1894–1956). This plane had been carried aloft by a B-29 and released at 30,000 feet (9,144 meters). Based on observational data of how the plane was handling, it is, however, highly likely that the sound barrier was broken on April 9, 1945, by Hans Guido Mutke (1921–2004) in a Messerschmitt Me262, the world's first operational jet aircraft. It is also probable that Chalmers Goodlin (1923–2005) broke the barrier in the Bell X-1 six months prior to Yeager's flight, and the barrier was broken again, by George Welch (1918–1954), in a North American XP-86 *Sabre* shortly before Yeager's flight. The Me262 and the XP-86 were both jet-powered aircraft. In 1949, the Douglas *Skyrocket* was credited as the first supersonic jet-powered aircraft to reach Mach 1 when Gene May (1904–1966) flew at Mach 1.03 at 26,000 feet (7,925 meters).

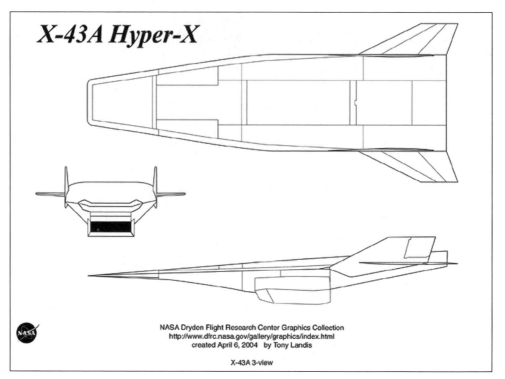

A NASA drawing of the X-43A, an unmanned scramjet capable of speeds nearing Mach 10.

What is the world's fastest aircraft?

NASA's X-43A, an unmanned scramjet, set the record for speed in November 2004 when it attained a speed of Mach 9.6 (approximately 7,000 miles per hour [11,265 kilometers per hour]).

The previous record was held by the North American X-15. The X-15 reached a peak altitude of 354,200 feet (108 kilometers). The X-15A-2 reached a speed of Mach 6.72 (4,534 miles per hour, 7,295 kilometers per hour). The X-15 is a rocket-powered aircraft that was launched/dropped by a modified B-52 bomber. It was designed in 1954 and first flown in 1959.

What is the world's fastest and highest flying jet aircraft?

The Lockheed SR-71, the Blackbird, is the fastest manned vehicle, achieving speeds of Mach 3.2. It holds three absolute world records for speed and altitude: speed in a straight line (2,193 miles per hour, 3,529 kilometers per hour); speed in a closed circuit (2,092 miles per hour, 3,366 kilometers per hour); and height in sustained horizontal flight (85,069 feet or 25,860 meters). The first flight of the Blackbird was on December 22, 1964. The final flight was on March 6, 1990. During the final flight, the SR-71 flew from Los Angeles to Washington, D.C., in 1 hour, 4 minutes, and 20 seconds at an average speed of 2,124 miles per hour (3,418 kilometers per hour). It is on display at the National Air and Space Museum's Steven F. Udvar-Hazy Center in Chantilly, Virginia.

What is the largest aircraft?

There are many ways to define the largest aircraft. Although length, height, wingspan, and empty weight are all valid criteria, maximum takeoff weight (MTOW) is most often used to identify the largest aircraft. The Russian Antonov An-225 is the largest airplane with a MTOW of 1.32 million pounds (600,000 kilograms). It can accommodate 46,000 cubic feet (1,300 cubic meters). The An-225 was built in the 1980s. It has a wingspan of 290 feet (88.4 meters) and is 275.58 feet (84 meters) long. It was built originally to carry the Russian space shuttle *Buran*.

How does the Airbus A380 compare to the Boeing 747-8 Intercontinental?

The Airbus A380 and Boeing 747-8 Intercontinental are two of the largest commercial passenger aircraft. Although the Boeing 747-8I is longer, the Airbus A380 has more capacity for passengers.

	Airbus A380	Boeing 747-8I
Length	238.6 feet (72.72 meters)	250.3 feet (76.3 meters)
Wingspan	261.6 feet (79.75 meters)	224.7 feet (68.5 meters)
Height	79.03 feet (24.09 meters)	63.6 feet (19.4 meters)
Seating capacity	525 in three-class configuration	467 in three-class configuration
Maximum takeoff weight	1,234,589 pounds (560,000 kilograms)	987,000 pounds (447,696 kilograms)
Cruising speed		Mach 0.85
Range	8,477 nautical miles (15,700 kilometers)	8,000 nautical miles (14,815 kilometers)

Who was the first to fly around the world in a balloon?

Steve Fossett (1944–c. 2007) was the first to fly solo around the world in a balloon. He departed western Australia on June 18, 2002, and returned on July 4, 2002, exactly 13 days, 23 hours, 16 minutes, and 13 seconds later. Fossett had previously made five attempts to circumnavigate the world in a balloon. Fossett disappeared while flying a small

Why don't tires on airplanes blow out when the airplane lands?

The Federal Aviation Agency requires airplane tires to meet rigid FAA standards to prevent accidental blowouts. A plug built into the tire will pop out when the air pressure gets too high, and the tire will deflate slowly rather than blow out.

plane over Nevada on September 3, 2007. His remains were not found until October 2008 amid the wreckage of his plane.

Who made the first parachute jump?

The first successful parachute jump from a great height was made by French aeronaut André-Jacques Garnerin (1769–1823) in 1797. He jumped 3,000 feet (914 meters) from a hot-air balloon.

What is avionics?

Avionics, a term derived by combining aviation and electronics, describes all of the electronic navigational, communications, and flight management aids with which airplanes are equipped today. In military aircraft it also covers electronically controlled weapons, reconnaissance, and detection systems. Until the 1940s, the systems involved in operating aircraft were purely mechanical, electric, or magnetic, with radio apparatus being the most sophisticated instrumentation. The advent of radar and the great advances made in airborne detection during World War II led to the general adoption of electronic distance-measuring and navigational aids. In military aircraft, such devices improve weapon delivery accuracy, and in commercial aircraft they provide greater safety in operation.

Steve Fossett, who made his fortune with his brokerage firms, traveled around the world in his balloon *The Spirit of Freedom* in 2002.

Where is the black box carried on an airplane?

Actually painted bright orange or red to make it more visible in an aircraft's wreckage, the black box is a tough metal-and-plastic case containing two recorders. Installed in the rear of the aircraft—the area most likely to survive a crash—the case has two shells of stainless steel with a heat-protective material between the shells. The case must be able to withstand a temperature of 2,000°F (1,100°C) for thirty minutes. Inside it, mounted in a shockproof base, is the aircraft's flight data and cockpit voice recorders. The flight data recorder provides information about airspeed, direction, altitude, acceleration, engine thrust, and rudder and spoiler positions from sensors that are located around the aircraft. The data are recorded as electronic pulses on stainless steel tape, which is about as thick as aluminum foil. When the tape is played back, it generates a computer printout. The cockpit voice recorder records the previous thirty minutes of the flight crew's conversation and radio transmission on a continuous tape loop. If a crash does not stop the recorder, vital information can be lost.

Who invented the black box?

The black box was invented by David Warren (1925–2010), an Australian aeronautical researcher. Warren began developing the black box after helping to investigate one of the world's first jet crashes in 1953. His research centered on developing a new use for a small recording device he had seen that was able to record voices for up to four hours. In 1954, he wrote a paper, "A Device for Assisting Investigation Into Aircraft Accidents." He completed building the prototype black box in 1957, naming it the ARL (for the Aeronautical Research Laboratories) Flight Memory Unit. The device was not accepted in Australia, but in 1958 a British aviation officer saw the device and immediately sent Warren and the ARL Flight Memory Unit to London where it was put into production. U.S. airlines began to install black boxes in the late 1950s.

When was the first full-scale wind tunnel for testing airplanes used?

It began operations on May 27, 1931, at the Langley Research Center of the National Advisory Committee for Aeronautics, Langley Field, Virginia. This tunnel, still in use, is 30 feet (9 meters) high and 60 feet (18 meters) wide. A wind tunnel is used to simulate air flow for aerodynamic measurement; it consists essentially of a closed tube, large enough to hold the airplane or other craft being tested, through which air is circulated by powerful fans.

What is the world's largest wind tunnel?

The largest wind tunnel in the world is the National Full-Scale Aerodynamics Complex at NASA Ames. The complex consists of two test areas, one 40 feet high by 80 feet wide (12 by 24 meters), the other 80 high by 120 feet wide (24 by 37 meters). The 80-by-120-foot area can generate wind speeds of up to 115 miles per hour (185 kilometers per hour).

What is a bird shot test?

Bird strikes—incidents where a bird collides with an aircraft—are not uncommon. To test components such as windshields, a small, dead bird, usually a chicken, is fired from a device into the windshield at an appropriate velocity.

Who designed the *Spruce Goose*?

American business tycoon, inventor, and aviator Howard Hughes (1905–1976) designed and built the all-wood H-4 Hercules flying boat, nicknamed the *Spruce Goose*. The aircraft had the greatest wingspan ever built and was powered by eight engines. It was flown only once—covering a distance of less than one mile at Los Angeles harbor on November 2, 1947, lifting only 33 feet (10.6 meters) off the surface of the water.

After the attack on Pearl Harbor on December 7, 1941, and the subsequent entry of the United States into World War II, the U.S. government needed a large, cargo-carrying airplane that could be made from noncritical wartime materials, such as wood. Henry J. Kaiser (1882–1967), whose shipyards were producing Liberty ships at the rate

The *Spruce Goose,* which flew a total of 33 feet in 1947, was designed by Howard Hughes. It was made too large to be practical.

of one per day, hired Hughes to build such a plane. Hughes eventually produced a plane that weighed 400,000 pounds (181,440 kilograms) and had a wingspan of 320 feet (97.5 meters). Unfortunately, the plane was so complicated that it was not finished by the end of the war. In 1947, Hughes flew the plane himself during its only time off the ground—supposedly just to prove that something that big could fly. The plane was on public display in Long Beach, California, but was sold in 1992 to aviation enthusiast Delford Smith (1930–2014) and shipped to McMinnville, Oregon, where it is now on display at the Evergreen Aviation Museum.

What is the difference between an amphibian plane and a seaplane?

The primary difference is that an amphibian plane has retractable wheels that enable it to operate from land as well as water, while a seaplane is limited to water takeoffs and landings, having only pontoons without wheels. Because its landing gear cannot retract, a seaplane is less aerodynamically efficient than an amphibian.

Who designed the first commercially successful helicopter?

Igor I. Sikorsky (1889–1972) built two rotating-wing machines in 1908, but they were not successful in leaving the ground. After emigrating to the United States from Russia in 1928, Sikorsky continued to design aircraft, including seaplanes and flying boats. During the 1930s, he again began to experiment with designing a helicopter. He flew his VS-300 on September 14, 1939, while it was tethered. Upon release, he discovered the VS-300 could fly in all directions except forward. He continued to modify the design until he was finally successful. The U.S. Army contracted to purchase its first helicopter in January 1941. Demand for helicopters continued to grow as their value for life-saving missions, first in the military and then in civilian life, became more recognized. Sikorsky was inducted into the National Hall of Fame in 1987 for the invention of the helicopter.

MILITARY VEHICLES

How did the military tank get its name?

During World War I, when the British were developing the tank, they called these first armored fighting vehicles "water tanks" to keep their real purpose a secret. This code word has remained in spite of early efforts to call them "combat cars" or "assault carriages."

Who invented the culin device on a tank?

In World War II, American tank man Sergeant Curtis G. Culin (1915–1963) devised a crossbar welded across the front of the tank with four protruding metal tusks. This device made it possible to break through the German hedgerow defenses. In the hedgerow country of Normandy, France, countless rows or stands of bushes or trees surrounded the fields, limiting tank movement. The culin device, also known as the "Rhinoceros" because its steel angled teeth formed a tusklike structure, cut into the base of the hedgerow and pushed a complete section ahead of it into the next field, burying any enemy troops dug in on the opposite side.

What is a Humvee?

The U.S. Army originally developed the HMMWV (High Mobility Multipurpose Wheeled Vehicle), or Humvee, in 1979 as a possible replacement for the M-151 or Jeep. Today, the military uses more than 100,000 "Hummers," which can operate in all weather extremes and are designed as troop transports, light-weapon platforms, ambulances, and mobile shelters.

Who was the Red Baron?

Manfred von Richthofen (1892–1918), a German fighter pilot during World War I, was nicknamed the "Red Baron" by the Allies because he flew a red-painted Albatros fighter. Although he became the top ace (aces are aviators credited with shooting usually at least five enemy aircraft during aerial combat) of the war by shooting down eighty Allied planes, only sixty of his kills were confirmed by both sides. The others are disputed and could have been joint kills by von Richthofen and his squadron, the Flying Circus (so named because of their brightly painted aircraft). Von Richthofen died on April 21, 1918, when he was attacked over the Semme River in France by Roy Brown (1893–1944), a Canadian ace, and Australian ground machine gunners. Both parties claimed responsibility for his death.

What is a Sopwith Camel and why is it so called?

The most successful British fighter plane of World War I, the Camel was a development of the earlier Sopwith Pup, with a much larger rotary engine. Its name "camel" was derived from the humped shape of the covering of its twin synchronized machine guns. The highly maneuverable Camel, credited with 1,294 enemy aircraft destroyed, proved far superior to all German types of fighter aircraft until the introduction of the Fokker D.VII in 1918. Altogether, 5,490 Camels were built by Sopwith Aircraft. Its top speed was 118 miles (189 kilometers) per hour and it had a maximum flying altitude of 24,000 feet (7,300 meters).

When was the B-17 Flying Fortress introduced?

A Fortress prototype first flew on July 28, 1935, and the first Y1B-17 was delivered to the Air Corps in March 1937, followed by an experimental Y1B-17A fitted with turbo-supercharged engines in January 1939. An order for thirty-nine planes was placed for this model under the designation B-17B. In addition to its bombing function, the B-17 was used for many experimental duties, including serving as a launching platform in the U.S. Army Air Force guided missile program and in radar and radio-control experiments. It was called a "Flying Fortress" because it was the best defended bomber of World War II. Altogether, it carried thirteen 50-caliber Browning M-2 machine guns, each having about 700 pounds (317.5 kilograms) of armor-piercing ammunition. Ironically, the weight of all its defensive armament and manpower severely restricted the space available for bombs.

The Sopwith Camel was a British fighter plane flown during World War I. A fun fact is that Snoopy from the comic strip "Peanuts" often pretended to fly one and have dog fights with the Red Baron.

Who were the Flying Tigers?

They were members of the American Volunteer Group, who were recruited early in 1941 by Major General Claire Lee Chennault (1890–1958) to serve in China as mercenaries. Some 90 veteran U.S. pilots and 150 support personnel served from December 1941 until June 1942 during World War II. The airplanes they flew were P-40 Warhawks, which had the mouths of tiger sharks painted on the planes' noses. It was from these painted-on images that the group got its nickname "Flying Tigers."

Why was the designation MiG chosen for the Soviet fighter plane used in World War II?

The MiG designation, formed from the initials of the plane's designers, Artem I. Mikoyan (1905–1970) and Mikhail I. Gurevich (1892–1976), sometimes is listed as the Mikoyan-Gurevich MiG. Appearing in 1940 with a maximum speed of 400 miles (644 kilometers) per hour, the MiG-3, a piston-engined fighter, was one of the few Soviet planes whose performance was comparable with Western types during World War II. One of the best-known fighters, the MiG-15, first flown in December 1947, was powered by a Soviet version of a Rolls-Royce turbo jet engine. This high performer saw action during the Korean War (1950–1953). In 1955 the MiG-19 became the first Soviet fighter capable of supersonic speed in level flight.

What is the name of the airplane that carried the first atomic bomb?

During World War II, the *Enola Gay*, a modified Boeing B-29 bomber, dropped the first atomic bomb on Hiroshima, Japan, at 8:15 A.M. on August 6, 1945. It was piloted by Col. Paul W. Tibbets Jr. (1915–2007) of Miami, Florida. The bombardier was Maj. Thomas W. Ferebee (1918–2000) of Mocksville, North Carolina. Bomb designer Capt. William S. Parsons (1901–1953) was aboard as an observer.

Three days later, another B-29, called *Bockscar*, dropped a second bomb on Nagasaki, Japan. The Japanese surrendered unconditionally on August 15, which confirmed the American belief that a costly and bloody invasion of Japan could be avoided at Japanese expense.

The *Enola Gay* was on display at the Smithsonian Institution's National Air and Space Museum in Washington, D.C., from 1995 to 1998. It went on permanent display at the National Air and Space Museum's Steven F. Udvar-Hazy Center in December 2003. *Bockscar* is on display at the U.S. Air Force Museum at Wright-Patterson Air Force Base in Dayton, Ohio.

Which aircraft was the first to use stealth technology?

The F-117A *Nighthawk* was first deployed in 1982. The goal of stealth technology is to make an airplane invisible to radar. There are two different ways to achieve invisibility; one is for the airplane to be shaped so that any radar signal it reflects is reflected away from the radar equipment, and the other is to cover the airplane in materials that absorb radar signals. Stealth aircraft have completely flat surfaces and very sharp edges to reflect away radar signals. Stealth aircraft are also treated so they absorb radar energy.

What are drones?

Drones are unmanned aerial vehicles (UAVs). While they are unmanned in the sense that there is no pilot or other crew aboard the aircraft while it is in flight, they are controlled by human operators. In some UAV systems, the operator remotely controls the vehicle with a stick and rud-

In addition to potential benefits, drones have become controversial because they can be hazards to commercial aircraft and, with cameras mounted to them, can pose privacy issues.

der control. In other systems, the operator enters way points and the aircraft autonomously determines how to reach that point. UAVs are often used for intelligence, surveillance, and reconnaissance—especially in situations that may put humans at risk. Newer designs of drones include those that resemble hummingbirds and insects equipped with video and audio equipment. The military has also investigated the use of unmanned combat air vehicles (UCAVs) for carrying weapons.

TRAINS AND TROLLEYS

What is a standard gauge railroad?

The first successful railroads in England used steam locomotives built by George Stephenson (1781–1848) to operate on tracks with a gauge of 4 feet, 8.5 inches (1.41 meters), probably because that was the wheel spacing common on the wagons and tramways of the time. Stephenson, a self-taught inventor and engineer, had developed in 1814 the steam-blast engine that made steam locomotives practical. His railroad rival, Isambard K. Brunel (1806–1859), laid out the line for the Great Western Railway at 7 feet, 0.25 inches (2.14 meters), and the famous "battle of the gauges" began. A commission appointed by the British Parliament decided in favor of Stephenson's narrower gauge, and the Gauge Act of 1846 prohibited using other gauges. This width eventually became accepted by the rest of the world. The distance is measured between the inner sides of the heads of the two rails of the track at a distance of 5/8 inch (16 millimeters) below the top of the rails.

What is the fastest train?

The French SNCF high-speed train TGV (*Train à Grande Vitesse*) *Atlantique* achieved the fastest speed—320.2 miles per hour (515.2 kilometers per hour)—recorded on any national rail system on May 18, 1990, between Courtalain and Tours.

The fastest passenger train in the Western Hemisphere is Amtrak's *Acela Express*. On the route between Boston and Washington, D.C., the top speed is generally 135 mph (217 kph) with a normal maximum speed of 150 miles per hour (241 kilometers per hour) on the 35-mile (56-kilometer) portion of the route between Boston and New Haven. The name *Acela* is a combination of the words "acceleration" and "excellence."

China has an extensive network of high-speed rail consisting of four North-South rail lines and four East-West rail lines. The total length of the grid is expected to be over 8,000 miles (13,000 kilometers) with plans to expand to over 31,000 miles (50,000 kilometers) by 2020. Average speeds on the lines are over 124 miles per hour (200 kilometers per hour) with some trains running at 217 miles per hour (350 kilometers per hour). The CRH380A has hit speeds of 302 miles per hour (486 kilometers per hour).

How does MAGLEV technology differ from standard railway design?

MAGLEV (*mag*netic *lev*itation) trains under development in Japan and Germany can travel 250 to 300 miles per hour (402 to 483 kilometers per hour) or more. These trains run on a bed of air produced from the repulsion or attraction of powerful magnetic fields (based on the principle that like poles of magnets repel and unlike poles [north and south] attract). The German Transrapid uses conventional magnets to levitate the train. The principle of attraction in magnetism, the employment of winglike flaps extending under the train to fold under a T-shaped guideway, and the use of electromagnets on board (that are attracted to the non-energized magnetic surface) are the guiding components. Interaction between the train's electromagnets and those built on top of the T-shaped track lift the vehicle 3/8 inch (1 centimeter) off the guideway. Another set of magnets along the rail sides provides lateral guidance. The train rides on electromagnetic waves. An alternating current in the magnet sets in the guideway changes their polarity to alternately push and pull the train along. Braking is done by reversing the direction of the magnetic field (caused by reversing the magnetic poles). To increase train speed, the frequency of the current is raised.

The Japanese MLV002 uses the same propulsion system, but the difference is in the levitation design, in which the train rests on wheels until it reaches a speed of 100 miles (161 kilometers) per hour. Then it levitates 4 inches (10 centimeters) above the guideway. The levitation depends on superconducting magnets and a repulsion system (rather than the attraction system that the German system uses).

What is the world's longest railway?

The Trans-Siberian Railway, from Moscow to Vladivostok, is 5,777 miles (9,297 kilometers) long. If the spur to Nakhodka is included, the distance becomes 5,865 miles (9,436 kilometers). It was opened in sections, and the first goods train reached Irkutsk on August 27, 1898. The Baikal-Amur Northern Main Line, begun in 1938, shortens the distance by about 310 miles (500 kilometers). The journey takes approximately seven days, two hours, and crosses seven time zones. There are 9 tunnels, 139 large bridges or viaducts, and 3,762 smaller bridges or culverts on the whole route. Nearly the entire line is electrified.

In comparison, the first American transcontinental railroad, completed on May 10, 1869, is 1,780 miles (2,864 kilometers) long. The Central Pacific Railroad was built eastward from Sacramento, California, and the Union Pacific Railroad was built westward to Promontory Point, Utah, where the two lines met to connect the line.

When was the first U.S. railroad chartered?

The first American railroad charter was obtained on February 6, 1815, by Colonel John Stevens (1749–1838) of Hoboken, New Jersey, to build and operate a railroad between the Delaware and Raritan rivers near Trenton and New Brunswick. However, lack of financial backing prevented its construction. The Granite Railway, built by Gridley Bryant (1789–1867), was chartered on October 7, 1826. It ran from Quincy, Massachusetts, to the Neponset River—a distance of 3 miles (4.8 kilometers). The main cargo was granite blocks used in building the Bunker Hill Monument.

When and why were cabooses eliminated from railroads?

The once familiar sight of a red caboose at the end of a train is mostly a historical memory now. Cabooses were home to conductors, brakemen, and flagmen on a train. In 1972, the Florida East Coast Railway uncoupled and eliminated its cabooses. In the early 1980s, the United Transportation Union agreed to the elimination of cabooses on many trains. Technology has replaced many of the functions of railroad employees. Computers have replaced the conductor's record keeping; an electronic "end-of-train device" monitors brake pressure, eliminating one of the rear brakeman's jobs; and trackside scanners have been installed to detect overhead axle bearings and report problems to the engineers.

What was the railroad velocipede?

In the nineteenth century, railroad track maintenance workers used a three-wheeled handcar to speed their way along the track. The handcar was used for inter-station express and package deliveries and for delivery of urgent messages between stations that could not wait until the next train. Also called an "Irish Mail," this 150-pound (68-kilogram) three-wheeler resembled a bicycle with a sidecar. The operator sat in the middle of the two-wheel section and pushed a crank back and forth, which propelled the triangle-shaped vehicle down the tracks. This manually powered handcar was replaced after World War

Train cabooses like this one are seen now only in museums or, sometimes, converted into restaurants.

I by a gasoline-powered track vehicle. This, in turn, was replaced by a conventional pickup truck fitted with an auxiliary set of flanged wheels.

What does the term "gandy dancer" mean?

A track laborer. The name derived from the special tools used for track work made by the Gandy Manufacturing Company of Chicago, Illinois. These tools were used during the nineteenth century almost universally by section gangs.

What was the route of the Orient Express?

This luxury train service was inaugurated in June 1883 to provide a connection between France and Turkey. It was not until 1889 that the complete journey could be made by train. The route left Paris and went via Chalons, Nancy, and Strasbourg into Germany (via Karlsruhe, Stuttgart, and Munich), then into Austria (via Salzburg, Linz, and Vienna), into Hungary (through Gyor and Budapest), south to Belgrade, Yugoslavia, through Sofia, Bulgaria, and finally to Istanbul (Constantinople), Turkey. Over the years, there were reductions in services until December 2009 when the Orient Express made its final trip. The Venice-Simplon Orient Express, a private luxury train, continues to operate.

How does a cable car, like those in San Francisco, move?

A cable runs continuously in a channel between the tracks located just below the street. The cable is controlled from a central station and usually moves about 9 miles (14.5 kilometers) per hour. Each cable car has an attachment, on the underside of the car, called a grip. When the car operator pulls the lever, the grip latches on to the moving cable and is pulled along by the moving cable. When the operator releases the lever, the grip disconnects from the cable and comes to a halt when the operator applies the brakes. Also called an endless ropeway, it was invented by Andrew S. Hallidie (1836–1900), who first operated his system in San Francisco in 1873.

What is a funicular railway?

A funicular railway is a type of railway used on steep grades, such as on a mountainside. Two counterbalanced cars or trains are linked by a cable, and when one moves down, the other moves up. Funicular railways in the United States are usually referred to as incline railways.

Where is the highest cable car in the world?

The Ba Na cable car in Da Nang, Vietnam, holds the record as the longest and highest nonstop cable car. The single-wire cable travels 3.6 miles (5.7 kilometers). The difference in height between the departure station and arrival station is 4,488 feet (1,369 meters). Prior to the opening of the Ba Na cable car, the highest cable car, which was also the longest, was the cable car in Merida in the Venezuelan Andes. It began in the mile-high (5,280 feet) city of Merida and ended after traversing 7.8 miles (12.5 kilometers) with

261

four station stops at Pico Espejo, Mirror Peak, at 15,633 feet (4765 meters). It took an hour to travel the entire distance. Service on the cable car in Merida was discontinued in 2008 after nearly 50 years of service due to deterioration of the infrastructure. It is being rebuilt with plans to reopen in 2015 as the Mukumbari cable car.

ENERGY

BASIC CONCEPTS

What is energy?

Physicists define energy as the capacity to do work. Work is defined as the force required to move an object some distance. There are many forms of energy, including chemical energy, heat (thermal) energy, light (radiant) energy, mechanical energy, electrical energy, and nuclear energy. The law of the conservation of energy states that within an isolated system, energy may be transformed from one form to another, but it can neither be created nor destroyed.

What is the role of energy in a modern, technological society?

Technology allows us to generate the energy we use in our daily lives. Energy use is divided into four economic sectors in our society: residential, commercial, transportation, and industrial. We use energy to heat, cool, and light our homes and offices, power our appliances and machines, provide power to our factories and manufacturing industries, and provide fuel for our transportation, including automobiles, buses, airplanes, and ships. Newer technologies allow us to generate and use energy more efficiently.

What are the two types of energy?

The two types of energy are kinetic energy and potential energy. Kinetic energy is moving energy. It is the energy possessed by an object as a result of its motion. Potential energy is stored energy. It is the energy possessed by an object as a result of its position. As an example, a ball sitting on top of a fence has potential energy. When the ball falls off the fence it has kinetic energy. The potential energy is transformed into kinetic energy.

263

What are the two groups of energy sources?

Energy sources are divided into two groups: renewable and nonrenewable. Renewable energy sources are sources that cannot be depleted. They include solar, wind, geothermal, biomass, and hydropower. Nonrenewable sources are sources that will be depleted with use over time. Fossil fuels, such as coal, natural gas, and oil, are nonrenewable energy sources. Renewable and nonrenewable energy sources are primary energy sources that are used to generate secondary energy sources such as electricity.

What is the difference between primary energy sources and secondary energy sources?

Primary energy sources may come from renewable or nonrenewable sources of energy, such as coal, solar, or nuclear sources. The energy from a primary source is converted for use as a secondary source. Secondary energy sources are derived from primary energy sources. For example, coal and wind are used to generate electricity. Electricity and hydrogen are examples of secondary energy sources. They are also referred to as energy carriers because they move energy in a useable form from one place to another place.

How is energy measured?

There are many different units used to measure energy. Two basic distinctions are units of energy that are not related to a particular fuel and ones that are "source-based" units that are related to the properties of a specific fuel. The basic unit used to measure energy in the English system is the British thermal unit (BTU). The basic unit used to measure energy in the metric system is the joule (J). The joule is named for James Prescott Joule (1818–1889), who discovered that heat is a type of energy. According to the International System of Units (SI), 1 BTU = 1055.06 J.

How did the historical definition of a calorie relate to the measurement of energy?

Historically, a calorie was defined as the amount of heat (energy) required to raise the temperature of 1 gram of water by 1°C, from 14.5°C to 15.5°C. Calorie is no longer used as a measurement of energy. The joule is the correct unit of measurement to use for en-

---

What is the relationship between energy and power?

Power is the rate at which energy is transferred. The unit of measure for power is the watt (W).

1 watt is equal to 1 joule per second

$$1 \text{ W} = 1 \text{ J/s}$$

The watt is named for Scottish engineer and inventor James Watt (1736–1819).

ergy measurement. The kilocalorie or "Calorie" was the unit of measurement for nutrition. The correct scientific measurement is the kilojoule.

Which economic sector—residential, commercial, transportation, or industrial— uses the most energy in the United States?

The industrial sector uses 32 percent of the total energy consumed in the United States. The transportation sector, including cars, trucks, buses, motorcycles, trains, boats, and aircraft, uses 28 percent of the total energy consumed in the United States. The remaining two sectors, residential and commercial, use 21 percent and 18 percent, respectively.

FOSSIL FUELS

Why are coal, oil, and natural gas called fossil fuels?

Coal, oil, and natural gas are composed of the remains of organisms that lived as long ago as 500 million years. These microscopic organisms (such as phytoplankton) became incorporated into the bottom sediments and then were converted, with time, to oil and gas. Coal is the remains of plants and trees (changing into peat and then lignite) that were buried and subjected to pressure, temperature, and chemical processes for millions of years. Fossil fuels are nonrenewable sources of energy. There is a finite supply of the resources for fossil fuels. Eventually, these resources will diminish to the point of being too expensive or too environmentally damaging to retrieve. Fossil fuels provide nearly 80 percent of all the energy consumed in the United States, including two-thirds of the electricity and nearly all of the transportation fuels.

What is the cleanest fossil fuel?

Natural gas, composed primarily of methane, is the cleanest of all fossil fuels, producing only carbon dioxide and water vapor when it is burned. Coal and oil produce higher levels of harmful emissions, including nitrogen oxides and sulfur dioxide when burned. Coal creates the most carbon dioxide when it is burned. Natural gas emits 25 to 50 percent less carbon dioxide than either oil or coal for each unit of energy produced.

Fossil Fuel Emission Levels

(Pounds per Billion BTU of Energy Input)

Pollutant	Natural Gas	Oil	Coal
Carbon Dioxide	117,000	164,000	208,000
Carbon Monoxide	40	33	208
Nitrogen Oxides	92	448	457
Sulfur Dioxide	1	1,122	2,591

How much of the world's energy use is fueled by coal?

Coal provides 29 percent of global primary energy needs and generates 40 percent of the world's electricity. In the United States, coal supplies more than half of the electricity consumed. Nearly 70 percent of total global steel production is dependent on coal.

How is coal used as an energy source?

Approximately 93 percent of the coal used in the United States is used in the electric power sector. It is used to generate approximately 39 percent of all the electricity generated in the United States. The

In the United States, coal still provides over half the energy for generating electricity.

remaining 7 percent of coal consumed in the United States is used as a basic energy source in other industries, including the steel, cement, and paper industries.

What countries produce the most oil?

The following countries produce the most oil as of 2014.

Country	Millions of Barrels	% of Total World Production*
United States	13.97	15%
Saudi Arabia	11.62	12%
Russia	10.85	12%
China	4.57	5%
Canada	4.38	5%
United Arab Emirates	3.47	4%
Iran	3.38	4%
Iraq	3.37	4%
Brazil	2.9	3%
Mexico	2.81	3%

*Source: U.S. Energy Information Administration.

When was the first oil well in the United States drilled?

The Drake well at Titusville, Pennsylvania, was completed on August 28, 1859 (some sources list the date as August 27). The driller, William "Uncle Billy" Smith (1812–1890), went down 69.5 feet (21 meters) to find oil for Edwin L. Drake (1819–1880), the well's operator. Within fifteen years, Pennsylvania oil field production reached over ten million 360-pound (163-kilogram) barrels a year. After 153 years, the well is still producing oil, although only about 1/10 of a barrel of oil each day. While not economically viable, it is a part of history. The oil is sold to a fuel company to make motor oil.

Where are the largest oil and gas fields in the world and in the United States?

The Ghawar field, discovered in 1948 in Saudi Arabia, is the largest in the world; it measures 150×22 miles (241×35 kilometers). The largest oil field in the United States is the Permian Basin, which covers approximately 100,000 square miles (160,934 square kilometers) in southeast New Mexico and western Texas.

Which country is the largest supplier of oil to the United States?

Canada is the largest supplier of crude oil to the United States. The five top suppliers of crude oil and petroleum products to the United States in 2012 were:

Top Crude Oil Suppliers	
Country	Percent of U.S. Imports
Canada	28%
Saudi Arabia	13%
Mexico	10%
Venezuela	9%
Russia	5%

Which U.S. state produces the most crude oil?

Texas is the largest producer of crude oil in the United States. In 2014, Texas produced 1,157,262 thousand barrels of oil. North Dakota is the second largest producer of crude oil in the United States, producing 396,886 thousand barrels of oil in 2014.

When was oil first produced in North Dakota?

Geologists began explorations for oil in North Dakota as early as 1923. Decades of exploration and drilling attempts failed to find oil. On April 4, 1951, the Clarence Iverson #1 in the Williston Basin began producing 240 barrels of oil per day.

What is shale?

Shale is a fine-grained sedimentary rock that forms from the compaction of silt and clay-sized mineral particles. Black shale contains organic material that can generate oil and natural gas. It also traps the generated oil and natural gas within its pores. It is only in the past decade that technologies of horizontal drilling and hydraulic fracturing have been developed so it is economically feasible to drill and reach the large volumes of oil and natural gas trapped in shale formations.

How much oil is estimated to be in the Bakken Formation?

The Bakken Formation is classified as a tight oil formation. Tight oil is produced from low-permeability sandstones, carbonates, such as limestone, and shale formations. The earliest wells of the Bakken Formation were drilled in the 1950s shortly after the Clarence Iverson #1 began producing oil. Estimates in the 1980s set the amount of recoverable oil in the Bakken Formation at 24.1 million barrels. By 1990, the estimate had risen to 61.5 million barrels. As recently as 1995, the U.S. Geological Survey (USGS) estimated there were 151 million barrels of recoverable oil in the Bakken Formation using the available technology. Due to advances in drilling technology, the USGS raised the estimate in the Bakken Formation to 3.0 from 4.3 billion barrels of recoverable oil. The adjacent Three Forks Formation in North Dakota, South Dakota, and Montana has an estimated 3.7 billion barrels of oil. Production of crude oil in the Bakken Formation surpassed one million barrels per day in 2014.

Which technologies have allowed North Dakota to become the second largest producer of crude oil in the United States?

The techniques of horizontal drilling and hydraulic fracturing have made recovery of crude oil in North Dakota more efficient and economically feasible. Traditional oil exploration uses vertical drilling to reach petroleum deposits. The advantage of horizontal drilling is that the well casing is in contact with a much greater percentage of the oil

Fracking—hydraulic fracturing—has become extremely controversial. Many people believe that energy companies are pumping poisons underground and endangering water supplies.

reservoir's surface oil. Horizontal drilling involves drilling straight down vertically until hitting shale. Once the drilling hits the shale layer, the drill string is bent to drill in parallel to the shale layer. Drilling sensors help the crew working above ground to steer the drill and make real-time adjustments based on directional data. Once the well is drilled, hydraulic fracturing is used to complete the well to boost oil production.

Is hydraulic fracturing a new technology?

The first commercial application of hydraulic fracturing, known colloquially as "fracking," was on March 17, 1949, in Oklahoma. Hydraulic fracturing is a well stimulation process used to maximize the extraction of underground resources—mostly oil and natural gas. Once the well is drilled and encased with steel and cement, fluid, oftentimes 99 percent water mixed with sand and less than 1 percent additives, is pumped into the well at high pressure to generate small fissures or fractures in the shale. Once the fractures are created, a propping agent is pumped into the fractures to keep them from closing when the pumping pressure is released. The fracturing fluids are collected at the surface for recycling or disposal. Hydraulic fracturing is applied to the majority of oil and gas wells in the United States.

How much oil is estimated to be in the Arctic National Wildlife Refuge?

Geologists are uncertain as to the exact amount of oil that is in the Arctic National Wildlife Refuge. Based on the one seismic survey that was done and the exploration of a few test wells drilled, it is estimated that there may be 16 billion barrels of oil under the tundra of the Arctic National Wildlife Refuge. Drilling for oil in the Arctic National Wildlife Refuge continues to pose many environmental issues and concerns.

Why is Pennsylvania crude oil so highly valued?

The waxy, sweet paraffinic oils found in Pennsylvania first became prominent because high-quality lubricating oils and greases could be made from them. Similar grade crude oil is also found in West Virginia, eastern Ohio, and southern New York. Different types of crude oil vary in thickness and color, ranging from a thin, clear oil to a thick, tarlike substance.

What are the steps in the refining process?

Once crude oil is recovered using drilling techniques, it is transferred either via pipelines or tankers or railroad cars to refineries. There are three basic steps in the refining process: 1) separation, 2) conversion, and 3) treatment. Modern technology for separating the

When was offshore drilling for oil first done?

The first successful offshore oil well was built off the coast at Summerland, Santa Barbara County, California, in 1896.

components or fractions of the crude oil involves piping oil through hot furnaces. The resulting liquids and vapors are discharged into distillation units. In the distillation units, the fractions are separated according to their weight and boiling point. The lighter fractions vaporize and rise to the top while the heavier ones condense back into solids and fall to the bottom. Once the fractions are separated they can go through the conversion process, often using the hydrocarbon cracking method. Other processes for conversion are alkylation and reforming. During the final treatment step, technicians adjust the octane level, vapor pressure ratings, and other settings for the final blend of the product.

What is the process known as hydrocarbon cracking?

Cracking is a process that uses heat to decompose complex substances. Hydrocarbon cracking is the decomposition by heat, with or without catalysts, of petroleum or heavy petroleum fractions (groupings) to give materials of lower boiling points. Thermal cracking, developed by William Burton (1865–1954) in 1913, uses heat and pressure to break some of the large, heavy hydrocarbon molecules into smaller gasoline-grade ones. The cracked hydrocarbons are then sent to a flash chamber where the various fractions are separated. Thermal cracking not only doubles the gasoline yield but has improved gasoline quality, producing gasoline components with good antiknock characteristics (no premature fuel ignition). Burton was inducted into the National Inventors Hall of Fame in 1984 for developing the cracking process.

How much does a barrel of oil weigh?

A barrel of oil weighs about 306 pounds (139 kilograms).

Why was lead added to gasoline and why is lead-free gasoline used in cars?

Tetraethyl lead was used for more than forty years to improve the combustion characteristics of gasoline. It reduces or eliminates "knocking" (pinging caused by premature ignition) in large, high-performance engines and in smaller, high-compression engines.

What does one barrel of crude oil produce when it is refined?

One barrel of crude oil contains 42 U.S. gallons. When one barrel of crude oil is refined, it produces approximately 45 gallons of petroleum products:

- 19 gallons of finished motor gasoline
- 12 gallons of diesel
- 6 gallons of other products
- 4 gallons of jet fuel
- 2 gallons of liquefied petroleum gases
- 1 gallon of heavy fuel oil
- 1 gallon of other distillates, such as heating oil

It provides lubrication to the extremely close-fitting engine parts, where oil has a tendency to wash away or burn off. However, lead will ruin and effectively destroy the catalyst presently used in emission control devices installed in new cars. Therefore, only lead-free gasoline must be used in modern vehicles. The sale of leaded gasoline for motor vehicles ended in 1996. All vehicles manufactured after July 1974 for sale in the United States were required to use unleaded gasoline.

What is a reformulated gasoline?

Oil companies are being required to offer new gasolines that burn more cleanly and have less impact on the environment. Typically, reformulated gasoline (RFG) contains lower concentrations of benzene, aromatics, and olefins; less sulfur; a lower Reid vapor pressure (RVP); and some percentage of an oxygenate (non-aromatic component), such as methyl tertiary butyl ether (MTBE). MTBE is a high-octane gasoline blending component produced by the reaction of isobutylene and methanol. It was developed to meet the ozone ambient air quality standards, but its unique characteristics as a water pollutant pose a challenge to the Environmental Protection Agency (EPA) in meeting the requirements of the Clean Air Act, the Safe Drinking Water Act, and the Underground Storage Tank Program. The Clean Air Act called for reformulated gasoline to be sold in the cities with the worst smog pollution beginning January 1, 1995. Reformulated gasoline is now used in seventeen cities and the District of Columbia.

What kinds of additives are in gasoline and why?

Gasoline Additives

Additive	Function
Antiknock compounds	Increase octane number
Scavengers	Remove combustion products of antiknock compounds
Combustion chamber	Suppress surface ignition and fouling of spark plug deposit modifiers
Antioxidants	Provide storage stability
Metal deactivators	Supplement storage stability
Antirust agents	Prevent rusting in gasoline-handling systems
Anti-icing agents	Suppress carburetor and fuel system freezing
Detergents	Control carburetor and induction system cleanliness
Upper-cylinder lubricants	Lubricate upper cylinder areas and control intake system deposits
Dyes	Indicate presence of antiknock compounds and identify makes and grades of gasoline

What do the octane numbers of gasoline mean?

The octane number is a measure of the gasoline's ability to resist engine knock. Two test fuels, normal heptane and isooctane, are blended for test results to determine the octane

271

number. Normal heptane has an octane number of zero and isooctane a value of 100. Gasolines are then compared with these test blends to find one that makes the same knock as the test fuel. The octane rating of the gasoline under testing is the percentage by volume of isooctane required to produce the same knock. For example, if the test blend has 85 percent isooctane, the gasoline has an octane rating of 85. The octane rating that appears on gasoline pumps is an average of research octane, determined in laboratory tests with engines running at low speeds, and motor octane, determined at higher speeds.

The numbers seen on gas fuel pumps indicate the percentage of isooctane.

What are the advantages and disadvantages of the alternative fuels to gasoline to power automobiles?

The emissions of gasoline are a major source of air pollution in most U.S. urban areas. Researchers are investigating alternative fuels that are cleaner, yet also economical. Alternative fuels include ethanol, methanol, biodiesel, compressed natural gas, electricity, and hydrogen. Currently, none of the alternatives deliver as much energy content as gasoline, so more of each of these fuels must be consumed to equal the distance that the energy of gasoline propels the automobile. Some predict that in the future, families may have different vehicles that use different fuels for different types of driving. For example, electric or hybrids for short trips, an ethanol/electric hybrid for longer trips, and a biodiesel vehicle for hauling large loads.

Types of Alternative Fuels

Alternative	Advantages	Disadvantages
Biodiesel from vegetable oils	Cleaner than petrodiesel; reduced emissions	Cost of pure biodiesel; blends (B30) turn into solids in low temperatures
Electricity from batteries	No vehicle emissions, good for stop-and-go driving	Short-lived, bulky batteries; limited trip range
Ethanol from corn, biomass, etc.	Relatively clean fuel	Costs; corrosive damage
Hydrogen from electrolysis, etc.	Plentiful supply; non-toxic emissions	High cost; highly flammable
Methanol from methanol gas, coal, biomass, wood	Cleaner combustion; less volatile	Corrosive; some irritant emissions
Natural gas from hydrocarbons and petroleum deposits	Cheaper on energy basis; relatively clean	Cost to adapt vehicle; bulky storage; sluggish performance

Alternative	Advantages	Disadvantages
Propane (liquefied petroleum gas)	Cleaner; good supply from domestic sources; less expensive, less readily available	Vehicles must be retrofitted

How do the costs of various fuels compare?

Fuel prices are reported in the units in which they are typically sold, for example, dollars per gallon. However, since the energy content per gallon of each fuel is different, the price paid per unit of energy is often different from the price paid per gallon. The following chart shows the average price, the average price in gasoline gallon equivalents, and the average price in dollars per million BTU as of 2014.

Average Fuel Prices (2014)

Fuel	Avg. Price/Gal.	Avg. Price in Gas Equivalents	Avg. Price in Dollars/Million BTU
Gasoline	$3.70	$3.70	$32.07
Compressed Natural Gas	$2.17 (per gas gallon equivalent)	$2.17	$18.80
Ethanol (E85)	$4.56	$5.09	$39.56
Propane	$4.24	$4.72	$36.71
Biodiesel (B20)	$3.63	$4.05	$31.48
Biodiesel (B99–B100)	$4.18	$4.66	$36.22

Interest in alternative fuels increases when the actual price differential per gallon increases, even if the savings is not as great on an energy-equivalent basis.

How is gasohol made?

Gasohol, a mixture of 90 percent unleaded gasoline and 10 percent ethyl alcohol (ethanol), has gained some acceptance as a fuel for motor vehicles. It is comparable in performance to 100 percent unleaded gasoline with the added benefit of superior anti-knock properties (no premature fuel ignition). No engine modifications are needed for the use of gasohol, and all auto manufacturers approve the use of gasohol (blends of 10 percent ethanol) in gasoline vehicles.

Since corn is the most abundant U.S. grain crop, it is predominantly used in producing ethanol. However, the fuel can be made from other organic raw materials, such as oats, barley, wheat, sorghum, sugar beets, or sugar cane. Potatoes, cassava (a starchy plant), and cellulose (if broken up into fermentable sugars) are possible other sources. The corn starch is processed through grinding and cooking. The process requires the conversion of a starch into a sugar, which in turn is converted into alcohol by reaction with yeast. The alcohol is distilled and any water is removed until it is 200 proof (100 percent alcohol).

Does natural gas have an odor?

Natural gas is odorless, colorless, and tasteless. The odor associated with natural gas is from mercaptan, a chemical that smells like sulfur—the characteristic "rotten egg" smell. It is added to natural gas prior to distribution for use by consumers as a safety measure. The odor allows it to be detected in the atmosphere in the event of a natural gas leak.

One acre of corn yields 250 gallons (946 liters) of ethanol; an acre of sugar beets yields 350 gallons (1,325 liters) while an acre of sugar can produce 630 gallons (2,385 liters). In the future, motor fuel could conceivably be produced almost exclusively from garbage, but currently its conversion remains an expensive process.

How does ethanol differ from gasohol?

Ethanol (E85) is a mixture of 85 percent ethyl alcohol and 15 percent gasoline. It can only be used in flexible fuel vehicles (FFVs). Flexible fuel vehicles are designed to run on gasoline, ethanol (E85), or any mixture of the two.

Which countries have the highest reserves of natural gas?

Worldwide reserves of natural gas are estimated at 6,972 trillion cubic feet as of 2014, with 40 percent of the total located in the Middle East and 33 percent of the total located in Europe and the former U.S.S.R.

Country	Natural Gas Reserves (Trillion Cubic Feet)
Russia	1,688
Iran	1,193
Qatar	885
United States	338
Saudi Arabia	291

How do advanced technologies assist scientists to locate natural gas deposits?

Geologists locate the type of rocks that are likely to contain oil and gas deposits. One technique they use is seismic surveys to find the best location to drill wells. Seismic surveys use echoes from a vibration source at the Earth's surface to collect information about the rocks beneath the surface. Oftentimes, dynamite is also used to create vibrations. Satellites and global positioning systems are also helpful to locate oil and gas deposits without excessive drilling. Once a natural gas deposit is located, drilling begins at the location. Natural gas flows up to the surface and into large pipelines to be transported to users.

A worker checks pressure gauges on natural gas pipelines. Natural gas supplies a quarter of the energy used in the United States.

How is natural gas transported from the processing plants to customers?

An extensive 2.4-million-mile (3.86-million-kilometer) underground pipeline system transports natural gas from the processing plants to customers throughout the United States. The pipeline system consists of small-diameter, low-pressure pipelines and wide-diameter, high-pressure pipelines that connect the drilling site to processing plants to the ultimate end-users in residential and commercial locations. Compressor (or pumping) stations keep the natural gas flowing forward along the pipeline system. While 1.2 million miles (1.9 million kilometers) were installed between 1950 and 1969, 225,000 miles (362,102 kilometers) of local distribution lines were added in the 1990s in response to the demand for access to natural gas for new commercial facilities and housing developments.

How is natural gas changed to a liquid?

Liquid natural gas (LNG) is natural gas that has been cooled to about −260°F (−162°C). The volume of LNG is about 600 times smaller than in the gaseous form, allowing it to be shipped or stored as a liquid. This allows LNG to be shipped to distribution centers via special tankers when pipelines are not an option, e.g., importing from distant countries. Once shipped to terminals, LNG is returned to the gaseous form and transported by the natural gas pipelines to distribution companies, industrial consumers, and power plants.

275

What are the uses of natural gas in the United States?

Natural gas accounts for approximately 25 percent of all the energy used in the United States. More than half of the homes in the United States use natural gas as the main heating fuel. One hundred cubic feet of natural gas can provide about 100,000 BTU (British thermal units) of heat. Some of the other appliances that use natural gas are stoves and ovens, water heaters, and clothes dryers. The single largest user of natural gas is the electric power generation industry.

How Natural Gas Is Used in the United States

Use	Amount used in trillion cubic feet (Tcf)	Percent of Total Used
Electric power generation	8.15	31
Industrial	7.46	29
Residential	4.94	19
Commercial	3.29	13
Lease and plant fuel consumption	1.41	5
Pipeline and distribution	0.74	3
Vehicle fuel	0.03	<1

What is the difference between "conventional" and "unconventional" natural gas?

"Conventional" gas and "unconventional" gas are identical, but the sources where they are found are different. "Conventional" gas describes natural gas accumulations found in high-permeability rock formations. The typical rock formations that produce conventional gas have high permeability such as sandstones or carbonates, including limestone. "Unconventional" gas describes natural gas accumulations found in rock formations that include some sort of trapping mechanism that prevents the gas from migrating and results in a high concentration of gas within the reservoir. Typical rock formations that trap "unconventional" gas are shale rock formations, low-permeability sandstones (also called tight gas), and coal seams.

When did shale gas become a commercial reality?

Shale gas production became a commercial reality around 2000. While geologists were aware of shale deposits since the early twentieth century, it was not economically feasible to drill for and extract the natural gas deposits in the shale. The combination of increased natural gas prices in the late 1990s and advances in drilling technologies in the early 2000s made it economically feasible to develop the shale gas.

Which states have significant quantities of shale gas?

According to the U.S. Energy Information Administration, Texas, Pennsylvania, Louisiana, and Arkansas produced nearly 80 percent of the shale gas in the United States in 2013. Together, these four states produced 26 billion cubic feet per day of shale gas.

Texas (Barnett, Eagle Ford, and Haynesville-Bossier)	11 billion cubic feet/day
Pennsylvania (Marcellus with a minimal amount from Utica)	8 billion cubic feet/day
Louisiana (Haynesville)	4 billion cubic feet/day
Arkansas (Fayetteville)	2.8 billion cubic feet/day

How much has shale gas production increased in recent years?

Since shale gas began being produced in the early 2000s, it has grown to representing 40 percent of total natural gas production and surpassed production from nonshale gas wells in 2013. That year the amount of shale gas production reached 33 billion cubic feet per day.

Gas Production Withdrawals (in million cubic feet)

Year	Production (conventional gas wells)	Production (shale gas wells)
2009	14,414,287	3,958,315
2010	13,247,498	5,817,122
2011	12,291,070	8,500,983
2012	12,504,227	10,532,858
2013	11,255,616	11,896,204

What are the environmental concerns associated with hydraulic fracturing?

The impact of hydraulic fracturing on air and water quality is an environmental concern. The Environmental Protection Agency is studying the environmental impact of the release of volatile organic compounds, hazardous air pollutants, and greenhouse gases. Since water plays a significant role in the process of hydraulic fracturing, the effects on groundwater, surface water, and drinking water from well construction and drilling are of concern.

What are synthetic fuels?

Synthetic fuels, commonly called synfuels, are gaseous and liquid fuels produced synthetically from coal and oil shale. Basically, coal is converted to gaseous or liquid forms in coal-based synfuels. These are easier to transport and burn more cleanly than coal itself. Synthetic natural gas is also produced from coal. Disadvantages of these processes are that they require a large amount of water and that the new fuels have 30–40 percent less fuel content than pure coal. Synfuels from oil shale are produced by extracting the oils from the rocky base. Gasoline and kerosene can be produced from oil shale. Synfuel may also be obtained from biomass from human and animal waste. The waste is converted to methane by the action of anaerobic bacteria in a digester.

How long will fossil fuel reserves last?

Determining how long fossil fuel reserves, the non-renewable sources of energy, will last is dependent on production rates and the rate of consumption. Other factors complicating the determination of how long fossil fuel reserves will last is whether additional supplies will be discovered, the economics of recovery of fossil fuels, and the growth of renewable sources of energy. Different models have been proposed, but researchers in the early twenty-first century predicted that oil and gas reserves would last another thirty-five and thirty-seven years, respectively. Coal reserves were predicted to last another 107 years. One study, published in 2009, predicted that coal will be the only fossil fuel available after 2042 and would last until 2112.

What is cogeneration?

Cogeneration is an energy production process involving the simultaneous generation of thermal (steam or hot water) and electric energy by using a single primary heat source. By producing two kinds of useful fuels in the same facility, the net energy yield from the primary fuel increases from 30–35 percent to 80–90 percent. Cogeneration can result in significant cost savings and can reduce any possible environmental effects conventional energy production may produce. Cogeneration facilities have been installed at a variety of sites, including oil refineries, chemical plants, paper mills, utility complexes, and mining operations.

NUCLEAR ENERGY

Is nuclear energy a renewable energy source?

Nuclear power plants require uranium for the process of fission. Uranium is a metal element found in rocks worldwide. It is mined and processed in order to be used in nuclear reactors. It is not a renewable resource, but it is relatively common metal. Nuclear power plants are a clean source of energy because they do not emit any of the greenhouse gases during production. There is the problem, however, of safely disposing nuclear waste.

What is the life of a nuclear power plant?

While there is some controversy about the subject, the working life of a nuclear power plant as defined by the industry and its regulators is approximately forty years, which is about the same as that of other types of power stations.

What are the two types of nuclear reactors in the United States?

Nuclear power plants contain a reactor unit that splits uranium atoms in a process called fission. During the process of fission, heat is released to turn water into steam, which

Nuclear power plants produce about twenty percent of America's energy. The good news is that, during operation, they emit very little pollution; the bad news is that the spent nuclear fuel (uranium or plutonium), which is no longer useful, must be disposed of safely for years.

then drives the turbine-generators that generate electricity. There are two types of reactors in the United States: boiling-water reactors and pressurized-water reactors. In a boiling-water reactor, the water heated by the reactor core turns directly into steam in the reactor vessel and is then used to power the turbine-generator. In a pressurized-water reactor, the water heated by the reactor core is kept under pressure and does not turn to steam. The hot, radioactive, liquid water then flows through and heats thousands of tubes in a steam generator. Outside the hot tubes in the steam generator is nonradioactive water, which boils and turns to steam. The radioactive water flows back into the reactor core where it is reheated and then returns to the steam generator.

When did the first nuclear power plant generate electricity for residential use?

The first nuclear power plant to generate electricity for residential use was a small facility in Obninsk, in the former Soviet Union, on June 27, 1954. The capacity of the power plant was only 5 megawatts. Due to the status of international relations during the Cold War, this accomplishment was not reported widely in the West. Two years later, in August 1956, the Calder Hall Unit 1 power plant became operational in Great Britain. Its initial capacity was 50 megawatts of electricity. Other units were added, and it continued to operate for nearly fifty years until March 2003. Calder Hall was the longest operating reactor in history.

A single uranium fuel pellet is about the size of a pencil eraser. It contains the same amount of energy as:

17,000 cubic feet of natural gas
1,780 pounds of coal
149 gallons of oil

When did the first commercial nuclear power plant in the United States become operational?

The first commercial electricity-generating nuclear power plant in the United States was located in Shippingport, Pennsylvania. It became operational on December 2, 1957, supplying energy to the Pittsburgh area. It continued to supply power until 1982. Although the Shippingport power plant was the first large-scale nuclear power plant in the United States, it was not the first civilian power plant. A few months earlier, in July 1957, the Santa Susana Sodium Reactor Experiment in California began providing power.

Where is the oldest operational nuclear power plant in the United States?

The oldest operational plant in the United States is the Oyster Creek, New Jersey, plant. It became operational on December 1, 1969. The original license for the Oyster Creek power plant was scheduled to expire on April 9, 2009. On April 8, 2009, the U.S. Nuclear Regulatory Commission approved a twenty-year license extension until April 9, 2029. The Nine Mile Point, Unit 1 also became operational on December 1, 1969. Its original license was to expire on August 22, 2009, but on October 31, 2006, the U.S. Nuclear Regulatory Commission granted a twenty-year extension to its license, extending it to August 22, 2029. Since the license for Oyster Creek was issued before the license for Nine Mile Point, Oyster Creek is officially the oldest operating U.S. reactor.

How many nuclear power plants are there worldwide?

As of 2015, 438 reactors were operational with 67 more under construction.

Nuclear Reactors Worldwide

Country	Number of Units	Number of Reactors Under Construction
Argentina	3	1
Armenia	1	
Belarus	0	2
Belgium	7	
Brazil	2	1
Bulgaria	2	
Canada	19	
China	27	24

Country	Number of Units	Number of Reactors Under Construction
Czech Republic	6	
Finland	4	1
France	58	1
Germany	9	
Hungary	4	
India	21	6
Iran	1	
Japan	43	2
Korea, South	24	4
Mexico	2	
Netherlands	1	
Pakistan	3	2
Romania	2	
Russia	34	9
Slovakia	4	2
Slovenia	1	
South Africa	2	
Spain	7	
Sweden	10	
Switzerland	5	
Taiwan	6	2
Ukraine	15	2
United Arab Emirates		1
United Kingdom	16	
United States	99	5

Which states in the United States have nuclear power plants?

As of 2015, the United States had 99 nuclear reactors at 61 power plants in operation in 30 states. Thirty-five of the power plants had two or more reactors. The plants generated a net of 797 billion kilowatt-hours of electricity, or 19.5 percent of domestic electricity, in 2014. In 2012, the U.S. Nuclear Regulatory Commission approved construction of two new nuclear reactors at the Vogtle nuclear power plant site in Georgia. They are expected to be operational in 2017 and 2018.

		Nuclear Plants per U.S. State (2013)			
State	# of Nuclear Reactors	# of Power Plants	Generating Capacity (megawatts)	Electricity (billion kilowatt-hours)	Percent of Electricity Generated for State
Alabama	5	2	5,044	33.7	27.1
Arizona	3	1	3,937	31.4	28.5
Arkansas	2	1	1,828	11.9	19.7
California	2	1	2,240	13.8	9.0
Connecticut	2	1	2,102	17.1	48.2
Florida	4	2	3,720	26.5	12.1
Georgia	4	2	4,061	32.9	27.2

State	# of Nuclear Reactors	# of Power Plants	Generating Capacity (megawatts)	Electricity (billion kilowatt-hours)	Percent of Electricity Generated for State
Illinois	11	6	9,722	97.1	47.9
Iowa	1	1	601	5.3	9.4
Kansas	1	1	876	7.1	14.7
Louisiana	2	2	2,134	17	16.7
Maryland	2	1	1,716	14.3	40.2
Massachusetts	1	1	677	4.3	12.8
Michigan	4	3	3,936	28.9	27.6
Minnesota	3	2	1,673	10.7	20.9
Mississippi	1	1	1,419	10.8	20.5
Missouri	1	1	1,190	8.4	9.1
Nebraska	2	2	1,245	6.9	18.5
New Hampshire	1	1	1,246	10.9	55.2
New Jersey	4	3	4,115	33.3	51.5
New York	6	4	5,264	44.8	33.1
North Carolina	5	3	5,056	40.2	32.2
Ohio	2	2	2,150	16.1	11.8
Pennsylvania	9	5	9,706	78.7	34.6
South Carolina	7	4	6,508	54.3	57.2
Tennessee	3	2	3,401	29	36.2
Texas	4	2	4,960	38.3	8.8
Vermont*	1	1	604	4.8	70
Washington	1	1	1,132	8.5	7.5
Wisconsin	2	1	1,182	9.9	17.8

*Vermont Yankee Nuclear Power Plant ceased operation at the end of 2014.

Which countries produce the most commercial nuclear power?

The United States produced the most commercial nuclear power in 2014. However, the percentage of electricity power by nuclear energy is greater for other countries.

World Nuclear Power

Country	Number of Reactors	Total Megawatts Electricity	Percent of Nation's Total Generated Electricity
France	58	63,130	76.9
Slovakia	4	1,814	56.8
Hungary	4	1,889	53.6
Ukraine	15	13,107	49.4
Belgium	7	5,921	47.5
Sweden	10	9,651	41.5
Switzerland	5	3,333	37.9
Slovenia	1	688	37.2
Czech Republic	6	3,904	35.8
Finland	4	2,752	34.6
Bulgaria	2	1,926	31.8

Country	Number of Reactors	Total Megawatts Electricity	Percent of Nation's Total Generated Electricity
Armenia	1	375	30.7
South Korea	24	21,667	30.4
United States	99	98,639	19.5

What actually happened at Three Mile Island?

The worst commercial nuclear accident in the United States occurred at Three Mile Island in March 1979. The Three Mile Island nuclear power plant in Pennsylvania experienced a partial meltdown of its reactor core and radiation leakage. On March 28, 1979, just after 4:00 A.M., a water pump in the secondary cooling system of the Unit 2 pressurized water reactor failed. A relief valve jammed open, flooding the containment vessel with radioactive water. A backup system for pumping water was down for maintenance.

Temperatures inside the reactor core rose, fuel rods ruptured, and a partial (52 percent) meltdown occurred because the radioactive uranium core was almost entirely uncovered by coolant for forty minutes. The thick steel-reinforced containment building prevented nearly all the radiation from escaping—the amount of radiation released into the atmosphere was one-millionth of that at Chernobyl. However, if the coolant had not been replaced, the molten fuel would have penetrated the reactor containment vessel, where it would have come into contact with the water, causing a steam explosion, breaching the reactor dome, and leading to radioactive contamination of the area similar to the Chernobyl accident.

What have been some notable nuclear accidents?

Nuclear Accidents

Date	Site	Description of Incident
October 7, 1957	Windscale plutonium production reactor (near Liverpool, England)	Fire in the reactor; released radioactive material; blamed for 39 cancer deaths
January 3, 1961	Idaho Falls, Idaho	Explosion in the reactor; radiation contained; 3 workers killed
October 5, 1966	Enrico Fermi demonstration breeder reactor (near Detroit, Michigan)	Partial core meltdown; radiation contained
January 21, 1969	Lucens Vad, Switzerland	Coolant malfunction; released radiation into a cavern; cavern sealed
March 22, 1975	Brown's Ferry Reactor (Decatur, Alabama)	Fire causes cooling water levels to become dangerously low
March 28, 1979	Three Mile Island, Pennsylvania	Partial core meltdown; minimal radiation released

Date	Site	Description of Incident
April 25, 1981	Tsunga, Japan	Workers exposed to radiation during repairs to nuclear plant
January 6, 1986	Kerr-McGee nuclear plant (Gore, Oklahoma)	Cylinder of nuclear material burst; 100 workers hospitalized; 1 death
April 26, 1986	Chernobyl, Ukraine (USSR in 1986)	Fire and explosions; radioactive material spread over much of Europe; at least 31 dead in the immediate aftermath; tens of thousands of cancer deaths and increased birth defects
September 30, 1999	Uranium-reprocessing facility (Tokaimura, Japan)	Container with uranium is overloaded; workers and nearby residents exposed to extremely high radiation levels
March 11, 2011	Fukushima Daiichi reactor (Fukushima, Japan)	Earthquake followed by a tsunami caused a loss of power; release of radioactive material; core meltdown; residents within 12 miles (20 kilometers) were evacuated

What is the Rasmussen report?

Dr. Norman Rasmussen (1927–2003) of the Massachusetts Institute of Technology (MIT) conducted a study of nuclear reactor safety for the U.S. Atomic Energy Commission. The 1975 study cost $4 million and took three years to complete. It concluded that the odds against a worst-case accident occurring were astronomically large—ten million to one. The worst-case accident projected about three thousand early deaths and $14 billion in property damage due to contamination. Cancers occurring later due to the event might number 1,500 per year. The study concluded that the safety features engineered into a plant are very likely to prevent serious consequences from a meltdown. Other groups criticized the Rasmussen report and declared that the estimates of risk were too low. After the Chernobyl disaster in 1986, some scientists estimated that a major nuclear accident might in fact happen every decade.

What is a meltdown, and what does it have to do with the "China Syndrome"?

A meltdown is a type of accident in a nuclear reactor in which the fuel core melts, resulting in the release of dangerous amounts of radiation. In most cases the large containment structure that houses a reactor would prevent the radioactivity from escaping. However, there is a small possibility that the molten core could become hot enough to burn through the floor of the containment structure and go deep into the earth. Nuclear engineers call this type of situation the "China Syndrome." The phrase derives from a discussion on the theoretical problems that could result from a meltdown, when a scientist commented that the molten core could bore a hole through the earth, coming out—if one happened to be standing in North America—in China. Although the scientist was grossly exaggerating, some took him seriously. In fact, the core would only bore a hole about 30 feet (10 meters) into the earth, but even this distance would have grave repercussions. All reactors are equipped with emergency systems to prevent such an accident from occurring.

What caused the Chernobyl accident?

The worst nuclear power accident in history, which occurred at the Chernobyl nuclear power plant in the Ukraine, will affect, in one form or another, 20 percent of the republic's population (2.2 million people). On April 26, 1986, at 1:23 A.M., during unauthorized experiments by the operators in which safety systems were deliberately circumvented in order to learn more about the plant's operation, one of the four reactors rapidly overheated, and its water coolant "flashed" into steam. The hydrogen formed from the steam reacted with the graphite moderator to cause two major explosions and a fire. The explosions blew apart the 1,000-ton (907-metric ton) lid of the reactor and released radioactive debris high into the atmosphere. It is estimated that 50 tons (45 metric tons) of fuel went into the atmosphere. An additional 70 tons (63.5 metric tons) of fuel and 700 tons (635 metric tons) of radioactive reactor graphite settled in the vicinity of the damaged unit. Human error and design features (such as a positive void coefficient type of reactor, use of graphite in construction, and lack of a containment building) are generally cited as the causes of the accident. Thirty-one people died from trying to stop the

The nuclear plant at Chernobyl, the Ukraine, remains abandoned to this day.

fires. More than 240 others sustained severe radiation sickness. Eventually, 150,000 people living near the reactor were relocated; some of whom may never be allowed to return home. Fallout from the explosions, containing radioactive isotope cesium-137, was carried by the winds westward across Europe.

The problems created by the Chernobyl disaster are overwhelming and continue today. Particularly troubling is the fact that by 1990–1991, a fivefold increase had occurred in the rate of thyroid cancers in children in Belarus. A significant rise in general morbidity has also taken place among children in the heaviest-hit areas of Gomel and Mogilev.

What was the distribution of radioactive fallout after the 1986 Chernobyl accident?

Radioactive fallout, containing the isotope cesium-137, and nuclear contamination covered an enormous area, including Byelorussia, Latvia, Lithuania, the central portion of the then Soviet Union, the Scandinavian countries, the Ukraine, Poland, Austria, Czechoslovakia, Germany, Switzerland, northern Italy, eastern France, Romania, Bulgaria, Greece, Yugoslavia, the Netherlands, and the United Kingdom. The fallout, extremely uneven because of the shifting wind patterns, extended 1,200 to 1,300 miles (1,930 to 2,090 kilometers) from the point of the accident. Estimates of the effects of this fallout range from 28,000 to 100,000 deaths from cancer and genetic defects within the next fifty years. In particular, livestock in high rainfall areas received unacceptable dosages of radiation.

Was the Fukushima nuclear accident in March 2011 caused by an earthquake?

The Tohoku (Great East Japan) Earthquake that occurred March 11, 2011, on the east coast of northern Honshu (the main island of Japan) had a magnitude of 9.0. It is believed to be one of the strongest earthquakes in recorded history. The nuclear power plants are designed with seismic sensors that are programmed to shut down the nuclear reactors when a significant earthquake occurs. The eleven nuclear reactors at four power plants that were operating at the time of the earthquake shut down automatically when they occurred.

Subsequent investigation revealed that the reactors functioned as designed in the event of an earthquake. When power was lost due to the earthquake, the emergency diesel generators started automatically to maintain the cooling of the reactors. However, the tsunami that occurred shortly after the earthquake caused significant problems. The tsunami waves that came ashore were about 49 feet (15 meters) high, submerging the turbines in

A sign warning of radioactivity is posted by a destroyed building in Fukushima, Japan.

more than 16 feet (5 meters) of seawater. All power, including the emergency power system, failed at the reactors. The cooling pumps were inoperable because of the flooding and lack of power. Without power for cooling, heat removal was difficult, and it took days until all the reactors reached cold shutdown.

How is nuclear waste stored?

Nuclear wastes consist either of fission products formed from the splitting of uranium, cesium, strontium, or krypton or from transuranic elements formed when uranium atoms absorb free neutrons. Wastes from transuranic elements are less radioactive than fission products; however, these elements remain radioactive far longer—hundreds of thousands of years. The types of waste are irradiated fuel (spent fuel) in the form of 12-foot (4-meter) long rods, high-level radioactive waste in the form of liquid or sludge, and low-level waste (non-transuranic or legally high-level) in the form of reactor hardware, piping, toxic resins, water from fuel pools, and other items that have become contaminated with radioactivity.

Currently, most spent nuclear fuel in the United States is safely stored in specially designed pools at individual reactor sites around the country. If pool capacity is reached, licensees may move toward use of above-ground dry storage casks. The three low-level radioactive waste disposal sites are Barnwell, located in South Carolina; Hanford, located in Washington; and Envirocare, located in Utah. Each site accepts low-level radioactive waste from specific regions of the country.

Most high-level nuclear waste has been stored in double-walled, stainless-steel tanks surrounded by 3 feet (1 meter) of concrete. The current best storage method, developed by the French in 1978, is to incorporate the waste into a special molten glass mixture, then enclose it in a steel container and bury it in a special pit. The Nuclear Waste Policy Act of 1982 specified that high-level radioactive waste would be disposed of underground in a deep geologic repository. Yucca Mountain, Nevada, was chosen as the single site to be developed for disposal of high-level radioactive waste. However, the Yucca Mountain site proved to be controversial due to dormant volcanoes in the vicinity and known earthquake fault lines. As of 2010, the Energy Department withdrew its application for a nuclear-waste repository at Yucca Mountain. Currently, the United States does not have a permanent underground repository for high-level nuclear waste. In 2012, the president's Blue Ribbon Commission on America's Nuclear Future recommended the development of one or more permanent deep geological facilities for the safe disposal of spent fuel.

What is vitrification?

Vitrification is a technique that transforms radioactive liquid waste into a solid. The nuclear material is mixed with sand or clay. The mixture is then heated to the melting point of the sand or to the point that the clay forms a ceramic. Although the resulting composite material is still radioactive, it is much easier and safer to handle, transport, and store the radioactive liquid waste as a solid.

RENEWABLE AND ALTERNATIVE ENERGY SOURCES

How do renewable energy resources differ from fossil fuels?

The main sources of renewable energy are biomass, such as wood, hydropower, geothermal, solar, and wind. Unlike fossil fuels, renewable energy resources are being replenished continuously and will never be depleted. Some sources of renewable energy, such as solar power and wind power, are perpetual, meaning the wind will blow and the sun will shine no matter how much energy is used. Sources of renewable energy that rely on agriculture, such as woods, are renewable as long as they are not depleted and exploited too rapidly.

How significant is the role of renewable energy resources in the United States?

Renewable energy resources accounted for approximately 9.6 quadrillion BTU or 10 percent of the total energy consumption in the United States in 2014. Biomass, including wood, biofuels, and biomass waste, accounted for 50 percent of the total renewable resources. Other renewable energy sources were: hydroelectric (28 percent), wind (18 percent), solar (4 percent), and geothermal (2 percent).

What are the advantages and disadvantages of solar power?

Solar energy is a clean, abundant, and safe energy source. More energy falls from the sun on Earth in one hour than is used by everyone in the world in one year. Over a two-week period, Earth gets as much energy from the sun as is stored in all known reserves of coal, oil, and natural gas. Solar energy can be used to heat water and spaces for homes and businesses or can be converted into electricity. The major advantages of solar power are that it does not produce air pollutants or emit greenhouse gases. Also, solar energy systems have minimal impact on the environment. Solar energy accounts for only about 3 percent of the total renewable energy resources, and less than 1 percent of electricity generation comes from solar power.

The disadvantages of solar energy are that sunlight is not constant. The amount of sunlight reaching the Earth varies with location, time of day, season, and weather. Also, there is not a large amount of energy from the sun delivered to any one place at one time. A large surface area is required to collect the energy at a useful rate.

What is the difference between passive solar energy systems and active solar energy systems?

Passive solar energy systems use the architectural design, the natural materials, or absorptive structures of the building as an energy-saving system. The building itself serves as a solar collector and storage device. An example would be thick-walled stone and adobe dwellings that slowly collect heat during the day and gradually release it at night. Passive systems require little or no investment of external equipment.

Active solar energy systems require a separate collector, a storage device, and controls linked to pumps or fans that draw heat from storage when it is available. Active

A mixed-use loft building in Hollywood, California, is designed to take advantage of passive solar energy. A system of screens and panels, positioned to allow light in or deflect it at certain times of the day, heat and cool the structure.

solar systems generally pump a heat-absorbing fluid medium (air, water, or an antifreeze solution) through a collector. Collectors, such as insulated water tanks, vary in size, depending on the number of sunless days in a locale. Another heat storage system uses eutectic (phase-changing) chemicals to store a large amount of energy in a small volume.

When were solar domestic hot water systems first developed?

Solar domestic hot water systems were developed in the 1960s in Australia, Japan, and Israel. They are considered a mature technology and are competitive with other technologies. Other countries that developed solar hot water systems were Spain, Cyprus, and, more recently, China.

How is solar energy converted into electricity?

Solar energy is converted into electricity using photovoltaic (PV) cells or concentrating solar power plants. Photovoltaic cells convert sunlight directly into electricity. Individual PV cells are combined in modules of about forty cells to form a solar panel. Ten to twenty solar panels are used to power a typical home. The panels are usually mounted on the home facing south or mounted onto a tracking device that follows the sun for the maximum exposure to sunlight. Power plants and other industrial locations combine more solar panels to generate electricity.

Concentrating solar power plants collect the heat (energy) from the sun to heat a fluid that produces steam that drives a generator to produce electricity. The three main types of concentrating solar power systems are parabolic trough, solar dish, and solar power tower, which describe the different types of collectors. Parabolic trough collectors have a long, rectangular, U-shaped reflector or mirror focused on the sun with a tube (receiver) along its length. A solar dish looks very much like a large satellite dish that concentrates the sunlight into a thermal receiver that absorbs and collects the heat and transfers it to the engine generator. The engine produces mechanical power that is used to run a generator converting mechanical power into electrical power. A solar tower uses a field of flat, sun-tracking mirrors, called heliostats, to collect and concentrate the sunlight onto a tower-mounted heat exchanger (receiver). A fluid is heated in the receiver to generate steam that is used in a generator to produce electricity.

How does a solar cell generate electricity?

A solar cell, also called a photovoltaic (PV) cell, consists of several layers of silicon-based material. When photons, particles of solar energy from sunlight, strike a photovoltaic cell, they are reflected, pass through, or are absorbed. Absorbed photons provide energy to generate electricity. The top, p-layer absorbs light energy. This energy frees electrons at the junction layer between the p-layer and the n-layer. The freed electrons collect at the bottom, n-layer. The loss of electrons from the top layer produces "holes" in the layer that are then filled by other electrons. When a connection, or circuit, is completed between the p-layer and n-layer, the flow of electrons creates an electric current. The photovoltaic effect, including the naming of the p-layer and n-layer, was discovered by Russell Ohl (1898–1987), a researcher at Bell Labs, in 1940.

When were photovoltaic cells developed?

A group of researchers at Bell Labs, Calvin Fuller (1902–1994), Daryl Chapin (1906–1995), and Gerald Pearson (1905–1987), developed the first practical silicon solar cell in 1954. They filed for a patent in 1954 and were granted U.S. Patent #2,780,765, "Solar Energy Converting Apparatus," in 1957. Fuller, Chapin, and Pearson were inducted into the National Inventors Hall of Fame in 2008 for the invention of the silicon solar cell. The earliest PV cells were used to power U.S. space satellites. The use of PV cells was then expanded to power small items such as calculators and watches.

Where is the largest solar generating plant located?

The largest solar generating plant in the world is the Ivanpah Solar Energy Generating System (ISEGS) located in California's Mojave Desert. The three-unit power system on 3,500 acres (14.2 square kilometers) produces 377 megawatts of electricity—enough power annually for 140,000 homes in California. It became operational in 2013. Similar to other electric generation plants, Ivanpah creates high-temperature steam to turn a conventional turbine. Over 300,000 software-controlled mirrors track the sunlight in two dimensions and reflect and focus the sunlight to boilers on top of three separate 459-foot-tall (140-meter-tall) towers. The concentrated sunlight strikes the boilers' pipes, heating the water to create superheated steam. The steam is piped from the boiler to a standard turbine to generate electricity.

Who invented the fuel cell?

The earliest fuel cell, known as a "gas battery," was invented by Sir William Grove (1811–1896) in 1839. Grove's fuel cell incorporated separate test tubes of hydrogen and oxygen, which he placed over strips of platinum. It was later modified by Francis Thomas

How much land is required to supply the United States with its total electric needs from solar energy?

Scientists estimate that a 10,000-square-mile area (100 miles × 100 miles, or 25,900 square kilometers) could generate 46,464,000 megawatt-hours of electricity per day or 17 trillion kilowatt-hours per year. In 2013, electricity consumption totaled nearly 3,831 billion kilowatt-hours in the United States. Renewable energy sources account for only 13 percent of the total electric generation, of which solar energy accounted for only 2 percent of the total in 2013.

Bacon (1904–1992) with nickel replacing the platinum. A fuel cell is equivalent to a generator—it converts a fuel's chemical energy directly into electricity.

What is biomass energy?

Biomass consists of organic material from plants and animals. Unlike fossil fuels, which were produced millions of years ago, biomass materials contain energy absorbed from the sun's rays through photosynthesis. Wood, crops and crop waste, and wastes of plant, mineral, and animal matter form the biomass. Much of it is in garbage, which can be burned for heat energy or allowed to decay and produce methane gas. However, some crops are grown specifically for energy, including sugar cane, sorghum, ocean kelp, water hyacinth, and various species of trees. It has been estimated that 90 percent of U.S. waste products could be burned to provide as much energy as 100 million tons of coal (20 percent will not burn but can be recycled). The use of biomass energy is significantly higher in developing countries where it is used as a primary energy source for cooking and heating. Biomass can be converted into biofuels such as biogas or methane, methanol, and ethanol. However, the process has been more costly than the conventional fossil fuel processes. Rubbish buried in the ground can provide methane gas through an aerobic decomposition. One ton of refuse can produce 8,000 cubic feet (227 cubic meters) of methane.

Is biomass energy a new technology?

Wood, a form of biomass energy, has been used as a fuel source for thousands of years. Until as recently as the mid-1800s, wood was the main source of energy—especially for heating and cooking—for most of the world, including the United States. In the twenty-first century, wood and wood products provide about 2 percent of the total amount of energy that is consumed. Nearly 80 percent of the energy from wood and wood products is used by industry, electric power producers, and commercial businesses. Many of the paper and wood manufacturing plants use the wood waste they generate to produce their own steam and electricity. The challenge of continuing to use wood and wood products as a source of energy is to manage it responsibly with sustainable practices so it continues to be a renewable source of energy. A second challenge is to use it in a way that minimizes the emissions from burning wood.

A pile of wood pulp waste serves as a fuel for this biomass plant.

How much wood is in a cord?

A cord of wood is a pile of logs 4 feet (1.2 meters) wide, 4 feet (1.2 meters) high, and 8 feet (2.4 meters) long. It may contain from 77 to 96 cubic feet of wood. The larger the unsplit logs, the larger the gaps, with fewer cubic feet of wood actually in the cord. Burning one full cord of wood produces the same amount of energy as one ton of coal.

Where are the largest biomass power plants in the world?

The largest biomass power plant is the Ironbridge power plant located in Severn Gorge, United Kingdom. Originally a coal-fired power station, two of the units at the plant were converted for biomass-based power production using wood pellets in 2013. It has a capacity of 740 megawatts. It is scheduled to close sometime in 2015 as this book goes to press.

The Polaniec biomass power plant in southeast Poland is the largest 100 percent biomass-fueled power plant in the world. It uses a mixture of 80 percent tree-farming products and 20 percent agricultural by-products as fuel to generate 205 megawatts of electricity. It uses a state-of-the-art technology circulating fluidized bed boilers to generate enough electricity for 600,000 households.

The largest biomass power plant in the United States is the New Hope Power Partnership located in South Bay, Florida. It burns sugar cane fiber and recycled urban wood to generate 140 megawatts of electricity—enough to power 60,000 homes along with the company's milling and refining operations. It is estimated that the facility saves approximately one million barrels of oil per year.

Which woods have the best heating quality in a wood-burning stove?

Wood accounts for 22 percent of the total of renewable energy resources in the United States. Woods that have high heat value, meaning that one cord equals 200 to 250 gallons (757 to 946 liters) of fuel oil or 250 to 300 cubic feet (7 to 8.5 cubic meters) of natural gas, are hickory, beech, oak, yellow birch, ash, hornbeam, sugar maple, and apple.

Woods that have medium heat value, meaning that one cord equals 150 to 200 gallons (567 to 757 liters) of fuel oil or 200 to 250 cubic feet (5.5 to 7 cubic meters) of natural gas, are white birch, Douglas fir, red maple, eastern larch, big leaf maple, and elm.

Woods that have a low heat value, meaning that one cord equals 100 to 150 gallons (378 to 567 liters) of fuel oil or 150 to 200 cubic feet (4 to 5.5 cubic meters) of natural gas, are aspen, red alder, white pine, redwood, western hemlock, eastern hemlock, Sitka spruce, cottonwood, western red cedar, and lodgepole pine.

Is wind energy a new technology?

Wind is a naturally occurring phenomenon caused by the difference in heating land and water areas on the earth's surface. Its force is captured and converted into mechanical energy. In ancient Egypt, wind propelled boats along the Nile River. Wind-powered water pumps were known in China in 200 B.C.E. Wind technology was brought to Europe from the Middle East by merchants and the Crusaders. The Dutch developed windpumps to drain lakes and marshes. Windmills were used in colonial America to pump water, grind grain, and cut wood at sawmills. During the late 1800s and early 1900s, small wind-electric generators were installed. However, in the late twentieth century, research was devoted to developing and harnessing wind energy to generate electricity. In 2014 wind energy in modern wind turbines generated about 4 percent of the total electricity generation in the United States.

How large are modern wind turbines?

The two basic types of modern wind turbines are horizontal-axis and vertical-axis turbines. Vertical-axis turbines look like a giant, two-bladed egg beater with blades attached to the top and bottom of a vertical rotor. Vertical-axis turbines may be as large as 100 feet (30.5 meters) tall and 50 feet (15 meters) wide.

Horizontal-axis turbines are much more frequently used since they out-perform vertical-axis turbines. The length of the blades on the turbine is a major factor in determining the electric generation capacity. Typical blades on modern turbines are made of laminated materials, such as composites, balsa wood, carbon fiber and fiberglass, and range from 112 feet (34 meters) to 180 feet (55 meters), with some as long as 260 feet (79 meters). The towers range in height from 262 feet (80 meters) to 328 feet (100 meters)—taller than the Statue of Liberty. Turbines are designed with a yaw drive system that rotates in response to wind direction changes.

How much electricity does one wind turbine generate?

The amount of electricity one wind turbine generates depends on the size of the turbine and the speed of the wind through the rotor. Small wind turbines used to power a single home or business may have a capacity of less than 100 kilowatts (100,000 watts). The largest turbines have a capacity of five to eight million watts.

How is the location of a wind farm determined?

Wind farms are clusters of wind turbines that generate electricity. Developers of wind farms collect data on how fast and how often the wind blows before developing a site as a wind farm. In general, the tops of smooth, rounded hills, open plains, shorelines, and mountain gaps that funnel and intensify wind are good locations for wind farms. Wind speeds of 14.3 miles per hour (23 kilometers per hour) at 164 feet (50 meters) above ground offer good choices for placement of wind turbines. Wind speeds vary on a daily and seasonal basis throughout the United States. Many times, the daily and seasonal variations correspond to seasonal variations in demand for electricity.

The San Gorgonio Pass Wind Farm near Palm Springs, California, is one of the largest in the world with over three thousand operational windmills.

Where are the largest wind farms?

The largest wind farms in the United States are in Texas, which has a total installed capacity of 35.9 million megawatt-hours. The Horse Hollow Wind Energy Center in Texas has 430 wind turbines distributed on 47,000 acres of land. There are 39 states with wind energy projects. The following chart lists the top five states with their capacity for electricity generated by wind:

State	Wind-generated Electricity
Texas	35.9 million megawatt-hours
Iowa	15.6 million megawatt-hours
California	13.2 million megawatt-hours
Oklahoma	10.9 million megawatt-hours
Illinois	9.6 million megawatt-hours

Which states generate the highest percentage of electricity from wind?

The United States generated 65,877 megawatts of electricity from the wind in 2014. This is enough to power over 16.7 million average American homes. The states that had the largest share of their total electricity generation from wind in 2014 were:

State	Power Needs from Wind
Iowa	28.5%
South Dakota	25.29%

State	Power Needs from Wind
Kansas	21.67%
Idaho	18.31%
North Dakota	17.58%

Which countries in the world generate the most electricity using wind energy?

The United States is the world's largest total producer of electricity from wind, but only 4.1 percent of total U.S. electricity generation was from wind in 2013. The five countries in the world that have the largest portion of their electricity generation from wind are:

Top Countries Generating Power from Wind

Country	Percent of Total Power
Denmark	34%
Portugal	23%
Spain	18%
Ireland	16%
Germany	8%

How significant is the offshore wind industry in Europe?

Studies have shown that wind speeds over water are often greater than those over land. As of 2014, there were 74 offshore wind farms in 11 European countries consisting of 2,488 wind turbines with an installed capacity of 8,045.3 megawatts of electricity. In a normal wind year, this produces 29.6 terawatt hours, which is approximately 1 percent of the European Union's total electricity consumption.

Are there any offshore wind farms in the United States?

There are no offshore wind farms currently operating in the United States, but the Cape Wind project is slated for construction on Horseshoe Shoal in Nantucket Sound off Cape Cod, Massachusetts. This wind farm is expected to be a 468-megawatt project with a projected cost of $2.5 billion. It is hoped that construction will begin in 2015, and the project will be up and running by 2016. If built, Cape Wind would include 130 wind turbines that could provide, under average winds, 75 percent of the energy needs for Cape Cod, Martha's Vineyard, and Nantucket.

When and where was the first offshore wind farm installed?

The first offshore wind farm was installed in 1991 in Vindeby, Denmark. It consists of eleven turbines and generates five MW of electricity. Thirty-four percent of Denmark's electricity is supplied by wind power.

What are the benefits and drawbacks of wind energy?

Wind turbines do not release emissions that can pollute the air or water. They also do not require water for cooling. Studies have shown that wind power is more efficient than other conventional power sources. In addition, wind turbine technology can range from a single turbine to wind farms consisting of hundreds or thousands of turbines. It takes less time to construct a wind farm than to construct coal-powered or natural gas power plants. One downside of wind energy is that it can be intermittent. Some individuals have objected to wind farms for aesthetic reasons. Also, some turbines can pose a threat to flying birds and bats.

How many tons of emissions are saved by the use of wind energy?

When the 10 million megawatt-hours of electricity is produced by wind farms, 6.7 million tons of carbon dioxide, 37,500 tons of sulfur dioxide, and 17,750 tons of nitrogen oxides are not released into the atmosphere. In addition, the amount of carbon pollution not entering the atmosphere is equivalent to a freight train of 50 cars, with each car holding 100 tons of solid carbon, every day.

What is geothermal energy?

Geothermal energy is heat contained beneath Earth's crust and brought to the surface in the form of steam or hot water. The five main sources of this geothermal reservoir are dry, super-heated steam from steam fields below Earth's surface; mixed hot water, wet steam, etc., from geysers; dry rocks (into which cold water is pumped to create steam); pressurized water fields of hot water and natural gas beneath ocean beds; and magma (molten rock in or near volcanoes and 5 to 30 miles [8 to 48 kilometers] below Earth's crust). Electric power production, industrial processing, and space heating are fed from geothermal sources.

Where is the largest geothermal power plant complex?

The California Geysers project, located 75 miles (121 kilometers) north of San Francisco, is the largest geothermal electric-generating complex in the world. The complex covers an area of 45 square miles (117 square kilometers) and has 15 power plants with 333 steam wells and 60 injection wells. It has a net generating capacity of 725 megawatts of electricity—approximately enough to power 725,000 homes or a city the size of San Francisco.

When was the first geothermal power plant built?

The first geothermal power station was built in 1904 in Larderello, Italy, by Prince Piero Conti (1865–1939). The prince attached a generator to a natural steam-driven engine to produce enough power to illuminate five light bulbs. In 1913 the first geothermal power plant with a capacity of 250 kilowatts was installed.

Which states have geothermal power plants?

Geothermal power plants in the United States produced 17 billion kilowatt-hours of electricity or 0.4 percent of the total U.S. electricity generation in 2014. Seven states—California, Nevada, Utah, Hawaii, Oregon, Idaho, and New Mexico—all have geothermal power plants. The geothermal power plants in California produced almost 76 percent of all geothermal-generated electricity in the United States.

What are the advantages and disadvantages of geothermal power?

Geothermal energy is a renewable resource that is mostly emission- and pollu-

A geothermal power plant in Svartsengi, Iceland. Iceland has a considerable amount of geothermal energy available to it due to its many natural hot springs and geysers.

tion-free. It is estimated that geothermal power plants release less than 1 percent of the carbon dioxide emissions released by a fossil fuel plant. Furthermore, geothermal power plants utilize scrubber systems to remove the hydrogen sulfide that is naturally found in the steam and hot water used to generate geothermal power. In addition, geothermal plants emit 97 percent less acid rain-causing sulfur compounds than are emitted by fossil fuel power plants. Costs to operate a geothermal power plant are relatively inexpensive. It has the potential to produce power consistently, unlike wind and solar power. Currently, drilling for geothermal energy is expensive. Since rocks lose heat over time, new sites and wells have to be drilled. A final disadvantage of geothermal energy is that it is restricted to the geographic area that has the natural resources to sustain a geothermal power plant.

How does hydropower produce electricity?

The water cycle is an integral part of how hydropower is used to generate electricity. The three steps of the water cycle are:

1. Solar energy heats water on the surface, causing it to evaporate.
2. Water vapor condenses into clouds and falls as precipitation.
3. Water flows through rivers back into the oceans, where it can evaporate and begin the cycle again.

The kinetic energy of moving water is used to turn turbines and create electricity. The two types of hydroelectric power plant systems are run-of-the-river systems and storage systems. Most hydropower plants have dams that collect and store a supply of water and release it to run through turbines when power is needed. In run-of-the-river systems, the force of the current of the river applies the needed force to run the turbines. The amount of power generated depends on the distance the water falls and the volume of water released.

Where was the first hydroelectric power plant in the United States built?

The Fox River in Appleton, Wisconsin, was the site of the first hydroelectric-generating plant in the United States. It began operation on September 30, 1882. It produced enough electricity to light the Appleton Paper and Pulp Company, Kimberly & Clark's nearby paper mill, and the home of Henry J. Rogers, who ran the Appleton Paper and Pulp Company.

How much electricity does the Niagara Power Plant produce?

The Niagara Power Plant consists of two main facilities—the Robert Moses Niagara Power Plant with 13 turbines and the Lewiston Pump-Generating Plant with 12 pump-turbines. The power plant diverts water from the Niagara River. The water flows through the turbines that power the generators. Together, they are capable of producing 2.4 million kilowatts of electricity, which is enough power to light 24 million 100-watt light bulbs at once.

Where is the largest hydroelectric dam in the world?

The Three Gorges Dam in China is the largest hydroelectric dam based on electricity production in the world. It is 594 feet (181 meters) high and has a length of about 7,770 feet (2,335 meters). It has a generating capacity of 22,500 megawatts. The dam forms the Three Gorges Reservoir, which has a surface area of about 400 square miles (1,045 square kilometers). It extends upstream from the dam about 370 miles (600 kilome-

China's Three Gorges Dam produces more electricity than any other dam in the world: 22,500 megawatts.

ters). The second largest hydroelectric dam in the world is the Itaipú hydroelectric power plant in Brazil and Paraguay. The Itaipú has a generating capacity of 14,000 megawatts. However, during the course of a year, both plants generate about the same amount of electricity. This is because seasonal variations in water availability on the Yangtze River in China limit the power generation capacity at the Three Gorges during several months of the year.

The Grand Coulee Dam on the Columbia River in the state of Washington is the largest hydroelectric dam in the United States with a generating capacity of 6,800 megawatts.

How can the ocean be used as a source of energy?

Tidal and wave energy contain enormous amounts of energy that can be harnessed. The idea is to harness the motion of wind-driven waves at the ocean's surface and convert this mechanical energy into electricity. The first tidal-powered mill was built in England in 1100; another in Woodbridge, England, built in 1170, has functioned for over 800 years. The Rance River Power Station in France, in operation since 1966, was the first large tidal electric generator plant, producing 160 megawatts. Differences in height between low and high tides are especially great in long, narrow bays, such as Alaska's Cook Inlet or the Bay of Fundy between New Brunswick and Nova Scotia. These locations are best for harnessing tidal energy, accomplished by erecting dams across the outlets of tidal basins. A tidal station works similar to a hydropower dam, with its turbines spinning as the tide flows through them. While tidal stations release few or no pollutants, they may have substantial impacts on the ecology of estuaries and tidal basins. Unfortunately, the tidal period of 13.5 hours causes problems in integrating the peak use with the peak generation ability.

What is the most commonly used renewable energy source for generation of electricity?

Renewable sources of energy accounted for 13 percent of the total generation of electricity in the United States in 2014. Hydroelectric power is the most common renewable energy source to generate electricity. Based on consumption, hydroelectric power accounted for nearly 6 percent of the total electricity generation in the United States and 48 percent of the total amount of renewable energy resources. Other renewable sources of energy used to generate electricity are:

1. Wind: 34 percent
2. Biomass wood: 8 percent
3. Biomass waste: 4 percent
4. Geothermal: 3 percent
5. Solar: 3 percent

GENERATION OF ELECTRICITY

What is a generator?

A generator is a device that converts a form of energy into electricity. Generators convert kinetic or mechanical energy into electrical energy. Most generators use an electromagnet to produce electricity. The principle of electromagnetic induction was discovered by Michael Faraday (1791–1867) in 1831. When an electric conductor, such as a copper wire, is moved through a magnetic field, electric current will flow ("be induced") in the conductor. The generator has coiled wire on its shaft surrounded by a giant magnet. When the shaft inside the turbine turns, an electric current is produced in the wire.

How does a power plant generate electricity?

A power plant consists of turbines and generators. Turbines may use steam, gas combustion, water, or wind as their energy source. The turbine rotates, turning the shaft surrounding the electromagnet. The mechanical energy of the spinning turbine is converted to electricity.

What are the fuel sources for the generation of electricity?

Although most electricity is produced by burning fossil fuels, nuclear and renewable energy sources also contribute to the generation of electricity. Specifically, the fuel sources for electricity generation are:

1. Coal: 39 percent
2. Natural gas: 27 percent
3. Nuclear: 19 percent
4. Renewable sources: 13 percent
5. Oil: 1 percent

How is electricity moved from a power plant to homes and industries?

Electricity is generated at many power plants throughout the country. Most of the electricity that is generated has a voltage of 25,000 watts. Electricity goes first to a transformer at the power plant that boosts the voltage to 400,000 volts. The boost in voltage is important, so electricity can be transferred more efficiently over long distances. Electricity is now moved over high-voltage power transmission lines throughout the country to substations near consumers. When the electricity reaches the substations, step-down transformers convert the power from very high voltage to lower voltages for use in homes and businesses. The distribution lines carry electricity via overhead or underground wires to homes and businesses. Transformers lower the voltage again before the electricity enters homes for use by consumers. Large appliances, such as some air conditioners, may require 220 volts, but smaller appliances require 110 volts.

How many power grids provide power to North America?

There are three distinct power grids, also called "interconnections," in North America. The Eastern Interconnection includes the eastern two-thirds of the United States and Canada from Saskatchewan east to the Maritime Provinces. The western third of the United States, along with Alberta and British Columbia and a portion of Baja California Norte, Mexico, are in the Western Interconnection. The third interconnection is most of the state of Texas. There are a few direct current (DC) ties that link the three distinct grids.

Electric power stations like this one use transformers to create high voltages in order to send the power over long distances.

When was one of the worst power outages in North America?

On August 14, 2003, parts of Ohio, Michigan, Pennsylvania, New York, Vermont, Massachusetts, Connecticut, New Jersey, and Ontario, Canada, experienced a power outage. An estimated 50 million people were affected by the power outage. Power was not restored for four days in parts of the United States, and rolling blackouts continued in Ontario for more than a week. It is estimated that the power outage resulted in a $6 billion economic loss.

How many miles of transmission lines form the power grid?

There are more than 450,000 miles (724,205 kilometers) of transmission lines in the grid. The power grid was first built in the 1890s with improvements during the subsequent years. There are more than 9,200 electric generating units capable of generating more than 1 million megawatts of electricity. The purpose of the power grid is to ensure that everyone in the United States has reasonable access to electricity service. Many local grids are interconnected for reliability. The grid transfers electricity from the producers to the users. It includes power generators, a system of substations, power lines, switches, and transformers to control and deliver the electricity to the users of electricity.

What is the Smart Grid?

The technical definition of the Smart Grid used by the National Institute of Standards and Technology is "a modernized grid that enables bidirectional flows of energy and uses two-way communication and control capabilities that will lead to an array of new functionalities and applications." Basically, the Smart Grid is a planned nationwide network that uses information technology to deliver electricity efficiently, reliably, and securely. It differs from the current grid because it will allow two-way flow of both electricity and information. It will be less centralized and more consumer-interactive. Smart grid technologies will be able to make real-time assessments of energy demand and meet the demand almost instantaneously. The goal is to have a more efficient, more reliable, and more secure power grid. The technologies for the Smart Grid are developing. It is anticipated that by 2030 the power grid will be very different from the current power grid infrastructure.

What are "smart" meters?

"Smart" meters are officially called advanced metering infrastructure (AMI) installations. They differ from standard electric meters because they measure and record electricity usage at a minimum of hourly intervals. They also provide data to both the electric utility company and the customer at least once per day. Some are real-time meters capable of transmitting instantaneous data. By 2013 electric utilities in the United States had installed 51,924,502 "smart" meters. The greatest use was for residential customers.

MEASUREMENT OF ENERGY

What is the weight per gallon of common fuels?

One Gallon of Fuel	Weight in Lbs/Kgs
Propane	4.23/1.9
Butane	4.86/2.2
Gasoline	6.00/2.7
Kerosene	6.75/3.1
Aviation gasoline	6.46–6.99/2.9–3.2

How do various energy sources compare?

Below are listed some comparisons (approximate equivalents) for the energy sources:

Energy Unit	Equivalent
1 BTU of energy	1 blue match tip
1 million BTU of energy	90 pounds of coal
	120 pounds of oven-dried hardwood
	8 gallons of motor gasoline
	11 gallons of propane
1 quadrillion BTU of energy	45 million short tons of coal
	60 million short tons of oven-dried hardwood
	1 trillion cubic feet of dry natural gas
	170 million barrels of crude oil
1 barrel of crude oil	5.6 thousand cubic feet of dry natural gas
	0.26 short tons (520 pounds) of coal
	1,700 kilowatt-hours of electricity

Energy Unit	Equivalent
1 short ton of coal	3.8 barrels of crude oil
	21 thousand cubic feet of dry natural gas
	6,500 kilowatt-hours of electricity
1,000 cubic feet of natural gas	0.18 barrels (7.4 gallons) of crude oil
	0.05 short tons (93 pounds) of coal
	300 kilowatt-hours of electricity
1,000 kilowatt-hours of electricity	0.59 barrels of crude oil
	0.15 short tons (310 pounds) of coal
	3,300 cubic feet of dry natural gas

Note: One quadrillion equals 1,000,000,000,000,000.

What are the approximate heating values of fuels?

Different types of energy are measured using different physical units. In the United States, the British thermal unit (BTU) is most commonly used to compare different types of fuel. A BTU is defined as the amount of energy required to raise the temperature of one pound of water by 1°F. The following chart lists the BTU content of common energy units, based on consumption in the United States in 2013:

Fuel	BTU	Unit of Measure
Oil	5,800,000	barrel (42 gallons)
Coal	19,480,000	short ton
Natural gas	1,025	cubic foot
Electricity	3,412	kilowatt-hour
Gasoline	124,262	gallon

What are the fuel equivalents to produce one quad of energy?

One quad (meaning one quadrillion) is equivalent to:

- 1×10^{15} BTU
- 252×10^{15} calories or 252×10^{12} K calories

In fossil fuels, one quad is equivalent to:

- 170 million gallons of crude oil
- 1 trillion cubic feet of natural gas
- 37.88 million tons of anthracite coal
- 38.46 million tons of bituminous coal

In nuclear fuels, one quad is equivalent to 2,500 tons of triuranium octaoxide if only uranium-235 is used.

In electrical output, one quad is equivalent to 2.93×10^{11} kilowatt-hours electric.

How is a heating degree day defined?

Early in the twenty-first century, engineers developed the concept of heating degree days as a useful index of heating fuel requirements. They found that when the daily mean temperature is lower than 65°F (18°C), most buildings require heat to maintain a 70°F (21°C) temperature. Each degree of mean temperature below 65°F (18°C) is counted as "one heating degree day." For every additional heating degree day, more fuel is needed to maintain a 70°F (21°C) indoor temperature. For example, a day with a mean temperature of 35°F (1.5°C) would be rated as 30 heating degree days and would require twice as much fuel as a day with a mean temperature of 50°F (10°C; 15 heating degree days). The heating degree concept has become a valuable tool for fuel companies for evaluation of fuel use rates and efficient scheduling of deliveries. Detailed daily, monthly, and seasonal totals are routinely computed for the stations of the National Weather Service.

What does the term "cooling degree day" mean?

It is a unit for estimating the energy needed for cooling a building. One unit is given for each degree Fahrenheit above the daily mean temperature when the mean temperature exceeds 75°F (24°C).

How many BTU are equivalent to one ton of cooling capacity?

1 ton = 288,000 BTU/24 hours or 12,000 BTU/hour.

CONSUMPTION AND CONSERVATION

Does the United States produce enough primary energy to meet its consumption needs?

No. Since 1958 the United States has consistently consumed more energy than it has produced. The energy that is not produced domestically is imported from foreign countries.

Year	Total U.S. Energy Production (quadrillion BTU)	Total U.S. Energy Consumption (quadrillion BTU)
1950	35.54	34.62
1960	42.80	45.09
1970	63.50	67.84
1980	67.18	78.07
1990	70.71	84.49
1995	71.17	91.03
2000	71.33	98.81
2005	69.43	100.28
2010	74.79	98.02
2011	77.99	97.46
2012	79.21	95.06
2013	81.94	97.85

Which fuel is used most by the transportation sector in the United States?

Gasoline is the dominant fuel used by cars, motorcycles, and light trucks in the United States. Diesel fuel is used more frequently by heavy trucks, buses, and trains.

Fuel	Percent of Total Use by Transportation Sector
Gasoline	57
Diesel	21
Jet fuel	11
Biofuels	4
Natural gas	3
Other	3

How does driving speed affect gas mileage for most automobiles?

Most automobiles reach the optimal fuel economy at different speeds, but it usually decreases at speeds above 50 miles per hour (80 kilometers per hour). The U.S. Department of Energy estimates that each 5 miles per hour (8 kilometers per hour) is comparable to paying an additional $0.16 per gallon for gasoline. Aggressive driving, such as speeding and rapid acceleration, can lower gas mileage by 33 percent on the highway and 5 percent for local driving.

When is it more economical to restart an automobile rather than let it idle?

Tests by the Environmental Protection Agency have shown that it is more economical to turn an engine off rather than let it idle if the idle time exceeds sixty seconds. Depending on the engine size of the vehicle and whether or not the air conditioning is also running, idling can use a quarter to a half gallon of fuel per hour while idling.

How much gasoline do underinflated tires waste?

Gas mileage can be improved by up to 3.3 percent by keeping the vehicle's tires inflated to the proper pressure. Underinflated tires can lower gas mileage by 0.3 percent for every 1 psi drop in pressure of all four tires. To save fuel, follow the automaker's guidelines regarding recommended air pressure levels for the tires. The proper pressure for a vehicle is usually found on a sticker in the driver's side door jamb or in the owner's manual.

Gas mileage improves by over 3 percent with properly inflated tires.

Which countries consume the most energy?

Top Energy-Consuming Countries (2012, in quadrillion BTU)

Country	Energy Consumption
China	110.60
United States	95.06
Russia	31.52
India	23.92
Japan	20.31
Germany	13.47
Canada	13.35
Brazil	12.10
South Korea	11.52
France	10.69

The total primary energy consumption in the world was 528.80 quadrillion Btu.

How much does it cost to generate one kilowatt-hour of electricity using different sources of energy?

The cost to generate one kilowatt-hour of electricity varies by the energy source.

Energy Source	Generating Cost (in cents per kilowatt-hour)
Hydropower	2–10¢
Geothermal	3–8¢
Wind	4–7¢
Natural gas	5–7¢
Solar thermal	5–13¢
Coal	6–8¢
Nuclear power	6–8¢
Biomass	6–9¢
Photovoltaics	15–25¢

How much energy does the average person in the United States use in a year?

In 2011, the United States, energy use per person was about 312 million BTU (British thermal units). Below is energy use per person for selected years:

Year	Energy Consumption per Person (Million BTU)
1950	227
1960	250
1970	331
1980	344
1990	339
2000	351
2005	340

Year	Energy Consumption per Person (Million BTU)
2008	326
2009	308
2010	316
2011	312

How much electricity does a typical American home use?

In 2013, the average monthly residential electricity consumption was 909 kilowatt-hours or 10,908 kilowatt-hours per year. More than half of the residential electricity consumption is for household appliances.

Has energy use in homes in the United States increased or decreased with the growth of electronic devices?

Since 1980, the number and average size of housing units has increased, and the number of homes with air conditioning and many more electronic devices has increased. However, due to improvements in the efficiency of space heating, air conditioning, and major appliances along with improved, better insulation of the housing structures and features such as double-pane windows, the total average per-household energy consumption has decreased since 1980.

Year*	Number of Housing Units	Average Energy Consumption per Housing Unit
1980	81.6 million	114.0 million BTU
1982	83.8 million	103.0 million BTU
1984	86.3 million	104.7 million BTU
1987	90.5 million	100.8 million BTU
1990	94.0 million	98.0 million BTU
1993	96.6 million	103.6 million BTU
1997	101.5 million	101.0 million BTU
2001	107.0 million	92.2 million BTU
2005	111.1 million	94.9 million BTU
2009	113.6 million	89.6 million BTU

*Data are from the U.S. Energy Information Administration Residential Energy Consumption Survey (survey for selected years).

How has residential energy end-use changed since 1980?

Total residential energy consumption has varied between 9.5 and 10.5 quadrillion BTU. Space heating and cooling used to account for more than 50 percent of the total residential energy consumption in the United States. Data released in 2011 and 2012 indicated that only 48 percent of total energy consumption in homes was for heating and cooling. Better insulation and more efficient heating and cooling equipment have led to the change in how energy is used by the residential sector.

| | U.S. Residential Energy Use | | | | |
Year	Total Energy Consumption	Space Heating (% of Total)	Appliances & Lighting (% of Total)	Water Heating (% of Total)	Air Conditioning (% of Total)
1980	9.5 quadrillion BTU	56	21	19	4
1993	10.01 quadrillion BTU	53	24	18	4.6
2009	10.18 quadrillion BTU	42	35	18	6

How is electricity used in homes in the United States?

In 2012 electricity was used to power appliances, air conditioning, heating, and the multitude of electronic devices in U.S. homes.

Use of Electricity	Quadrillion BTU	Billion Kilowatt-Hours	Percent of Total Electric Use
Space cooling	0.85	250	18
Lighting	0.64	186	14
Water heating	0.45	130	9
Refrigeration	0.38	111	8
Televisions, home theater systems, DVD players, video game consoles	0.33	98	7
Space heating	0.29	84	6
Clothes dryers	0.20	59	4
Computers, monitors, and related equipment	0.12	37	3
Cooking	0.11	31	2
Dishwashers	0.10	29	2
Furnace fans and boiler circulation pumps	0.09	28	2
Freezers	0.08	24	2
Clothes washers	0.03	9	1
Small electric devices, heating elements, and other uses	1.02	299	22
TOTAL	4.69	1,375	

How does the Energy Star program promote energy efficiency?

Energy Star is a dynamic government/industry partnership that offers businesses and consumers energy-efficient solutions, making it easy to save money while protecting the environment for future generations. In 1992 the U.S. Environmental Protection Agency (EPA) introduced Energy Star as a voluntary labeling program designed to identify and promote energy-efficient products to reduce greenhouse gas emissions. The

Energy Star label is now on major appliances, office equipment, lighting, consumer electronics, and more. The EPA has also extended the label to cover new homes and commercial and industrial buildings. Through its partnerships with more than 18,000 private and public sector organizations, Energy Star delivers the technical information and tools that organizations and consumers need to choose energy-efficient solutions and best management practices. Energy Star has successfully delivered energy and cost savings across the country, saving businesses, organizations, and consumers $24 billion in 2012. Energy Star has been instrumental in promoting the widespread use of such technological innovations as LED traffic lights, efficient fluorescent lighting, power management systems for office equipment, and low standby energy use.

Setting the thermostat at 68°F (20°C) in the winter and 78°F (25°C) in the summer when you are home will save you a lot on your utility bill.

Which products were the first to be Energy Star qualified products?

The first products to receive the Energy Star label were personal computers and monitors in June 1992. In January 1993, qualified printers were introduced, followed by fax machines in October 1994.

How much money can be saved by lowering the setting on a home thermostat?

Studies have shown that, when heating a home, a 10° to 15°F (5° to 8°C) reduction in the home thermostat setting for approximately eight hours a day will save 5 to 15 percent in fuel costs. This amounts to a savings of 1 percent for each degree if the setback period is eight hours long.

How much energy is saved by raising the setting for a house air conditioner?

In general, for every 1°F the inside temperature is increased, the energy needed for air conditioning is reduced by 3 percent. If all consumers raised the settings on their air conditioners by 6°F, for example, 190,000 barrels of oil could be saved each day.

How much fuel can be saved when a home is properly insulated?

Insulation of a single-family house with EPS or XPS (extruded polystyrene) over a 50-year period has the potential to save 80 metric tons of heating oil.

How much energy is required to use various electrical appliances?

The formula to estimate the amount of energy a specific appliance consumes is:

> ## Is it true that the savings gained from turning down the thermostat in winter is lost when the temperature is raised because the heating system has to work harder to rewarm the living area?
>
> No! In truth, once the temperature in the house drops, it loses energy more slowly to the surrounding environment. The lower the interior temperature, the slower the heat loss (the difference between the exterior and interior temperatures will be less). The longer the house stays at a lower temperature, the greater the savings—both in energy consumed and financially.

(wattage × hours used per day)/1,000 = daily kilowatt-hour (kWh) consumption.

The formula to calculate the annual consumption is:

Daily kilowatt-hour (kWh) × number of days per year the appliance is used = annual energy consumption.

The annual cost to run an appliance is calculated by multiplying the kWh per year by the local utility's rate per kWh.

The table below indicates the wattage for various household electrical products.

Appliance	Wattage
Clock radio	10
Coffeemaker	900–1,200
Clothes washer	350–500
Clothes dryer	1,800–5,000
Dishwasher (using the drying feature greatly increases energy consumption)	1,200–2,400
Ceiling fan	65–175
Window fan	55–250
Furnace fan	400
Whole house fan	240–750
Hair dryer	1,200–1,875
Microwave oven	750–1100
PC CPU—awake/asleep	120/30 or less
PC Monitor—awake/asleep	150/30 or less
Laptop	50
Refrigerator (frost-free, 16 cubic feet)	725
less than 31" LED TV	50
31–49" LED TV	75
50" or larger LED TV	100

Appliance	Wattage
50" or larger plasma TV	200
Toaster oven	1,225
DVD player	20–25
Vacuum cleaner	630
Water heater (40 gal.)	4,500–5,500

Is it possible to compare the energy efficiency of different brands of appliances?

In 1980, the Federal Trade Commission's Appliance Labeling Rule became effective. It requires EnergyGuide labels to be placed on all new refrigerators, freezers, water heaters, dishwashers, clothes washers, room air conditioners, heat pumps, furnaces, and boilers. The bright yellow labels with black lettering identify energy consumption and operating costs for each of the various household appliances. EnergyGuide labels show the estimated yearly electricity consumption to operate the product along with a scale of comparison among similar products. The comparison scale shows the least and most energy used by comparable models.

Who invented the incandescent light bulb?

The concept of an incandescent light bulb dates back to at least 1802 when Humphry Davy (1778–1829) demonstrated an arc lamp. His electric light had two wires connected to a battery, and he attached a charcoal strip between the other two ends of the wires.

While other inventors experimented with different filaments and whether the bulb should be filled with a gas or the air should be vacuumed out of the bulb, Thomas Edison (1847–1931) is credited with developing a long-burning, incandescent light bulb. Edison received U.S. Patent #223,898 for the electric lamp. Edison was inducted into the National Inventors Hall of Fame in 1973 for the invention of the electric lamp.

How is lighting brightness measured?

Lighting brightness is measured in lumens. The higher the number of lumens, the brighter the light bulb. In the past, light bulbs were sold based on their wattage, a measure of how much energy they consumed, not the brightness of the light they produced. As a comparison, when replacing a 100W incandescent bulb

It wasn't Thomas Edison but, rather, Humphry Davy (shown here) who invented the incandescent light bulb.

When was gas lighting invented?

In 1799, Philippe Lebon (1767–1804) patented a method of distilling gas from wood for use in a "Thermolamp," a type of lamp. By 1802, William Murdock (1754–1839) installed gas lighting in a factory in Birmingham, England. The introduction of widespread, reliable interior illumination enabled dramatic changes in commerce and manufacturing.

with an energy-efficient bulb, use one with a brightness of about 1600 lumens. Similarly, replace a 75W bulb with one that gives about 1100 lumens, a 60W with one that gives about 800 lumens, and a 40W with one that gives about 450 lumens.

What are the different types of newer, energy-efficient light bulbs?

While incandescent light bulbs were a revolutionary technology, they are very inefficient since more than approximately 90 percent of the energy they produce is heat and not light. Three of the newer, modern light bulbs are halogen incandescents, compact fluorescent lamps (CFLs), and light-emitting diodes (LEDs). Halogen incandescents differ from regular incandescent light bulbs because they have a capsule inside that holds gas around the filament to increase the bulb efficiency. Compact fluorescent lamps are curly versions of traditional fluorescent bulbs. They last up to ten times longer, use about a quarter of the energy, and produce 90 percent less heat, while producing more light per watt. Light-emitting diodes (LEDs) are a type of solid-state lighting. They are semiconductors that convert electricity into light. Although more expensive than traditional incandescent light bulbs, they use only 20 percent to 25 percent of the energy as an incandescent and last up to 25 times longer than traditional incandescents.

Comparison between Traditional Incandescents, Halogen Incandescents, CFLs, and LEDs

	60W Incandescent	43W Halogen Incandescent	CFL	LED
Energy $ Saved (%)		~25%	~75% (60W Traditional) ~65% (43W Halogen)	~75-80% (60W Traditional) ~72% (43W Halogen)
Annual Energy Cost (based on 2 hours of usage/day; 11 cents/kilowatt-hour)	$4.80	$3.50	$1.20	$1.00
Expected Life of Bulb	1,000 hours	1,000 to 3,000 hours	10,000 hours	25,000 hours

According to the Department of Energy, replacing 15 traditional incandescent light bulbs with energy-saving light bulbs could save the average household at least $50 per year.

Did the Energy Independence and Security Act of 2007 (EISA 2007) ban incandescent light bulbs?

The Energy Independence and Security Act of 2007 (EISA 2007) did not ban the use of incandescent light bulbs. Instead, it set standards requiring light bulbs to be approximately 25 percent more efficient, beginning January 1, 2012, for 100W light bulbs, January 1, 2013, for 75W bulbs, and January 1, 2014, for 60W and 40W bulbs. Specifically, the act limits the import or manufacture of inefficient light bulbs. Stores are permitted to sell remaining inventory, and consumers are permitted to continue using inefficient light bulbs.

When should a fluorescent light be turned off to save energy?

Fluorescent lights use a lot of electric current getting started, and frequently switching the light on and off will shorten the lamp's life and efficiency. It is energy efficient to turn off a fluorescent light only if it will not be used again within fifteen minutes.

Who invented the first visible-spectrum LED?

The first visible-spectrum LED was invented by Nick Holonyak Jr. (1928–) in 1962. Holonyak's diode emitted only red light. Today, LEDs are used on instrument panels, such as bicycle tail lights. The technology led the way for lasers used for CD and DVD players. Holonyak was inducted into the National Inventors Hall of Fame in 2008 for the invention of LEDs. He received the National Medal of Science in 1990 and National Medal of Technology in 2002 for the development of LED technology and its applications for digital displays, consumer electronics, automotive lighting, traffic signals, and general illumination.

When were blue light-emitting diodes invented?

Blue light-emitting diodes were not invented until the 1990s. The invention of blue light-emitting diodes was essential for the development of white light LED lamps. The Nobel Prize in Physics was awarded to Isamu Akasaki (1929–), Hiroshi Amano (1960–), and Shuji Nakamura (1954–) for the invention of blue light-emitting diodes in 2014.

INTRODUCTION, HISTORICAL BACKGROUND, AND ENVIRONMENTAL MILESTONES

How does technology impact the environment?

Technology can provide the tools to create a sustainable future, or it can destroy the environment with dire results for future generations. Technological development and advancement such as the use of chlorofluorocarbons, which have led to ozone depletion, has negative impacts on the environment. However, the use of technology to develop renewable sources of energy may ensure the sustainability of the planet. In the 1970s, biologist Paul R. Ehrlich (1932–) and physicist John P. Holdren (1944–) proposed a mathematical model to show the relationship between environmental impacts and the forces that propel them. The equation of their model is:

$$I = P \times A \times T$$

... where I is the impact of humans on the environment, P is the population, A is the demand for resources (affluence), and T is the technological factor.

Can the scientific method be applied to the study of technology and the environment?

The scientific method can be applied to study the impact of technology on the environment. For example, when scientists consider the problem of how global warming affects a specific ecosystem, such as a coral reef, field scientists will make observations that will be collected and analyzed. A hypothesis will be developed based on these observations. One hypothesis may be that a slight increase in water temperature can cause the

315

animals living in the coral to die. The hypothesis is then tested in a laboratory setting. The results of the laboratory study are compared with the data collected in the field. Based on a careful analysis of all data, the hypothesis can be proved/disproved. The results can then be shared with other scientists through publications.

How are environmental problems investigated?

When changes in the environment are suspected of having adverse effects on ecosystems or populations (including human) in the natural world, scientists will gather data to determine the cause of the problem. One part of the process will be a study of risk analysis. Risk analysis or assessment identifies a hazard, explores and assesses the relationship between the hazard and the adverse effect, and determines the probability of an adverse effect. Once a risk assessment has identified a problem, scientists, the public, and legislators will work to develop ways to solve and control the problem. A final component of the process will be to monitor the implementation of the plan.

What is the ecological footprint?

The ecological footprint is a measurement of how fast humans consume resources and generate waste compared to how fast nature can absorb the waste and generate new resources. The measurement helps scientists, government agencies, and institutions assess the impact of human activity on the planet. Since the late 1970s, humanity has been in ecological overshoot; i.e., the annual demands on nature exceed what the earth can generate in a year. It currently takes the earth one year and six months to regenerate what is used in one year. Mathis Wackernagel (1962–) and William Rees (1943–) developed the concept of the ecological footprint in 1990.

What are some important milestones in the history of technology and the environment?

Milestones in the history of technology and the environment are significant either immediately or become significant with the passage of time.

Technology and the Environment Milestones

Date	Event
1661	John Evelyn (1620–1706) publishes *The Smoke of London*, attacking air pollution.
1691	Edmond Halley (1656–1742) describes the water cycle.
1798	Thomas Robert Malthus (1766–1834) publishes *An Essay on the Principle of Population*, which warned about the dangers of unchecked population growth.
Late 18th century– early 19th century	Industrial Revolution.
1849	U.S. Department of the Interior established to protect the nation's resources and outdoors.

Date	Event
1854	Henry David Thoreau (1817–1862) publishes *Walden*, discussing the loss of wilderness due to the rise of industrialization. Readers are inspired to live simply.
1862	Homestead Act speeds the settlement of the Plains states and destruction of the prairie ecosystem.
1864	George Perkins Marsh (1801–1882) publishes *Man and Nature*, which initiates the conservation movement.
1872	Yellowstone in Wyoming is named the first national park.
1875	American Forestry Association founded to encourage wise forest management.
1892	John Muir (1838–1914) founds the Sierra Club.
1908	Chlorination is introduced in U.S. water treatment plants.
1913	Charles Fabry (1867–1945) discovers the presence of ozone in the upper atmosphere. Later research shows the importance of ozone as a screen in preventing damaging UV radiation from reaching the Earth's surface.
1916	National Park Service Act creates the National Park Service, the first such agency in the world.
1933	Tennessee Valley Authority created with the goal of developing hydropower using new technologies.
1949	Aldo Leopold (1887–1948) publishes *A Sand County Almanac*, which establishes guidelines for the conservation movement and suggests the concept of a land ethic.
1952	Oregon is the first state to endorse a program to control air pollution.
1962	Rachel Carson (1907–1964) publishes *Silent Spring*, which establishes the link between pesticides and environment. This book and her work, in general, is generally considered the beginning of the modern environmental movement.
1970	Environmental Protection Agency (EPA) created.
1970	First Earth Day in the United States.
1972	Oregon is first state to mandate bottle recycling.
1972	The use of the pesticide DDT is phased out in the United States.
1974	Chlorofluorocarbons are hypothesized to cause ozone thinning.
1978	Oil tanker *Amoco Cadiz* runs aground, leaking 220,000 tons of oil.
1978	Residents of Love Canal, New York, are evacuated after investigations show the deadly effects of chemical wastes in the area.
1979	Partial meltdown of Three Mile Island nuclear power plant accident in Pennsylvania.
1982	Convention on the Law of the Sea (international treaty) developed to protect ocean resources.
1984	Union Carbide's pesticide plant accident in Bhopal, India.
1984	Scientists discover ozone hole over Antarctica.
1986	Chernobyl Nuclear Power Station experiences a nuclear core meltdown and releases radioactive material over large areas of the Soviet Union and northern Europe.

Date	Event
1987	Montreal Protocol signed requiring countries to phase out ozone-depleting chemicals.
1989	Tanker *Exxon Valdez* oil spill off the coast of Alaska.
1991	Oil spill during Gulf War.
1997	Kyoto Protocol (never ratified by the United States).
2004	Google Earth uses satellite imagery and other new technologies to make the world available to the public.
2005	United Nations report indicates that the hole in the ozone layer is increasing, despite efforts over twenty years to curb the emission of greenhouse gases.
2008	World leaders meet at a Group of Eight (G8) industrialized nations summit and agree to cut emissions of greenhouse gases by 50 percent by 2050.
2009	World leaders meet at a Group of Eight (G8) industrialized nations summit and agree to cut emissions of greenhouse gases by 80 percent by 2050 by using newer technological advances and techniques.
2010	Deepwater Horizon explosion in the Gulf of Mexico kills 11 and severely injures 17 and spills 206 million gallons (780 million liters) of oil from Louisiana to Florida.
2010	EPA issues rules on automotive fuel efficiency and, for the first time, regulates greenhouse gas emissions.
2010	The previous decade, 2000–2009, was the warmest on record according to the World Meteorological Organization.
2010	EPA celebrates its 40th anniversary.
2011	Arctic sea ice reaches an historic low.
2011	EPA approves use of 15 percent ethanol blended fuels in gasoline, sparking renewed debate over environmental benefits and drawbacks of corn ethanol.
2011	Nuclear reactor meltdowns, explosions, and spent fuel fires at the Fukushima power complex.
2011	Renewable energy could power 80 percent of the world by 2050 according to the UN-affiliated International Renewable Energy Agency by incorporating newer technologies.
2012	50th anniversary of the publication of Rachel Carson's *Silent Spring*.
2012	UN Conference on Sustainability in Rio de Janeiro states that climate change is the greatest challenge facing the world.
2014	UN Intergovernmental Panel on Climate Change releases alarming report on the state of the environment.

What was the environmental significance of Rachel Carson's *Silent Spring*?

In the book *Silent Spring*, published in 1962, Rachel Carson (1907–1964) exposed the dangers of pesticides, particularly DDT, to the reproduction of species that prey upon the insects for whom the pesticide was intended. *Silent Spring* raised the public awareness and is considered a pivotal point at the beginning of the environmental movement.

When was the first Earth Day celebration?

The first Earth Day, April 22, 1970, was coordinated by Denis Hayes (1944–) at the request of Gaylord Nelson (1916–2005), U.S. senator from Wisconsin. Nelson is sometimes called the father of Earth Day.

Rachel Carson famously wrote *Silent Spring* about the impact of pesticides and other chemicals on the environment.

His main objective was to organize a nationwide public demonstration so large it would attract the attention of politicians and force the environmental issue into the political dialogue of the nation. Important official actions that began soon after the celebration of the first Earth Day were the establishment of the Environmental Protection Agency (EPA); the creation of the President's Council on Environmental Quality; and the passage of the Clean Air Act, establishing national air quality standards. In 1995 Nelson received the Presidential Medal of Freedom for his contributions to the environmental protection movement. Earth Day continues to be celebrated each spring, raising the awareness of our responsibility to the environment.

When was the Environmental Protection Agency (EPA) created and what does it do?

In 1970 President Richard M. Nixon (1913–1994) signed an executive order that created the Environmental Protection Agency (EPA) as an independent agency of the U.S. government. The creation of a federal agency by executive order rather than by an act of the legislative branch is somewhat uncommon. The EPA was established in response to public concern about unhealthy air, polluted rivers and groundwater, unsafe drinking water, endangered species, and hazardous waste disposal. Responsibilities of the EPA include environmental research, monitoring, and enforcement of legislation regulating environmental activities. The EPA also manages the cleanup of toxic chemical sites as part of a program known as Superfund.

What is some of the major environmental legislation in the United States and when was it passed?

Environmental legislation in the United States dates back to 1862 and the passage of the Homestead Act, which made free homesteads available on unappropriated lands in the Prairie states available on a large-scale basis. This led to rapid settlement and destruction of the prairie ecosystem. Although in the late nineteenth century legislation was passed to develop natural resources, in later decades and throughout the twentieth century, legislation was passed to protect the environment, often regulating industrial development.

Important U.S. Environmental Legislation

Date	Act	Purpose
1872	Mining Law	Allows any person finding mineral deposits on public land to file a claim granting him or her free access to the site for mining or other development
1891	Forest Reserve Act	Authorizes the president to set aside forest reserves from public land
1898	Rivers and Harbors Act	Passed to control pollution on navigable waterways
1902	Reclamation Act	Allows the "reclamation" of dry lands in the Western states through irrigation and damming of rivers
1916	National Park Service Act	Creates the National Park Service, the first such agency in the world
1920	Mineral Leasing Act	Regulates the exploitation of fuel and fertilizer minerals on public lands
1935	Soil Conservation Act	Establishes the Soil Conservation Service to deal with soil erosion after the Dust Bowl
1938	Food, Drug, and Cosmetics Act	Establishes standards for the quality and labeling of foods, drugs, and cosmetics
1948	Water Pollution Control Act	Provides state and local governments with funding to address water pollution
1955	Air Pollution Control Act	Provides funding to states for air pollution control activities
1956	Federal Water Pollution Control Act	Increases federal funding to states and local governments for water pollution control activities; initiates development of water quality standards
1963	Clean Air Act (amended 1970, 1977, 1990)	Initial legislation that aims to control air pollution
1964	Wilderness Act	Allows for the preservation of wilderness areas for present and future generations
1965	Solid Waste Disposal Act (amended 1970, 1976)	Addresses the problem of disposing of solid wastes
1965	Water Quality Act	Establishes federal water quality standards
1969	Coal Mine Health and Safety Act	Focuses on safety concerns in the mining industry

Date	Act	Purpose
1970	National Environmental Policy Act (NEPA)	Requires all federal agencies to take into account the environmental consequences of their plans and actions
1970	National Mining and Minerals Act	Encourages the mining industry to pursue a financially viable industry while searching for ways to be wise and efficient with the use of the minerals
1970	Resources Recovery Act	An amendment to the Solid Waste Recovery Act, which finances recycling programs and authorizes a major assessment of solid waste disposal
1972	Clean Water Act	Sets a goal of attaining "fishable and swimmable" water quality in all U.S. surface waters
1974	Safe Drinking Water Act (amended 1996)	Sets and requires minimum standards for water quality
1976	Energy Policy and Conservation Act	Sets fuel economy standards for cars and light trucks
1976	Resource Conservation and Recovery Act	Regulates the storage, shipping, processing, and disposal of hazardous substances
1976	Toxic Substances Control Act	Categorizes toxic and hazardous substances and regulates the use and disposal of poisonous chemicals
1977	Surface Mining Control and Reclamation Act	Limits the scarring of the landscape, erosion, and water pollution caused by surface mining
1980	Comprehensive Environmental Response, Compensation, and Liability Act (Superfund)	Establishes a Superfund to finance government cleanup activities often due to the improper disposal of hazardous waste
1988	Alternative Motor Fuels Act	Encourages auto manufacturers to design and build cars that run on alternative fuels, such as methanol and ethanol
1989	Montreal Protocol on Substances That Deplete the Ozone Layer	Bans propellants and other substances that damage the ozone layer
1996	Food Quality Protection Act	Requires that all pesticides used on food products be of no harm to human populations
1997	Kyoto Protocol	Sets binding greenhouse gas emissions targets for participating nations; the United States signs but does not ratify the Kyoto Protocol
2000	Water Resources Development Act	U.S. Senate approves $7.8 billion fund to restore the Florida Everglades
2002	Small Business Liability Relief and Brownfields Revitalization Act	Provides funds to assess and clean up brownfields
2005	U.S. Energy Policy Act proposed	Kyoto Protocol officially goes into effect
2007	Environmental Justice Act	Legislation approved that allows environmental justice for hurricane victims

How do environmental impact statements protect the environment?

The National Environmental Policy Act (NEPA) requires federal agencies to consider the environmental impacts of proposed actions, such as federal highway construction and other public works, flood and erosion control, and military projects. Environmental impact statements (EISs) are detailed documents describing the proposed project or action, long- and short-term effects on the environment, and possible alternatives that would lessen the adverse effects on the environment.

What is sustainable materials management (SMM)?

Sustainable materials management is an approach to resource management that attempts to use and reuse materials more productively over their entire life cycles. Adopting a sustainable materials management approach encompasses using materials in the most productive way possible with an emphasis on using less. It includes an objective of reducing toxic chemicals and environmental impacts throughout the material life cycle with an ultimate goal of assuring sufficient resources for today's needs and those of the future. Products designed following a SMM philosophy will reduce environmental impacts beginning with the raw material acquisition for the product, using fewer, less toxic, and more durable materials. The product will be designed so that at the end of its useful life, it can be easily disassembled and the materials can be recycled or disposed of in an environmentally friendly manner.

LAND USE

What is a biogeochemical cycle?

The elements that organisms need most (carbon, nitrogen, phosphorus, and sulfur) cycle through the physical environment, the organism, and then back to the environment. Each element has a distinctive cycle that depends on the physical and chemical properties of the element. Examples of biogeochemical cycles include the carbon and nitrogen cycles, both of which have a prominent gaseous phase. Examples of biogeochemical cycles with a prominent geologic phase include phosphorus and sulfur, where a large portion of the element may be stored in ocean sediments. Examples of cycles with a prominent atmospheric phase include carbon and nitrogen. These biogeochemical cycles involve biological, geologic, and chemical interactions.

What is the hydrologic cycle?

The hydrologic cycle takes place in the hydrosphere, which is the region containing all the water in the atmosphere and Earth's surface. It involves five phases: condensation, infiltration, runoff, evaporation, and precipitation. Rain, and other precipitation, is part of the hydrologic cycle.

What is the carbon cycle?

To survive, every organism must have access to carbon atoms. Carbon makes up about 49 percent of the dry weight of organisms. The carbon cycle includes movement of carbon from the gaseous phase (carbon dioxide in the atmosphere) to solid phase (carbon-containing compounds in living organisms) and then back to the atmosphere via decomposers. The atmosphere is the largest reservoir of carbon, containing 32 percent carbon dioxide. Biological processes on land shuttle carbon between atmospheric and terrestrial compartments, with photosynthesis removing carbon dioxide from the atmosphere and cell respiration returning carbon dioxide to the atmosphere.

What percentage of Earth's surface is forest area?

The world's total forest area is just over 15 million square miles (4 billion hectares) or 31 percent of the total land area. More than half of the world's forest areas are in the five most forest-rich countries: the Russian Federation, Brazil, Canada, the United States, and China.

How rapidly is deforestation occurring?

Agriculture, excessive logging, and fires are major causes of deforestation. Afforestation and the natural expansion of forests help to decrease the rate of deforestation. The rate of deforestation has decreased from 16 million hectares (61,776 square miles) per year during the 1990s to 13 million hectares (50,193 square miles) per year in the decade 2000–2010. The net change in forest area for 2000–2010 is estimated at −5.2 million

Clearcutting large areas of forest not only results in a decline in wildlife populations but also has an impact on everything from soil quality for farming to the weather. The loss of trees results in more carbon dioxide in the atmosphere, contributing to global warming.

What percentage of Earth's surface is tropical rain forest?

Rain forests account for approximately 7 percent of Earth's surface, or about 3 million square miles (7.7 million square kilometers). The Amazon Basin is the world's largest continuous tropical rain forest. It covers about 2.7 million square miles (6.9 million square kilometers).

hectares (20,007 square miles) per year (an area about the size of Costa Rica), down from −8.3 million hectares (32,046 square miles) per year in the 1990s. Unfortunately, in Brazil deforestation has risen again after dropping from 2005 to 2013. Changes in the government there have seen the destruction of the Amazon rain forest increase in some areas by over 400 percent.

What is the importance of the rain forest?

Half of all medicines prescribed worldwide are derived originally from wild products, and the U.S. National Cancer Institute has identified more than two thousand tropical rain forest plants with the potential to fight cancer. Rubber, timber, gums, resins and waxes, pesticides, lubricants, nuts and fruits, flavorings and dyestuffs, steroids, latexes, essential and edible oils, and bamboo are among the products that would be drastically affected by the depletion of the tropical forests. In addition, rain forests greatly influence patterns of rain deposition in tropical areas; smaller rain forests mean less rain. Large groups of plants, like those found in rain forests, also help control levels of carbon dioxide in the atmosphere.

What products come from tropical forests?

Houseplants	Spices	Foods	Oils, Medicines, Alkaloids
Anthurium	Allspice	Avocado	Camphor oil
Croton	Black pepper	Banana	Cascarilla oil
Dieffenbachia	Cardamom	Coconut	Coconut oil
Dracaena	Cayenne	Grapefruit	Eucalyptus oil
Fiddle-leaf fig	Chili	Lemon	Oil of star anise
Mother-in-law's tongue	Cinnamon	Lime	Palm oil
Parlor ivy	Cloves	Mango	Patchouli oil
Philodendron	Ginger	Orange	Rosewood oil
Rubber tree plant	Mace	Papaya	Tolu balsam oil
Schefflera	Nutmeg	Passion fruit	Annatto
Silver vase bromeliad	Paprika	Pineapple	Curare
Spathiphyllum	Sesame seeds	Plantain	Diosgenin
Swiss cheese plant	Turmeric	Tangerine	Quinine
Zebra plant	Vanilla bean	Brazil nuts	Reserpine
		Cane sugar	Strophanthus
		Cashew nuts	Strychnine

Houseplants	Spices	Foods	Oils, Medicines, Alkaloids
Zebra plant		Chocolate	Yang-Yang
		Coffee	
		Cola Beans	
		Cucumber	
		Hearts of palm	
		Macadamia nuts	
		Manioc/tapioca	
		Okra	
		Peanuts	
		Peppers	
		Tea	

Gums, Resins	Fibers	Woods
Chicle latex	Bamboo	Balsa
Copaiba	Jute/Kenaf	Mahogany
Copal	Kapok	Rosewood
Gutta percha	Raffia	Sandalwood
Rubber latex	Ramie	Teak
Tung oil	Rattan	

What is the importance of wetlands?

A wetland is an area that is covered by water for at least part of the year and has characteristic soils and water-tolerant plants. Examples of wetlands include swamps, marshes, bogs, and estuaries. Wetlands may be freshwater, saltwater (marine), or deepwater (lakes and other deepwater habitats). Wetlands protect and improve water quality, provide habitats for fish and wildlife, store floodwaters, and maintain surface water flow during dry periods. Technological advances in the quality of remotely sensed imagery and computerized mapping techniques have enhanced the ability to collect more detailed information about the nation's wetlands. Scientific research strives to improve, protect, and maintain the water quality.

Type of Wetland	Typical Features
Swamp	Tree species such as willow, cypress, and mangrove
Marsh	Grasses such as cattails, reeds, and wild rice
Bog	Floating vegetation, including mosses and cranberries
Estuary	Specially adapted flora and fauna, such as crustaceans, grasses, and certain types of fishes

How many acres of wetlands have been lost in the United States?

The United States lost approximately 100 million acres (40.5 million hectares) of wetland areas between colonial times and the 1970s. The most recent Wetlands Status and

Trends report (2004–2009) published in 2011 by the U.S. Fish and Wildlife Service estimates there were 110.1 million acres (44.6 million hectares) of wetlands in the United States. Wetlands composed 5.5 percent of the surface area of the United States. Ninety-five percent of all wetlands were freshwater, and the remaining 5 percent were in salt-water (marine) systems. In the five-year period from 2004 to 2009, the total wetland area declined by an estimated 62,300 acres (25,200 hectares), or an average annual loss of 13,800 acres (5,590 hectares), which is considered statistically insignificant. While freshwater wetlands increased in area slightly, marine areas saw a slight decrease. The reasons for wetland area increase or decrease reflect economic conditions, land use trends, changing regulation, and enforcement and climatic changes.

How did technology contribute to the loss of wetlands?

The earliest reasons for wetland drainage were for settlement and development, especially for agricultural growth. The Swamp Lands Act of 1849 allowed the reclamation of swamp and overflow lands in Louisiana. In 1850 and 1860, the act became applicable to fourteen other states, clearly indicating that the federal government promoted land reclamation and drainage projects for settlement and development.

The expansion of the railroads in the mid- to late-nineteenth cleared many forest wetlands to introduce new transportation routes and for the timber to build railroad lines. The advent of the industrial age brought new machinery that allowed the conversion of wetlands to farmlands to occur more rapidly. Concurrently, other machinery made it possible to cultivate larger areas of land more easily.

Wetlands serve as habitat for wildlife, but they can serve other useful purposes, such as helping to keep pollutants out of rivers and streams and mitigating the extent of floods.

Engineering projects to drain wetlands continued through most of the twentieth century. For example, by the 1920s, approximately 70 percent of the original wetland acreage of the Central Valley in California had been modified with levees, drainage, and water-diversion projects. The U.S. government continued to encourage drainage projects until the 1970s. As knowledge of the importance of wetlands increased beginning in the late 1970s through the end of the twentieth century and early twenty-first century, the focus has shifted to wetland reclamation projects.

What is a watershed?

A watershed is an area of land that contains a common set of streams and rivers that all drain into a single, larger body of water such as a larger river, a lake, or an ocean. For example, the Mississippi River watershed is an enormous watershed in which all the tributaries to the Mississippi that collect rainwater eventually drain into the Mississippi, which eventually drains into the Gulf of Mexico.

What is an estuary?

Estuaries are places where freshwater streams and rivers meet the sea. The salinity of such areas is less than that of the open ocean but greater than that of a typical river, so organisms living in or near estuaries have special adaptations. Estuaries are rich sources of invertebrates, such as clams, shrimps, and crabs, as well as fishes such as striped bass, mullet, and menhaden. Unfortunately, estuaries are also popular locations for human habitation and businesses. Contamination from shipping, household pollutants, and power plants are carried to the sea by rivers and streams and threaten the ecological health of many estuaries.

What is bioremediation?

Bioremediation is the degradation, decomposition, or stabilization of pollutants by microorganisms such as bacteria, fungi, and cyanobacteria. Oxygen and organisms are injected into contaminated soil and/or water (e.g., oil spills). The microorganisms feed on and eliminate the pollutants. When the pollutants are gone, the organisms die. The massive cleanup in Alaska following the Exxon *Valdez* oil spill in 1989 relied on bioremediation. The superficial layer of oil was removed by suction and filtration, but the oil-soaked beach was cleaned by bacteria that could use oil as an energy source.

What is eutrophication?

Eutrophication is a process in which the supply of plant nutrients in a lake or pond is increased. In time, the result of natural eutrophication may be dry land where water once flowed, caused by plant overgrowth. Natural fertilizers, washed from the soil, result in an accelerated growth of plants, producing overcrowding. As the plants die off, the dead and decaying vegetation depletes the lake's oxygen supply, causing fish to die. The accumulated dead plant and animal material eventually changes a deep lake to a shallow one, then to a swamp, and finally, it becomes dry land. While the process of eutrophication is a natural one, it has been accelerated enormously by human activities. Fertilizers from farms, sewage, industrial wastes, and some detergents all contribute to the problem.

What does it mean when a lake is brown or blue?

When a lake is brown, it usually indicates that eutrophication is occurring. This process refers to the premature "aging" of a lake when nutrients are added to the water, usually due to runoff, which may be either agricultural or industrial in origin. Due to this rich supply of nutrients, blue-green algae begin to take over the green algae in the lake, and food webs within the lake are disturbed, leading to an eventual loss of fish. When a lake is blue, this usually means that the lake has been damaged by acid precipitation. The gradual drop in pH caused by exposure to acid rain causes disruption of the food webs, eventually killing most organisms. The end result is clear water, which reflects the low productivity of the lake.

What is phytoremediation?

Phytoremediation, a newer green technology, is the use of green plants to remove pollutants from the soil or render them harmless. Certain plants, such as alpine pennycress (*Thlaspi caerulescens*), have been identified as metal hyperaccumulators. Researchers have found that these plants can grow in soils contaminated with toxic heavy metals, including cadmium, zinc, and nickel. The plants extract the toxic heavy metals from the soil and concentrate them in the stems, shoots, and leaves. These plant tissues may then be collected and disposed of in a hazardous waste landfill. Researchers are investigating ways to recover the metals by extracting them from the plants. Although phytoremediation is limited by how deep the roots of the plants grow in the soil, it is a less costly alternative way to clean up hazardous waste sites.

AIR POLLUTION

What is air pollution?

Air pollution is the contamination of Earth's atmosphere at levels high enough to harm humans, other organisms, or other materials. The major air pollutants are particulate matter, ozone, carbon monoxide, nitrogen oxides, sulfur dioxide, and lead. Primary air pollutants, including carbon monoxide, nitrogen oxides, sulfur dioxide, and particulate matter, enter the atmosphere directly. Secondary air pollutants are harmful chemicals that form from other substances released into the atmosphere. Ozone and sulfur trioxide are examples of secondary air pollutants.

What are the sources of air pollution?

Most air pollutants originate from human-made sources, such as forms of transportation (cars, trucks, and buses), industrial processes, such as refineries, iron and steel mills, paper mills, and chemical plants. Another significant source of air pollution is fuel combustion as a result of burning fossil fuels.

What is the difference between primary air pollutants and secondary air pollutants?

Primary air pollutants are harmful chemicals that enter directly into the atmosphere due to either human activities or natural processes. Examples are carbon oxides, nitrogen oxides, sulfur dioxide, particulate matter, and hydrocarbons. Secondary air pollutants are harmful chemicals that form in the atmosphere when primary air pollutants react with one another or with natural components of the atmosphere. Examples are ozone, sulfur trioxide, and several acids, including sulfuric acid.

Pollution in many Chinese cities, such as Beijing, has gotten so bad that many people living there wear masks when outside.

What are the health effects of the major air pollutants?

Common health effects of exposure to even low levels of air pollutants are irritated eyes and inflammation of the respiratory tract. There is some evidence that exposure to air pollutants suppresses the immune system, increasing susceptibility to infections.

Pollutant	Source	Health Effects
Particulate matter	Industries, electric power plants, motor vehicles, construction, agriculture	Aggravates respiratory illnesses; long-term exposure may cause increased incidence of chronic conditions such as bronchitis; suppresses immune system; heavy metals and organic chemicals may cause cancer
Nitrogen oxides	Motor vehicles, industries, heavily fertilized farmland	Irritate respiratory tract; aggravate asthma and chronic bronchitis
Sulfur oxides	Electric power plants and other industries	Irritate respiratory tract; long-term exposure may cause increased incidence of chronic conditions such as bronchitis; suppress immune system; heavy metals and organic chemicals may cause cancer
Carbon monoxide	Motor vehicles, industries, fireplaces	Reduces blood's ability to transport oxygen; low levels cause headaches and fatigue; higher levels cause mental impairment or death
Ozone	Formed in atmosphere	Irritates eyes and respiratory tract; produces chest discomfort; aggravates asthma and chronic bronchitis

What are the main components of motor vehicle exhaust?

The main components of exhaust gas are nitrogen, carbon dioxide, and water. Smaller amounts of nitrogen oxides, carbon monoxide, hydrocarbons, aldehydes, and other products of incomplete combustion are also present. The most important air pollutants, in order of amount produced, are carbon monoxide, nitrogen oxides, and hydrocarbons.

What invention played a major role in reducing air pollution?

One of the major technological developments that reduced air pollution was the invention of the catalytic converter. Once the emissions from the automobile engine were identified as a major source of air pollution, researchers began to search for ways to eliminate the toxic emissions. Carl D. Keith (1920–2008), John J. Mooney (1929–), and a team of engineers at Engelhard Corporation invented the three-way catalytic converter in the early 1970s. It neutralized hydrocarbons, carbon monoxide, and nitrogen oxides, rendering them harmless to humans and the environment. The invention of the three-way catalytic converter allowed the automobile industry to meet the standards of the Clean Air Act. The EPA has estimated that because of the catalytic converter, cars are now 98 percent cleaner in terms of nitrogen oxide emissions than those built in the 1970s. Demands for ultralow emissions vehicles (ULEVs) and super ULEV vehicles have forced engineers to develop vehicles with even fewer emissions, reducing emissions during engine startup and electric vehicles.

What is the Air Quality Index (AQI)?

The Air Quality Index (AQI) is an index for reporting daily air quality. The AQI value indicates how clean or polluted the air is in a given location and whether it will cause any health effects. The U.S. Environmental Protection Agency calculates the AQI for five major air pollutants regulated by the Clean Air Act. The five major pollutants are ground-level ozone, particle pollution (also known as particulate matter), carbon monoxide, sulfur dioxide, and nitrogen dioxide. There are corresponding national air quality standards established by the EPA. The AQI is divided into six categories:

Air Quality Index (AQI) Values	Color	Air Quality Index Level of Health Concern	Health Precautions
0–50	Green	Good	Air quality is considered satisfactory, and air pollution poses little or no risk.

Air Quality Index (AQI) Values	Color	Level of Health Concern	Health Precautions
51–100	Yellow	Moderate	Air quality is acceptable; however, for small numbers of the population who are extremely sensitive to air pollution, there may be moderate health concerns.
101–150	Orange	Unhealthy for sensitive groups	Members of sensitive groups may experience health effects. The general public is not likely to be affected.
151–200	Red	Unhealthy	Everyone may begin to experience health effects with members of sensitive groups experiencing more serious health effects.
201–300	Purple	Very unhealthy	Health warnings of emergency conditions. The entire population is more likely to be affected.
301–500	Maroon	Hazardous	Health alert; everyone may experience more serious health effects.

What are the components of smog?

Smog, the most widespread pollutant in the United States, is a photochemical reaction resulting in ground-level ozone. Ozone, an odorless, tasteless gas in the presence of light, can initiate a chain of chemical reactions. Ozone is a desirable gas in the stratospheric layer of the atmosphere, but it can be hazardous to health when found near Earth's surface in the troposphere. The hydrocarbons, hydrocarbon derivations, and nitric oxides emitted from such sources as automobiles are the raw materials for photochemical reactions. In the presence of oxygen and sunlight, the nitric oxides combine with organic compounds, such as the hydrocarbons from unburned gasoline, to produce a whitish haze, sometimes tinged with a yellow-brown color. In this process, a large number of new hydrocarbons and oxyhydrocarbons are produced. These sec-

At top is a photo taken of lower Manhattan and the Twin Towers in 1970, before the Clean Air Act; below, a photo taken in 1998.

331

ondary hydrocarbon products may comprise as much as 95 percent of the total organics in a severe smog episode.

What are some of the accomplishments achieved in reducing air pollution since the Clean Air Act was passed in 1970?

Since the passage of the Clean Air Act in 1970, the economy has grown dramatically due to advances in technology, such as manufacturing, while the levels of air pollution have decreased dramatically. One of the goals of the EPA was to set national air quality standards for six common air pollutants—carbon monoxide, lead, nitrogen dioxide, ozone, particulate matter, and sulfur dioxide. Since passage of the Clean Air Act in 1970 through 2012, the amount of these six pollutants in the air has decreased an average 72 percent while the gross domestic product grew by 219 percent. Other accomplishments are:

- Large industrial sources (chemical plants, petroleum refineries, and paper mills) emit 1.5 million tons (1361 metric tons) less of toxic air pollutants than in 1990
- New cars and trucks are 99 percent cleaner
- New locomotives are 90 percent cleaner than pre-regulation locomotives
- Sulfur in gasoline has been reduced by 90 percent while in diesel fuel, it has been reduced by 99 percent
- Power plants have cut emissions that cause acid rain
- Fewer serious health effects in the American population caused by air pollution
- The use of ozone-depleting substances, such as CFCs and halons, has been phased out

What are some technological innovations that were developed to meet the requirements of the Clean Air Act?

Once the problems of air pollution caused by industrial development were acknowledged, research focused on developing new technologies that would reduce the impact on the environment and health of all humans. Examples of these newer technologies are:

- Catalytic converters
- Air scrubbers
- Low- or zero-volatile organic compound (VOC) paints, cleaners, caulk, adhesives, and other consumer products
- Chlorofluorocarbon (CFC)- and hydrochlorofluorocarbon (HCFC)-free air conditioners, refrigerators, aerosol sprays, and cleaning solvents
- Electric and hybrid vehicles

What is acid rain?

Acid deposition is the fallout of acidic compounds in solid, liquid, or gaseous forms. Wet deposition occurs as precipitation while dry deposition is the fallout of particulate matter.

Acid rain is the best known form of acid deposition. The term "acid rain" was coined by British chemist Robert Angus Smith (1817–1884) who, in 1872, published *Air & Rain: The Beginnings of a Chemical Climatology*. Since then, acid rain has become an increasingly used term for rain, snow, sleet, or other precipitation that has been polluted by acids such as sulfuric and nitric acids. When gasoline, coal, or oil is burned, its waste products of sulfur dioxide and nitrogen dioxide combine in complex chemical reactions with water vapor in clouds to form acids. Acid rain may be either wet deposition or dry deposition from the atmosphere. Wet deposition refers to acidic rain, fog, and snow. Dry deposition occurs in dry weather when the chemicals are incorporated into dust and/or smoke. It falls and sticks to the ground, buildings, cars, and trees. Dry deposited particles are washed off in rain, leading to increased runoff. The runoff makes the mixture more acidic. In the course of the hydrological cycle, about half of the acidity in the atmosphere falls back to the earth through dry deposition. Natural emissions of sulfur and nitrogen compounds combined with those in air pollution have resulted in severe ecological damage to lakes, streams, and forests. Acid rain also effects the composition of building materials, including metals (for example, bronze used in statues), marble, and limestone.

How much have emissions of sulfur dioxide and nitrogen oxide decreased since the 1990 amendments to the Clean Air Act?

Sulfur dioxide and nitrogen oxide are two of the main components of acid rain. In 1990, emissions of sulfur dioxide and nitrogen oxides totaled nearly 50 million tons.

Year	Sulfur Dioxide Emissions (million tons)	Nitrogen Oxide Emissions (million tons)
1990	23.1	25.5
1995	18.6	25.0
2000	16.3	22.6
2005	14.5	20.3
2010	7.8	15.0
2011	6.6	14.8
2012	5.3	13.7
2013	5.2	13.1

How acidic is acid rain?

Acidity or alkalinity is measured by a scale known as the pH (potential for Hydrogen) scale. It runs from zero to 14. Since it is logarithmic, a change in one unit equals a tenfold increase or decrease. Therefore, a solution at pH 2 is 10 times more acidic than one at pH 3 and 100 times as acidic as a solution at pH 4. Zero is extremely acid, 7 is neutral, and 14 is very alkaline. Any rain measuring below 5.0 is considered acid rain. Normal rain and snow containing dissolved carbon dioxide (a weak acid) measure about pH 5.6. Most acidic rain in the United States has a pH of 4.2–4.4. The area centered around Lake Erie and Lake Ontario has acid rain with the highest pH in North America.

What technological development has been successful in reducing acid rain?

The most successful technology that reduced acid rain is the introduction of "scrubbers" that filter pollutants, especially sulfur dioxide, in smokestacks. The use of such technology has reduced the pollutants from factory emissions significantly.

What are the types of scrubbers used to control sulfur emissions?

Scrubbers, or more technically, flue gas desulfurization (FGD) technology, are installed to remove the sulfur dioxide from the flue gases. The three basic types of scrubbers are wet scrubbers, spray dry scrubbers, and

An air-separating factory uses scrubbers to remove sulfur dioxide from waste gases.

dry scrubbers. The most commonly used type is the wet scrubber, which is capable of removing more than 90 percent of the sulfur dioxide. Wet scrubber technology involves injecting a slurry of sorbent material, usually limestone or lime, into the flue gas. The sulfur dioxide dissolves into the slurry droplets and falls to the bottom of the spray tower. The slurry is sent to a reaction tank where the reaction is completed and a neutral salt is formed.

In spray dry systems, also called semi-dry systems, the slurry has a higher sorbent concentration. When the hot flue gas mixes with the slurry solution, water from the slurry is evaporated. The process forms a dry waste product that is collected and can be disposed or recycled to the slurry.

In dry systems, a powdered sorbent, typically a calcium- or sodium-based alkaline reagent, is pneumatically injected directly into the furnace or ductwork. The dry waste product is then removed.

Which pollutants lead to indoor air pollution?

Indoor air pollution, which may cause "sick building syndrome," results from conditions in modern, high-energy-efficiency buildings, which have reduced outside air exchange or have inadequate ventilation, chemical contamination, and microbial contamination. Indoor air pollution can produce various symptoms, such as headaches, nausea, and eye, nose, and throat irritation. In addition, houses are affected by indoor air pollution emanating from consumer and building products and from tobacco smoke. Below are listed some pollutants found in houses:

Pollutant	Sources	Effects
Asbestos	Old or damaged insulation, fireproofing, or acoustical tiles	Many years later, chest and abdominal cancers and lung diseases

Pollutant	Sources	Effects
Biological pollutants	Bacteria, mold, and mildew; viruses; animal dander and cat saliva; mites; cockroaches; and pollen	Eye, nose, and throat irritation; shortness of breath; dizziness; lethargy; fever; digestive problems; asthma; influenza and other infectious diseases
Carbon monoxide	Unvented kerosene and gas heaters; leaking chimneys and furnaces; wood stoves and fireplaces; gas stoves; automobile exhaust from attached garages; tobacco smoke	At low levels, fatigue; at higher levels, impaired vision and coordination; headaches; dizziness; confusion; nausea. Fatal at very high concentrations
Formaldehyde	Plywood, wall paneling, particle board, and fiberboard; foam insulation; fire and tobacco smoke; textiles and glues	Eye, nose, and throat irritations; wheezing and coughing; fatigue; skin rash; severe allergic reactions; may cause cancer
Lead	Automobile exhaust; sanding or burning of lead paint; soldering	Impaired mental and physical development in children; decreased coordination and mental abilities; kidneys, nervous system, and red blood cell damage
Mercury	Some latex paints	Vapors can cause kidney damage; long-term exposure can cause brain damage
Nitrogen dioxide	Kerosene heaters and unvented gas stoves and heaters; tobacco smoke	Eye, nose, and throat irritation; may impair lung function and increase respiratory infections in young children
Organic gases	Paints, paint strippers, solvents, and wood preservatives; aerosol sprays; cleansers and disinfectants; moth repellents; air fresheners; stored fuels; hobby supplies; dry-cleaned clothing	Eye, nose, and throat irritation; headaches; loss of coordination; nausea; damage to liver, kidney, and nervous system; some organics cause cancer in animals and are suspected of causing cancer in humans
Pesticides	Products used to kill household pests and products used on lawns or gardens that drift or are tracked inside the house	Irritation to eye, nose, and throat; damage to nervous systems and kidneys; cancer
Radon	Earth and rock beneath the home; well water, building materials	No immediate symptoms; estimated to cause about 10 percent of lung cancer deaths; smokers at higher risk

Why is exposure to asbestos a health hazard?

Asbestos fibers were used in building materials between 1900 and the early 1970s as insulation for walls and pipes, as fireproofing for walls and fireplaces, in soundproofing and acoustic ceiling tiles, as a strengthener for vinyl flooring and joint compounds, and as a paint texturizer. Asbestos poses a health hazard only if the tiny fibers are released into the air, but this can happen with any normal fraying or cracking. Asbestos removal aggravates this normal process and multiplies the danger level—it should only be handled by a contractor trained in handling asbestos. Once released, the particles can hang suspended in the air for more than twenty hours.

Exposure to asbestos has long been known to cause asbestosis. This is a chronic, restrictive lung disease caused by the inhalation of tiny mineral asbestos fibers that scar lung tissues. Asbestos has also been linked to cancers of the larynx, pharynx, oral cavity, pancreas, kidneys, ovaries, and gastrointestinal tract. The American Lung Association reports that prolonged exposure doubles the likelihood that a smoker will develop lung cancer. It takes cancer fifteen to thirty years to develop from asbestos. Mesothelioma is a rare cancer affecting the surface lining of the pleura (lung) or peritoneum (abdomen) that generally spreads rapidly over large surfaces of either the thoracic or abdominal cavities. Current treatment methods include surgery, radiation, and chemotherapy, although mesothelioma continues to be difficult to control.

What causes formaldehyde contamination in homes?

Formaldehyde contamination is related to the widespread construction use of wood products bonded with urea-formaldehyde resins and products containing formaldehyde. Major formaldehyde sources include subflooring of particle board; wall paneling made from hardwood plywood or particle board; and cabinets and furniture made from particle board, medium density fiberboard, hardwood plywood, or solid wood. Urea-formaldehyde foam insulation (UFFI) has received the most media notoriety and regulatory attention. Formaldehyde is also used in drapes, upholstery, carpeting, wallpaper adhesives, milk cartons, car bodies, household disinfectants, permanent-press clothing, and paper towels. In particular, mobile homes seem to have higher formaldehyde levels than other houses. The release of formaldehyde into the air by these products (called outgassing or offgassing) can develop poisoning symptoms in humans. The EPA classifies formaldehyde as a potential human carcinogen (cancer-causing agent).

How does the release of volatile organic compounds (VOCs) differ outdoors and indoors?

Volatile organic compounds (VOCs) are found both outdoors and indoors. They are regulated by the EPA for outdoor use to prevent the formation of ground-level ozone, a component of photochemical smog. Indoors, the main concern regarding VOCs is the adverse health effects on individuals exposed to them. The EPA began to regulate the manufacture of paint and other coatings in the late 1990s to improve indoor air quality. Manufacturers developed new paints and coatings with low- or zero-VOC content.

Why is radon a health hazard?

Radon is a colorless, odorless, tasteless, radioactive gaseous element produced by the decay of radium. It has three naturally occurring isotopes found in many natural materials, such as soil, rocks, well water, and building materials. Because the gas is continually released into the air, it makes up the largest source of radiation that humans receive. A National Academy of Sciences (NAS) report noted that radon was the second-leading cause of lung cancer. It has been estimated that it may cause as much as 12 percent, or about 15,000 to 22,000 cases, of lung cancer deaths annually. Smokers seem to be at a higher risk than non-smokers. The U.S. Environmental Protection Agency (EPA) recommends that in radon testing the level should not be more than four picocuries per liter. The estimated national average is 1.3 picocuries per liter (pCi/L). Because the EPA's "safe level" is equivalent to 200 chest X-rays per year, some experts believe that lower levels are appropriate. The American Society of Heating, Refrigeration, and Air-Conditioning Engineers (ASHRAE) recommends two picocuries/liter. The EPA estimates that nationally, 1 in 15 or 6 percent of all homes have radon levels that are above the four picocuries/liter limit. The EPA recommends that homes with values of 2–4 pCi/L install a radon-reduction system in the home. A common radon-reduction system is a vent pipe system and a fan that pulls the radon from beneath the house and vents it to the outside.

GLOBAL CLIMATIC CHANGE

What is the greenhouse effect?

The greenhouse effect is a warming near the earth's surface that results when the earth's atmosphere traps the sun's heat. The atmosphere acts much like the glass walls and roof of a greenhouse. The effect was described by John Tyndall (1820–1893) in 1861. It was given the greenhouse analogy much later in 1896 by Swedish chemist Svante Arrhenius (1859–1927). The greenhouse effect is what makes the earth habitable. Without the presence of water vapor, carbon dioxide, and other gases in the atmosphere, too much heat would escape and the earth would be too cold to sustain

Greenhouse gases encircle the earth, preventing heat from escaping and raising the average temperature of the planet.

life. Carbon dioxide, methane, nitrous oxide, and other greenhouse gases absorb the infrared radiation rising from the earth and hold this heat in the atmosphere instead of reflecting it back into space.

In the twentieth century, the increased buildup of carbon dioxide, caused by the burning of fossil fuels, has been a matter of concern. There is some controversy concerning whether the increase noted in the earth's average temperature is due to the increased amount of carbon dioxide and other gases or is due to other causes. Volcanic activity, destruction of the rain forests, use of aerosols, and increased agricultural activity may also be contributing factors.

What are the greenhouse gases?

Scientists recognize carbon dioxide, methane, chlorofluorocarbons, nitrous oxide, and water vapor as significant greenhouse gases. Greenhouse gases account for less than 1 percent of the earth's atmosphere. These gases trap heat in Earth's atmosphere, preventing the heat from escaping back into space. Human activities, such as burning fossil fuels for gasoline in automobiles, electricity, and heat account for the dramatic increase in the amount of greenhouse gas emissions in the United States in the last 150 years.

Emissions of Greenhouse Gases in the United States (Million Metric Tons of Gas)

Gas	1990	1995	2000	2005	2010	2011	2012	2013*
Carbon dioxide	5,124	5,451	6,002	6,134	5,705	5,569	5,358	5,505
Methane	745	750	716	708	667	661	648	636
Nitrous oxide	330	371	335	356	360	372	366	355
Fluorinated gases	102	122	159	153	167	175	174	176
Total	6,301	6,695	7,213	7,350	6,899	6,777	6,545	6,673

*Latest data available as of August 2015 (source: U.S. Environmental Protection Agency).

What is the main greenhouse gas responsible for climate change?

Carbon dioxide is the primary greenhouse gas emitted through human activities. Since the late 1700s, carbon dioxide concentration has increased from about 280 ppm to 395 ppm in 2013, resulting in the highest concentration in at least 400,000 years. This relatively rapid increase is due to the movement of carbon from ancient sediments (coal, gas, oil) to the atmosphere in the form of carbon dioxide. Activities such as clearing and

How much carbon dioxide is produced per gallon of gasoline?

A gallon of gasoline produces 19.4 pounds (8.8 kilograms) of carbon dioxide. The average passenger vehicle produces 5.2 metric tons (11,464 pounds) of carbon dioxide annually.

burning of forests, increased mining of fossil fuels, and inefficiency of the burning of fossil fuels have accelerated the release of carbon dioxide.

What are the primary sources of greenhouse gas emissions in the United States?

Technological development has increased the greenhouse gas emissions in the United States.

Source of Greenhouse Gas Emissions	Percent of Greenhouse Gas Emissions in the United States (2013)
Electricity production (burning fossil fuels, in particular coal and natural gas)	31%
Transportation (burning gasoline and diesel [fossil fuels] for cars, trucks, buses, ships, trains, and planes)	27%
Industry (burning fossil fuels for energy; also from certain chemical reactions)	21%
Commercial and residential (burning fossil fuels for heat)	12%
Agriculture (livestock, in particular cows, soils, and rice production)	9%

Source: Environmental Protection Agency

What are some clean coal technologies that are being researched?

Researchers are studying methods and techniques that will increase the efficiency of generating electricity with coal while at the same time reducing the amount of carbon dioxide emissions when it is burned. One method is to capture and store the carbon dioxide. Another technology is coal gasification. Coal gasification is a process in which coal is converted to a gas, called syngas, before it is burned. It is then easier to separate the carbon dioxide as a relatively pure gas before power is generated. Another technique is to burn coal in oxygen instead of in air. This reduces the amount of flue gas that must be processed to isolate carbon dioxide. Once the carbon dioxide is captured, it must be stored permanently. It may be stored in oil and gas reservoirs, coal seams that cannot be mined, or deep saline aquifers. All of these geological formations are geologically sealed so the carbon dioxide will not be able to escape. Research continues to find an economically feasible option to reduce carbon dioxide emissions while using coal.

What is the enhanced greenhouse effect?

The buildup of carbon dioxide and other greenhouse gases warms the atmosphere by absorbing some of the infrared (heat) radiation. Some of the heat in the warmed atmosphere is transferred back to Earth's surface, warming the land and ocean even more.

What technical methods are used to infer and measure climate change?

Drilling into the Earth's ice caps and glaciers allows scientists to extract ice cores, which can provide information about atmospheric gas concentration, temperature changes, snowfall, solar activity, and, if soot has been trapped, the frequency of forest fires. Drilling cores into sedimentary beds underneath bodies of water provides information about vegetation and climatic changes. Perhaps the most used and reliable method involves computer modeling of the vast amount of data generated by charting of weather and atmospheric patterns.

What is a carbon sink?

Land areas that absorb carbon dioxide from the atmosphere are known as "sinks." Land use and forestry offset 15 percent of the greenhouse gas emissions in the United States. Thus, managed forests and other land uses are a net sink, absorbing more carbon dioxide than they emit.

What is carbon management and what are some ways to deal with the increasing quantities of carbon dioxide?

Carbon management refers to the separation and capture of carbon dioxide produced during the combustion of fossil fuels and then sequestering or storing it. Some of the ways to reduce carbon dioxide include:

1. planting trees
2. increasing the efficiency of coal-fired power plants
3. replacing coal-fired power plants with nuclear power, hydropower, wind power, or natural gas
4. increasing the fuel economy of motor vehicles
5. redesigning cities to decrease reliance on single-occupant vehicles
6. insulating buildings to reduce winter heating and summer cooling

Which countries emit the most carbon dioxide into the air?

Carbon dioxide is one of the greenhouse gases. In 2011 the ten countries that emitted the most carbon dioxide from the consumption of energy were:

Country	Carbon Dioxide Emissions (in million metric tons)	Percentage of Global Total
China	8,547.7	26.1
United States	5,270.4	16.1
India	1,830.9	5.6
Russia	1,781.7	5.4
Japan	1,259.1	3.8
Germany	788.3	2.4
South Korea	657.1	1.8
Iran	603.6	1.8
Saudi Arabia	582.7	1.8
Canada	550.8	1.7
World	32,723.1	

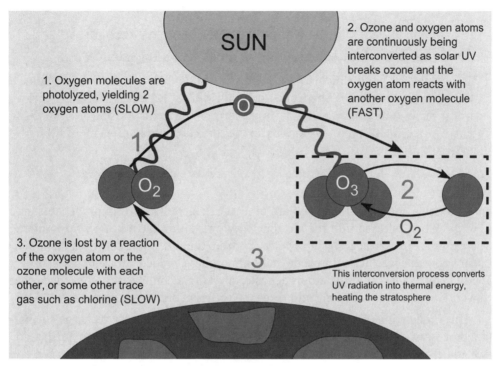

1. Oxygen molecules are photolyzed, yielding 2 oxygen atoms (SLOW)

2. Ozone and oxygen atoms are continuously being interconverted as solar UV breaks ozone and the oxygen atom reacts with another oxygen molecule (FAST)

3. Ozone is lost by a reaction of the oxygen atom or the ozone molecule with each other, or some other trace gas such as chlorine (SLOW)

This interconversion process converts UV radiation into thermal energy, heating the stratosphere

Ozone is created when chemical reactions change O_2 to O_3.

What one single pollutant most frequently escapes EPA standards in urban areas?

Ozone in the troposphere, the level of the atmosphere closest to the earth, is the one pollutant that most frequently exceeds EPA standards.

Is ozone beneficial or harmful to life on Earth?

Ozone, a form of oxygen with three atoms instead of the normal two, is highly toxic; less than one part per million of this blue-tinged gas is poisonous to humans. In Earth's upper atmosphere (stratosphere), it is a major factor in making life on Earth possible. About 90 percent of the planet's ozone is in the ozone layer. The ozone belt shields the Earth from and filters the excessive ultraviolet (UV) radiation generated by the sun. Scientists predict that a diminished or depleted ozone layer could lead to increased health problems for humans, such as skin cancer, cataracts, and weakened immune systems. Increased UV radiation can also lead to reduced crop yields and disruption of aquatic ecosystems, including the marine food chain. While beneficial in the stratosphere, near ground level it is a pollutant that helps form photochemical smog and acid rain.

How do chlorofluorocarbons affect the Earth's ozone layer?

Chlorofluorocarbons (CFCs) are hydrocarbons, such as freon, in which part or all of the hydrogen atoms have been replaced by fluorine atoms. These can be liquids or gases, are nonflammable and heat-stable, and are used as refrigerants, aerosol propellants, and solvents. When released into the air, they slowly rise into Earth's upper atmosphere, where they are broken apart by ultraviolet rays from the sun. Some of the resultant molecular fragments react with the ozone in the atmosphere, reducing the amount of ozone. The CFC molecules' chlorine atoms act as catalysts in a complex set of reactions that convert two molecules of ozone into three molecules of ordinary oxygen.

This is depleting the beneficial ozone layer faster than it can be recharged by natural processes. The resultant "hole" lets through more ultraviolet light to Earth's surface and creates health problems for humans, such as cataracts and skin cancer, and disturbs delicate ecosystems (for example, making plants produce fewer seeds). In 1978 the U.S. government banned the use of fluorocarbon aerosols, and currently aerosol propellants have been changed from fluorocarbons to hydrocarbons, such as butane.

How large is the Antarctic ozone hole?

The term "hole" is widely used in popular media when reporting on ozone. However, the concept is more correctly described as a low concentration of ozone that occurs in August–October (springtime in the Southern Hemisphere). It was not observed until 1979. The first scientific article on ozone depletion in the Antarctic was published in *Nature* in 1985. The largest ozone hole ever observed, 11.4 million square miles (29.6 million square kilometers), occurred on September 24, 2006. The average size of the ozone hole was 8.1 million square miles (21.0 million square kilometers) during September–October 2014. The single-day maximum ozone hole area for 2014 was 9.3 million square miles (24 million square kilometers). The 2014 ozone hole was comparable in size to the years 2010, 2012, and 2013. It was smaller than the large holes between 1998 and 2006.

Who discovered the hole in the ozone layer?

The hole in the ozone layer was discovered by Joseph Farman (1930–2013). Farman was a British geophysicist who began collecting ozone readings in Antarctica in 1957. During the course of the first twenty-five years after he began collecting ozone readings, there was little change in the readings. In 1982, however, the Antarctica readings were so drastically different that at first Farman thought his recording devices were faulty. He ordered new devices and found the readings to be even more dramatically different. The data were double- and triple-checked and found to be correct. Farman and his colleagues, Brian Gardiner and Jonathan Shanklin (1953–), published an article in *Nature* in 1985 announcing that ozone levels over Antarctica had fallen by about 40 percent from 1975 to 1984.

What are some specific physical indicators of climate change?

According to the Third Assessment Report issued by the Intergovernmental Panel on Climate Change (IPCC) in 2001, some of the physical indicators of climate changes include:

- Increase in sea level by 4–8 inches (10–20 centimeters) during the twentieth century
- A decrease in duration of river and lake ice in the Northern Hemisphere by 2 weeks
- Thinning of Arctic sea ice by 10–40 percent
- Retreat of mountain glaciers
- Thawing and warming of permafrost
- El Niño events magnified in the past 40 years
- Lengthening of growing season by 1–4 days per decade

What were the goals of the Montreal Protocol?

The Montreal Protocol was signed in 1987 by members of the United Nations to phase out and reduce production of CFCs and other ozone-damaging chemicals by 50 percent by 1998. Amendments and revisions to the Montreal Protocol were made in 1990, 1992, and 1997 to include other chemicals and accelerate the phase-out of certain chemicals. Production of CFCs, carbon tetrachloride, and methyl chlororform was phased out by 1996 in the United States and other highly developed countries.

What is the Kyoto Protocol?

The Kyoto Protocol was an international summit held in Kyoto, Japan, in December 1997. Its goal was for governments around the world to reach an agreement regarding emissions of carbon dioxide and other greenhouse gases. The Kyoto Protocol called for the industrialized nations to reduce national emissions over the period 2008–2012 to 5 percent below the 1990 levels. The protocol included these greenhouse gases: carbon dioxide, methane, nitrous oxide hydrofluorocarbons, perfluorocarbons, and sulfur dioxides. In December 2012, the Doha Amendment to the Kyoto Protocol was adopted. This amendment provides for a second commitment period from January 1, 2013, to December 31, 2020, for countries to reduce the greenhouse gas emissions by at least 18 percent below 1990 levels.

WATER POLLUTION

What are the major types of water pollution?

Water pollution is a physical or chemical change in water that adversely affects the health of humans or other organisms. There are eight major types of water pollution:

Water Pollution

Type of Pollution	Source	Examples
Sewage	Wastewater from drains or sewers	Human wastes, soaps, detergents
Disease-causing agents	Wastes of infected individuals	Bacteria, viruses, protozoa, parasitic worms
Sediment pollution	Erosion of agricultural lands, forest soils, overgrazed rangelands, strip mines, construction	Clay, silt, sand, and gravel suspended in water and settling out
Inorganic plant and algal nutrients	Human and animal wastes, plant residues, fertilizer runoff, atmospheric deposition	Nitrogen and phosphorus
Organic compounds	Landfills, industrial waste, agricultural runoff	Synthetic chemicals, pesticides, plastics, industrial chemicals, cleaning solvents
Inorganic chemicals	Industries, mines, oil drilling, urban runoff	Acids, salts, heavy metals, including lead, mercury, and arsenic
Radioactive substances	Nuclear power plants, nuclear weapons industry, medical and scientific research facilities	Unstable isotopes of radioactive materials such as uranium and thorium
Thermal pollution	Industrial runoff	Heated water produced during industrial processes and then released into waterways

How did the Clean Water Act of 1972 improve regulation of water pollution?

The first major federal legislation addressing the concerns of water pollution was the Federal Water Pollution Control Act of 1948. Although it recognized a growing concern for clean, safe water, the amendments of the Clean Water Act of 1972 strengthened the regulation of pollutants into the nation's waters. The key points of the 1972 Clean Water Act were:

• Established the basic structure for regulating pollutants discharged into U.S. waters.

Industrial waste is, of course, one of the major contributors to water pollution across the globe. Farms and residential runoff (such as fertilizers from lawns) also play a part in contamination.

> ## What is gray water?
>
> **G**ray water is water that has already been used for washing dishes—by hand or in dishwashers—taking showers, or used in washing machines. Gray water is recycled to flush toilets, wash cars, or water lawns. In these instances, clean water is not required.

- Provided the EPA with the authority to implement pollution control programs, such as setting wastewater standards for industry.
- Continued to set water quality standards for all contaminants in surface water.
- Made it unlawful for any person to discharge any pollutant from a point source into navigable waters without obtaining a permit.
- Recognized the need to address the problems of nonpoint source water pollution.

What is nonpoint source water pollution?

Nonpoint source (NPS) water pollution comes from many different sources. A major source of NPS is from rainfall and snowmelt. As the water from the precipitation runs off various surfaces, it picks up pollutants—mostly human-made—and carries them to streams, rivers, lakes, wetlands, and oceans.

How does the National Pollution Discharge Elimination System (NPDES) permit program control water pollution?

The NPDES permit program controls water pollution by regulating point sources that discharge pollutants into the water. Point sources are discrete conveyances such as pipes or man-made ditches that release water. All industrial and municipal facilities that discharge water must obtain a permit it they discharge water directly into surface waters.

Can advances in technology help solve the problem of freshwater depletion |and shortage?

Desalination, also known as desalinization, generates freshwater by removing salt from seawater. One method is to mimic the hydrologic cycle by hastening evaporation from allotments of ocean water with heat and then condensing the vapor—that is, distilling freshwater. Another method involves forcing water through membranes to filter out salts. More than 17,000 desalination facilities were in operation worldwide as of 2013.

What are the sources of oil pollution in the oceans?

Most oil pollution results from minor spillage from tankers and accidental discharges of oil when oil tankers are loaded and unloaded. Other sources of oil pollution include improper disposal of used motor oil, oil leaks from motor vehicles, routine ship maintenance, leaks in pipelines that transport oil, and accidents at storage facilities and refineries.

Source	Percent of Total Oil Spillage
Natural seepage	46
Discharges from consumption of oils (operational discharges from ships and discharges from land-based sources)	37
Accidental spills from ships	12
Extraction of oil	3

Where did the first major oil spill occur?

The first major commercial oil spill occurred on March 18, 1967, when the tanker *Torrey Canyon* grounded on the Seven Stones Shoal off the coast of Cornwall, England, spilling 830,000 barrels (119,000 tons) of Kuwaiti oil into the sea. This was the first major tanker accident. However, during World War II, German U-boat attacks on tankers between January and June of 1942 off the United States East Coast spilled 590,000 tons of oil.

Although the Exxon *Valdez* was widely publicized as a major spill of 35,000 tons in 1989, it is dwarfed by the deliberate dumping of oil from Sea Island into the Persian Gulf on January 25, 1991. It is estimated that the spill equaled almost 1.5 million tons of oil. A major spill also occurred in Russia in October 1994 in the Komi region of the Arctic. The size of the spill was reported to be as high as 2 million barrels (286,000 tons).

In addition to the large disasters, day-to-day pollution occurs from drilling platforms where waste generated from platform life, including human waste, and oils, chemicals, mud, and rock from drilling are discharged into the water.

Date	Cause	Tons Spilled
1/42–6/42	German U-boat attacks on tankers off the East Coast of U.S. during World War II	590,000
3/18/67	Tanker *Torrey Canyon* grounds off Land's End in the English Channel	119,000
3/20/70	Tanker *Othello* collides with another ship in Tralhavet Bay, Sweden	60–100,000
12/19/72	Tanker *Sea Star* collides with another ship in Gulf of Oman	115,000
5/12/76	*Urquiola* grounds at La Coruna, Spain	100,000
3/16/78	Tanker *Amoco Cadiz* grounds off Northwest France	223,000
6/3/79	Itox I oil well blows in southern Gulf of Mexico	600,000

Date	Cause	Tons Spilled
7/79	Tankers *Atlantic Express* and *Aegean Captain* collide off Trinidad and Tobago	300,000
2/19/83	Blowout in Norwuz oil field in the Persian Gulf	600,000
8/6/83	Fire aboard *Castillo de Beliver* off Cape Town, South Africa	250,000
3/24/89	Exxon *Valdez* accident in Alaska	35,000
1/25/91	Iraq begins deliberately dumping oil into Persian Gulf from Sea Island, Kuwait	1,450,000
2/94	A structure to prevent pipeline leaks fails, spilling oil in the Komi Republic in northern Russia	almost 102,000,000
2/4/99	Tanker *New Carissa* spills some of its oil in Coos Bay, Oregon	238
11/13/02	Tanker *Prestige* splits in half off Galicia, Spain	67,000
4/20/10	Deepwater Horizon offshore oil rig (BP) spill after an explosion	736,880

What are the most commonly collected debris items found along ocean coasts?

Debris found along ocean coasts is categorized as land-based, ocean-based, or general-source debris. Land-based debris blows, washes, or is discharged into the water from

This satellite photo taken on May 24, 2010, shows the extent of the Deepwater Horizon spill in the Caribbean.

How much oil was spilled into the Caribbean Sea when the British Petroleum offshore oil rig exploded in 2010?

The largest accidental oil spill in the world began April 20, 2010, when the BP Deepwater Horizon rig exploded in the Gulf of Mexico. Initial estimates reported that only 1,000 barrels of oil per day were being released into the Gulf. These estimates quickly changed to 5,000 barrels per day and then 12,000–19,000 barrels per day by late May. As researchers continued to evaluate the situation, the estimate continued to rise until they determined that an average of 53,000 barrels of oil were released per day between April 20 and July 15, 2010. The final estimate was that 4.9 million barrels of oil flowed into the Gulf of Mexico, making it the largest oil spill ever.

land areas. Land-based debris originates with beachgoers, fishermen, shore-based manufacturers, processing facilities, and waste-management facilities. Boating and fishing activities are often sources of ocean-based debris. During a five-year study from 2001 to 2006, a total of 238,103 items of debris were collected from various sites along the nation's coasts. Land-based debris accounted for 52 percent of the total, general-source debris accounted for 34 percent of the total, and ocean-based debris accounted for 14 percent of the total. The most frequently collected items were:

Debris along Ocean Coasts during 2001–2006 Study

Source	Number of Items Collected	Percent Total
Land-based		
Straws	65,384	27.5
Balloons	18,509	7.8
Metal beverage cans	17,705	7.4
General sources		
Plastic beverage bottles	30,858	13
Plastic bags with seam <1 meter	21,477	9
Plastic food bottles	8,355	3.5
Ocean-based		
Rope >1 meter	13,023	5.5
Fishing line	8,032	3.4
Floats/buoys	3,488	1.5

Why is ocean trash a serious problem?

Trash in the oceans and along the shores can harm swimmers and beachgoers. It also harms wildlife either by entangling the wildlife or if it is eaten. During the annual international coastal waterway cleanup in 2014, there were 561,895 volunteers, who collected 16,186,759 pounds (7,342,190 kilograms) of trash along 13,360 miles (21,500 kilometers) of coast worldwide.

Top 10 Trash Items Found*

Item	Number Collected
Cigarette butts	2,248,065
Food wrappers	1,376,133
Plastic beverage bottles	988,965
Plastic bottle caps	811,871
Straws	519,911
Other plastic bags	489,968
Plastic grocery bags	485,204
Glass beverage bottles	396,121
Beverage cans	382,608
Plastic cups and plates	376,409

*Source: Ocean Conservancy

SOLID AND HAZARDOUS WASTE

How are hazardous waste materials classified?

Hazardous wastes may be listed by the EPA as being hazardous, identified by certain characteristics; universal hazardous wastes, such as batteries; and mixed wastes, which contain radioactive and hazardous waste components. There are four types of hazardous waste materials identified by their characteristics—corrosive, ignitable, reactive, and toxic. Hazardous wastes may be liquids, solids, gases, or sludges.

- *Corrosive* materials can wear away or destroy a substance. Most acids are corrosive and can destroy metal, burn skin, and give off vapors that burn the eyes.
- *Ignitable* materials can burst into flames easily. These materials pose a fire hazard and can irritate the skin, eyes, and lungs. Gasoline, paint, and furniture polish are ignitable.
- *Reactive* materials can explode or create poisonous gas when combined with other chemicals. Combining chlorine bleach and ammonia, for example, creates a poisonous gas.
- *Toxic* materials or substances can poison humans and other life. They can cause illness or death if swallowed or absorbed through the skin. Pesticides and household cleaning agents are toxic.

What is the Toxic Release Inventory (TRI)?

TRI is a government-mandated, publicly available compilation of information on the release of nearly 650 individual toxic chemicals by manufacturing facilities in the United

States. These chemicals are used in the production of products our society consumes, such as pharmaceuticals, automobiles, clothing, and electronics. The chemicals on the list have been identified to cause significant adverse health effects, including cancer and other chronic conditions or significant adverse environmental effects. The law requires manufacturers to state the amounts of chemicals they release directly to air, land, or water, or that they transfer to off-site facilities that treat or dispose of wastes. The U.S. Environmental Protection Agency compiles these reports into an annual inventory and makes the information available in a computerized database available for analysis.

In 2013, 21,598 facilities managed 25.63 billion pounds (11.63 billion kilograms) of toxic chemical waste. Thirty-six percent was recycled, 11 percent was used for energy recovery, 37 percent was treated, and 16 percent or 4.14 billion pounds (1.8 billion kilograms) of toxic chemicals were released into the environment. The majority of these releases, 2.75 billion pounds (1.25 billion kilograms), were disposed on-site to land (including landfills and underground injection). The remaining of the total on-site releases were 0.77 billion pounds (.35 billion kilograms) into the air and 0.21 billion pounds (.095 kilograms) into water. The total amount of toxic chemicals released in 2013 was 15 percent higher than the amount released in 2012 and 7 percent lower than the amount released in 2003. A main reason for the increase was due to on-site land disposal by the metal mining sector.

Which industries release the greatest percentage of chemicals?

The metal mining industry released the greatest quantity of toxic chemicals for the year 2013, accounting for 47 percent of total chemical releases.

Industry	Total Releases (pounds)	Percent of Total
Metal mining	1,448.8 million	47
Chemicals	544.6 million	12
Electric utilities	519.3 million	13
Primary metals	440,462,686	8
Paper	186,356,909	5
Hazardous waste management	168,601,834	4
Food	167,223,768	3
All others	538,718,584	8

How do source reduction activities impact the release of toxic chemicals?

Source reduction activities are practices that reduce the total quantity of chemical waste generated at the source. Some activities adopted by manufacturing plants that reduce the generation of chemical waste are good operating practices (for example, spill and leak prevention at the plants), process modifications (for example, instituting re-circulation within a process), and raw materials modifications (for example, increasing the purity of raw materials).

What are the POPs?

POPs is an acronym for Persistent Organic Pollutants (POPs). The Stockholm Convention on Persistent Organic Pollutants banned a dozen chemicals known as the "dirty dozen." All of these chemicals possess toxic properties, resist degradation, and are transported across international boundaries via air, water, and migratory species. The ban became effective in 2004. Health effects of these compounds include disruption of the endocrine system, cancer, and adverse effects in the developmental processes of organisms.

The Original "Dirty Dozen"

Persistent Organic Pollutant (POP)	Use
Aldrin	Insecticide
Chlordane	Insecticide
DDT (dichlorodiphenyl-trichloroethane)	Insecticide
Dieldrin	Insecticide
Endrin	Rodentcide and insecticide
Heptachlor	Fungicide
Hexachlorobenzene	Insecticide and fire retardant
Mirex™	Insecticide
Toxaphene™	Insecticide
PCBs (polychlorinated biphenyls)	Industrial chemical
Dioxins	By-product of certain manufacturing processes
Furans (dibenzofurans)	By-product of certain manufacturing processes

In 2009, an additional nine chemicals were added to the list. The ban on this group became effective in 2010. At the fifth meeting of the Stockholm Convention, held in 2011, one more chemical was added to the list. The new POPs are:

Persistent Organic Pollutant (POP)	Use
Alpha hexachlorocylohexane	Insecticide
Beta hexachlorocyclohexane	Insecticide
Chlordecone	Pesticide
Hexabromobiphenyl	Industrial chemical used as a flame retardant
Hexabromodiphenyl ether and heptabromodiphenyl ether (commercial octabromodiphenyl ether)	Industrial chemical
Lindane	Insecticide (still acceptable for use as a pharmaceutical for control of head lice and scabies as second line treatment)
Pentachlorobenzene	Fungicide, used in PCB products and a flame retardant
Perfluorooctane sulfonic acid, its salts, and perfluorooctane sulfonyl fluoride	Industrial chemical

Persistent Organic Pollutant (POP)	Use
Tetrabromodiphenyl ether and pentabromodiphenyl ether (commercial pentabromodiphenyl ether)	Industrial chemical
Endosulfan	Insecticide

How did DDT affect the environment?

Although DDT was synthesized as early as 1874 by Othmar Zeidler (1859–1911), it was Swiss chemist Paul Müller (1899–1965) who recognized its insecticidal properties in 1939. He was awarded the 1948 Nobel Prize in Physiology or Medicine for his development of dichloro-diphenyl-trichloro-ethene, or DDT. Unlike the arsenic-based compounds then in use, DDT was effective in killing insects and seemed not to harm plants and animals. In the following twenty years, it proved to be effective in controlling disease-carrying insects (mosquitoes that carry malaria and yellow fever and lice that carry typhus) and in killing many plant crop destroyers. Publication of Rachel Carson's (1907–1964) *Silent Spring* in 1962 alerted scientists to the detrimental effects of DDT. Increasingly, DDT-resistant insect species and the accumulative hazardous effects of DDT on plant and animal life cycles led to its disuse in many countries during the 1970s.

What are PCBs?

Polychlorinated biphenyls (PCBs) are a group of chemicals that were widely used before 1970 in the electrical industry as coolants for transformers and in capacitors and other electrical devices. They caused environmental problems because they do not break down and can spread through the water, soil, and air. They have been linked by some scientists to cancer and reproductive disorders and have been shown to cause liver function abnormalities. Government action has resulted in the control of the use, disposal, and production of PCBs in nearly all areas of the world, including the United States.

What problems may be encountered when polyvinyl chloride (PVC) plastics are burned?

Chlorinated plastics, such as PVC, contribute to the formation of hydrochloric acid gases. They also may be a part of a mix of substances containing chlorine that form a precursor to dioxin in the burning

DDT decreased bald eagle populations because the chemical, when ingested by egg-laying females, caused the shells to soften and break before the chicks could hatch.

process. Polystyrene, polyethylene, and polyethylene terephthalate (PET) do not produce these pollutants.

What is the Toxic Substances Control Act (TSCA)?

In 1976, the U.S. Congress passed the Toxic Substances Control Act (TSCA). This act requires the premarket testing of toxic substances. When a chemical substance is planned to be manufactured, the producer must notify the Environmental Protection Agency (EPA), and, if the data presented are determined to be inadequate to approve its use, the EPA will require the manufacturer to conduct further tests. Or, if it is later determined that a chemical is present at a level that presents an unreasonable public or environmental risk or if there are insufficient data to know the chemical's effects, manufacturers have the burden of evaluating the chemical's characteristics and risks. If testing does not convince the EPA of the chemical's safety, the chemical's manufacturing, sale, or use can be limited or prohibited. Food, drugs, cosmetics, and pesticides are generally excluded from the TSCA.

What is the Superfund Act?

The discovery of toxic waste dumps, such as Love Canal in New York and Times Beach in Missouri, prompted Congress to develop a program to clean up abandoned hazardous waste sites. In 1980, the U.S. Congress passed the Comprehensive Environmental Response, Compensation, and Liability Act, commonly known as the Superfund program. This law (along with amendments in 1986 and 1990) established a $16.3-billion Superfund financed jointly by federal and state governments and by special taxes on chemical and petrochemical industries (which provide 86 percent of the funding). The purpose of the Superfund is to identify and clean up abandoned hazardous-waste dump sites and leaking underground tanks that threaten human health and the environment. To keep taxpayers from footing most of the bill, cleanups are based on the polluter-pays principle. The EPA is charged with locating dangerous dump sites, finding the potentially liable culprits, ordering them to pay for the entire cleanup, and suing them if they don't. When the EPA can find no responsible party, it draws money out of the Superfund for cleanup.

An abandoned street in the Love Canal neighborhood of Niagara Falls, New York. The Hooker Chemical Company was found negligent of properly disposing the chemicals that made the area uninhabitable.

Which states have the most hazardous waste (Superfund) sites?

As of 2014, there were 1,318 hazardous waste sites and 51 proposed sites on the

National Priorities List. There are hazardous waste sites in all of the fifty states except North Dakota. The states with the most hazardous waste sites are:

- New Jersey: 113
- Pennsylvania: 94
- California: 97
- New York: 86
- Michigan: 65
- Florida: 53
- Texas: 50
- Washington: 50
- Illinois: 44
- Wisconsin: 37

The average cost to clean up a site is $25 million. Over 300 sites have been cleaned up and deleted from the National Priorities List.

Where is one of the largest Superfund sites in the nation?

One of the largest Superfund sites in the nation is near Butte, Montana, and includes the former Berkeley Pit mine. Once operations at the pit ceased, the pumps that were used to remove water from the bottom of the pit were removed, and the pit began to fill with groundwater seeping in from the surrounding aquifers. The water is highly contaminated due to the presence of toxic minerals, including zinc, cadmium, sulfuric acid, and arsenic. The pH of the water has been measured at 2.5. As of 2014, the water level in the pit had risen to 5,319.78 feet (1,621.5 meters) above sea level. When the water level reaches 5,410 feet (1,649 meters), pumping and treating of the water will begin in accordance with the Superfund remedial plan of action. In the meantime, the area is monitored closely as a Superfund site with other remedial actions in place in the surrounding areas.

What are brownfields?

The Environmental Protection Agency defines brownfields as abandoned, idled, or underused industrial or commercial sites where expansion or redevelopment is complicated by real or perceived environmental contamination. Real estate developers perceive brownfields as inappropriate sites for redevelopment. There are approximately 450,000 brownfields in the United States, with the heaviest concentrations being in the Northeast and Midwest.

How much garbage does the average American generate?

According to the Environmental Protection Agency, nearly 251 million tons of municipal waste was generated in 2012. This is equivalent to 4.38 pounds (almost 2.0 kilograms) per person per day, or approximately 1,599 pounds per person (725 kilograms) per year.

How much municipal solid waste is generated globally?

The growth of solid waste generation is a consequence of urbanization and economic growth. Globally, nearly 1.4 billion tons (1.3 metric tons) of municipal solid waste are generated annually, or 2.6 pounds (1.2 kilograms) per person per day. Waste generation is greater in urban areas where there is easier access to purchased products and less recycling and reuse.

Country	Total Municipal Solid Waste Generation per Day (Tons/Metric Tons)	Per Capita Municipal Solid Waste Generation (Pounds/Kilograms)
United States	688,614/624,700	5.7/2.6
Germany	140,893/127,816	4.7/2.1
Japan	159,247/144,466	3.8/1.7
Brazil	164,350/149,096	2.2/1.0
China	573,806/520,548	2.2/1.0

Based on data published in 2012, although data are collected at different times by different countries.

What are the components of municipal solid waste?

Municipal solid waste consists of things we commonly use and then throw away, such as paper and packaging, food scraps, yard waste, tires, and large household items, including old sofas, appliances, and computers. It does not include industrial, hazardous, or construction waste. The distribution of the municipal solid waste generated in 2012 is illustrated below:

Waste Product	Weight (millions of tons)	Percent of Total
Paper and paperboard	68.62	27.4
Yard wastes	33.96	13.5
Food scraps	36.43	14.5
Plastics	31.75	12.7
Metals	22.38	8.9
Rubber, leather, and textiles	21.86	8.7
Wood	15.82	6.3
Glass	11.57	4.6
Other materials	4.6	3.4

When was the first garbage incinerator built?

The first garbage incinerator in the United States was built in 1885 on Governor's Island in New York Harbor.

How much methane fuel does a ton of garbage make?

Over a period of 10 to 15 years, a ton of garbage will make 14,126 cubic feet (400 cubic meters) of fuel, although a landfill site will generate smaller amounts for 50 to 100 years. One ton of garbage can produce more than 100 times its volume in methane over a decade. Landfill operators tend not to maximize methane production.

How is municipal waste managed?

More than half of the municipal waste generated in the United States is discarded in landfills. The balance is either recovered through recycling programs or combusted with energy recovery.

Management of Municipal Waste (2012)

Method of Disposal	Amount (millions of tons)	Percent of Total
Discarded in landfills	135	54
Recovered through recycling and composting	87	35
Combustion with energy recovery	29	12
Total	251	100

How has disposal of solid waste to landfill facilities changed?

Since there are large amounts of available land in the United States, landfilling has been an essential component of waste management for several decades. In areas where land is less available, combustion has a more significant role in waste management. However, combustion requires proper air emission control equipment to minimize the impact on air pollution. In 1960, 94 percent of all garbage was sent to landfills. During the following decades, although the amount of municipal solid waste increased, the amount going to landfills decreased. During the years 1990 to 2012, the total amount of waste going to landfills decreased from 145.3 million tons, 69.8 percent of the total municipal solid waste generated, to 135 million tons, representing 53.8 percent of the total municipal solid waste generated.

Despite shrinking space, the United States still dumps half of its trash into landfills.

Year	Percent of Municipal Solid Waste in Landfills
1960	93.6%
1970	93.1%
1980	88.6%
1990	69.8%
2000	57.6%
2005	56.1%
2010	54.3%
2012	53.8%

Much of this decrease was due to an increased amount of waste being recycled. As of 2012, there were 1,908 landfills in the continental United States.

What are some innovative technologies for managing landfills?

Two innovative technologies for landfills are bioreactors and projects that recover and use the landfill gas emissions. A bioreactor landfill accelerates the decomposition and stabilization of waste in the landfill. Decomposition in a bioreactor is accomplished in years rather than decades in a traditional landfill. Landfill gas emissions projects focus on how to capture, convert, and use the landfill gas emissions as an energy resource.

Why does the trash in a bioreactor landfill decompose more quickly than in a traditional landfill?

Bioreactor landfills may be aerobic, anaerobic, or a hybrid of aerobic and anaerobic. In aerobic bioreactor landfills, leachate is removed and then recirculated through the landfill. Air is injected into the waste mass to promote aerobic activity and accelerate waste stabilization. In an anaerobic bioreactor landfill, water may be added to the leachate to increase the rate of decomposition. A hybrid facility will use both aerobic and anaerobic methods to degrade waste in the upper levels of the landfill and collect gas from the lower levels. The Yolo County, California, landfill was one of the first experimental bioreactor landfill projects.

How is landfill gas (LFG) extracted from landfills?

Landfill gas consists of approximately 50 percent methane and 50 percent carbon dioxide. It accounted for 18 percent of the total methane emissions in the United States in 2012. Rather than allowing LFG escape into the atmosphere, contributing to the total greenhouse gas emissions, it is extracted using a series of wells and a blower/flare (or vacuum) system. The system collects the gas to a central point where it can be processed. It may be used to generate electricity or upgraded to pipeline-quality gas to be either used or processed into an alternative vehicle fuel.

RECYCLING AND CONSERVATION

What is the significance of the recycling symbol?

The familiar recycling symbol consists of three arrows representing the three components of recycling. The three steps are:

1. Collection and processing of recyclable materials

2. Use of recyclables by industry to manufacture new products

3. Consumer purchase of products made from recycled materials

What is the Resource Conservation and Recovery Act?

In 1976, the U.S. Congress passed the Resource Conservation and Recovery Act (RCRA), which was amended in 1984 and 1986. This law requires the EPA to identify hazardous wastes and set standards for their management, including generation, transportation, treatment, storage, and disposal of hazardous waste. The law requires all firms that store, treat, or dispose of more than 220 pounds (100 kilograms) of hazardous wastes per month to have a permit stating how such wastes are managed.

How much has recycling of municipal solid waste grown since 1960?

In 1960 the recycling rate in the United States was 6.4 percent. It has grown to 34.5 percent in 2012. On average, Americans recycled and composted 1.5 pounds (0.7 kilograms) of our individual waste generation of 4.38 pounds (2 kilograms) per person per day.

Recycling, 1960–2012*

Year	Weight (thousands of tons)	Percent of Total Solid Waste
1960	5,610	6.4
1970	8,020	6.6
1980	14,520	9.6
1990	33,240	16.0
2000	69,460	28.5
2010	85,160	34.0
2012	86,620	34.5

*Most recent data available as of August 2015.

When was the first curbside recycling program established in the United States?

The first municipal-wide curbside recycling program was started in 1974 in University City, Missouri, where city officials designed and distributed a container for collecting newspapers. As of 2011, there were almost 10,000 curbside recycling programs, serving 73 percent of the total population in the United States.

Which materials have the highest recycling rates as a percent of the total municipal solid waste generation?

The recycling (recovery) rates in 2012 for various materials are listed below:

Product	Percent Recycled (2012)
Paper and paperboard	64.6
Yard trimmings	57.7
Metals	34.0
Glass	27.7
Plastics	8.8

When was the first metal recycling project in the United States?

The first metal recycling in the United States occurred in 1776 when patriots in New York City toppled a statue of King George III (1738–1820), which was melted down and made into 42,088 bullets.

When did aluminum recycling begin?

Aluminum recycling dates back to 1888 when the smelting process was invented. Since aluminum is so valuable, companies involved in manufacturing aluminum were motivated to discover ways to make aluminum from aluminum. Recycling aluminum saves more than 90 percent of the energy required to make new aluminum since it eliminates the mining of new ore. It is estimated that 75 percent of all aluminum ever produced in the United States is still in use today. Nearly 40 percent of the aluminum in North America is created through recycling.

How can plastics be made biodegradable?

Most plastics are made from oil- and petroleum-based products. As non-renewable resources, oil and petroleum will one day no longer be available. In addition, although plastics do not rust or rot, there is no environmentally friendly way to dispose of them. Researchers and scientists are developing biodegradable plastics that are economical to produce. Degradable plastic has starch in it so that it can be attacked by starch-eating bacteria to eventually disintegrate the plastic into bits. Chemically degradable plastic can be broken up with a chemical solution that dissolves it. Used in

Blocks of crushed aluminum cans await recycling. Recycling aluminum has proven to be a very successful money-saving program in the United States and the world.

359

surgery, biodegradable plastic stitches slowly dissolve in the body fluids. Photodegradable plastic contains chemicals that disintegrate over a period of one to three years when exposed to light. One-quarter of the plastic yokes used to package beverages are made from a plastic called Ecolyte, which is photodegradable.

What do the numbers inside the recycling symbol on plastic containers mean?

The Society of the Plastics Industry developed a voluntary coding system for plastic containers to assist recyclers in sorting plastic containers. The symbol is designed to be imprinted on the bottom of the plastic containers. The numerical code appears inside three arrows that form a triangle. A guide to what the numbers mean is listed below. The most commonly recycled plastics are polyethylene terephthalate (PET) and high-density polyethylene (HDPE).

Code	Material	Examples
1	Polyethylene terephthalate (PET/PETE)	2-liter soft drink bottle
2	High-density polyethylene (HDPE)	Milk and water jugs
3	Vinyl (PVC)	Plastic pipes, shampoo bottles
4	Low-density polyethylene (LDPE)	Produce bags, food storage containers
5	Polypropylene (PP)	Squeeze bottles, drinking straws
6	Polystyrene (PS)	Fast-food packaging, other packaging
7	Other	Food containers

What products are made from recycled plastic?

Resin	Common Uses	Products Made from Recycled Resin
HDPE	Beverage bottles, milk jugs, detergent bottles, milk and soft drink crates, pipe, cable, film	Motor oil bottles, pipes, and pails
LDPE	Film bags such as trash bags, coatings, plastic bottles	New trash bags, pallets, carpets, fiberfill
PET	Soft drink, detergent, and juice bottles	Bottles/containers

Resin	Common Uses	Products Made from Recycled Resin
PP	Auto battery cases, screw-on caps and lids; some yogurt and margarine tubs, plastic film	Auto parts, batteries, carpets
PS	Housewares, electronics, fast food carry-out packaging, plastic utensils	Insulation board, office equipment, reusable cafeteria trays
PVC	Sporting goods, luggage, pipes, auto parts; packaging for shampoo bottles, blister packaging, and films	Drainage pipes, fencing, house siding

A new clothing fiber called Fortrel EcoSpun is made from recycled plastic soda bottles. The fiber is knit or woven into garments such as fleece for outerwear or long underwear. The processor estimates that every pound of Fortrel EcoSpun fiber results in ten plastic bottles being kept out of landfills.

When offered a choice between plastic or paper bags for your groceries, which should you choose?

The answer is neither. Both are environmentally harmful, and the question of which is the more damaging has no clear-cut answer. Twelve million barrels of oil (a nonrenewable resource) are required to produce 100 billion plastic bags. Plastic bags degrade slowly in landfills and can harm wildlife if swallowed, and producing them pollutes the environment. In contrast, 35 million trees are cut down to produce 25 million brown paper bags accompanied by air and water pollution during the manufacturing process. Although each can be recycled, the EPA estimates that only 1 percent of plastic bags and 20 percent of paper bags are recycled. Instead of choosing between paper and plastic bags, bring your own reusable canvas or string containers to the store, and save and reuse any paper or plastic bags you acquire.

What natural resources are saved by recycling paper?

One ton (907 kilograms) of recycled waste paper would save an average of 7,000 gallons (26,460 liters) of water, 3.3 cubic yards (2.5 cubic meters) of landfill space, 3 barrels of

How much landfill space is saved by recycling one ton of paper?

Each ton (907 kilograms) of office paper saves more than 3 cubic yards of landfill space while a ton of newsprint saves 4.6 cubic yards of landfill space. In addition, recycling one ton (907 kilograms) of cardboard saves 9 cubic yards of landfill space.

oil, 17 trees, and 4,000 kilowatt-hours of electricity—enough energy to power the average home for six months. It would also reduce air pollution by 74 percent.

When was paper recycling started?

Paper recycling was actually born in 1690 in the United States when the first paper mill was established by the Rittenhouse family on the banks of Wissahickon Creek, near Philadelphia. The paper at this mill was made from recycled rags.

How much newspaper must be recycled to save one tree?

Seventeen trees are saved with each ton of recycled paper, as well as a lot of energy and landfill space!

One 35-to-40-foot (10.6-to-12-meter) tree produces a stack of newspapers 4 feet (1.2 meters) thick; this much newspaper must be recycled to save a tree.

How much paper is recycled in the United States?

In 2012, 44 million tons of paper were recycled, amounting to 65 percent of all paper that was generated as waste.

What is the average per capita domestic water usage in the United States?

Domestic water usage includes water used for drinking, food preparation, washing clothes and dishes, flushing toilets, and showers. It also includes outdoor water used for watering lawns and gardens and washing cars. Water may be either self-supplied through fresh groundwater or supplied by public water companies. It is estimated that the domestic per capita water usage in the United States is 98 gallons (371 liters) per day or approximately 400 gallons (1,514 liters) per day for a family of four. Domestic water use accounts for only about 1 percent of the total water usage in the United States. Our water usage is divided into these categories:

Average Daily per Capita Water Usage (U.S.)			
Use/Activity	Gallons per Capita	Liters per Capita	Percent Total Use
Toilets	18.5	70	26.7
Clothes washers	15.0	57	21.7
Showers	11.6	44	16.8
Faucets	10.9	41	15.7
Leaks	9.5	36	13.7
Baths	1.2	4.5	1.7
Dishwashers	1.0	3.8	1.4
Other household uses	1.6	6	2.2

How is technology used to increase the conservation of water resources?

Water efficiency is defined as using improved technologies and practices that deliver equal or better service with less water. WaterSense is an Environmental Protection Agency (EPA) partnership program promoting greater water efficiency by encouraging manufacturers to design products that are more efficient and providing consumers with a way to identify efficient products with a simple product label. Products that have the WaterSense label are 20 percent more efficient than traditional products without sacrificing performance. Products that meet the WaterSense requirements include faucets that reduce excessive flow volumes without sacrificing performance, showerheads, and toilets that use 20 percent less water per flush but perform as well or better than older models. The improved technology of WaterSense products has saved 757 billion gallons (2,865 billion liters) of water between 2006 and 2013.

Comparison of Older Fixtures and Appliances and WaterSense Fixtures and Appliances

Fixture/Appliance	Water Usage in Older Models	Water Usage in WaterSense Models
Toilet	3.5–7 gallons (13 to 26 liters)/flush	1.6 gallons (6 liters)/flush
Showerheads	5.5 gallons (21 liters)/minute	2.5 gallons (9.5 liters)/minute
Faucets	2.2 gallons (8 liters)/minute	1.5 gallons (5.6 liters)/minute
Washing Machines	27–54 gallons(102–204)/load	<27 gallons (<102 liters)/load

When did the manufacture of low flow toilets become mandatory?

Beginning in 1995, the EPA mandated that all toilets manufactured in the United States could use no more than 1.6 gallons (6.1 liters) of water per flush. In the late 1990s, these low flow toilets did not meet consumers' expectations. They were prone to clogging, required more than one flush, or caused other problems with home plumbing. However, with newer designs and technological advances, manufacturers have improved the product so even the higher efficiency, 1.28-gallons (4.8-liters) per-flush toilets toilet perform well.

Is washing dishes by hand better for the environment than using an automatic dishwasher?

Dishwashers often save energy and water compared to hand washing. Depending on the brand, an Energy Star-efficient dishwasher typically consumes less than 5.5 gallons (20.8

How much water could be saved if all the inefficient toilets in the United States were converted to WaterSense labeled models?

Converting all the inefficient, older toilets in the United States to WaterSense labeled efficient models could save more than 640 billion gallons (2,423 billion liters) of water per day or the equivalent of 15 days of water flow over Niagara Falls.

liters) of water per normal wash. Newer models of dishwashers have been designed with sensors that automatically adjust the length of the wash cycle and water temperature, depending on how soiled the dishes are in the load. It is possible to save nearly 5,000 gallons (18,927 liters) of water per year using an Energy Star-efficient dishwasher. Older models of dishwashers may use 8 to 15 gallons (30 to 57 liters) of water per load. Hand-washing a day's worth of dishes may use up to 27 gallons (102 liters) of water. It is possible to increase the efficiency of hand washing by using a faucet with an aerator, turning off the water while washing, soaking the dishes first in a basin of soapy water, and scraping off excess food.

What are some innovative uses of discarded tires?

The Rubber Manufacturers Association calculated 3,781 thousand tons (3,430 thousand metric tons), or nearly 231 million, tires were scrapped in the United States in 2011. Rather than sending these tires to landfills, advances in technology have created new markets and innovative uses for discarded tires. The three major markets for scrap tires are tire-derived fuel, ground rubber applications, and civil engineering projects. Nearly 87 million tires were diverted to the tire-derived fuel (TDF) market, accounting for 37.7 percent or 1,427 thousand tons (1,295 metric tons) of the scrap tires generated. TDF is used in a variety of combustion technologies, including cement kilns, pulp and paper mill boilers, and utility and industrial boilers. Ground rubber applications, including new rubber products, playground and other sports surfacing, and rubber-modified asphalt, consumed 57 million tires, weighing 928.5 thousand tons (842 thousand metric tons, 24.5 percent of the total) of scrapped tires. Another 295 thousand tons (268 thousand metric tons, 7.8 percent of the total) of tires was used in civil engineering applications. These include tire shreds used in road and landfill construction, septic tank leach fields, and other construction projects.

How has the growth of consumer electronic products in recent decades affected the environment?

Consumer electronics include computers (desktops and portable laptops and tablets), computer monitors, computer peripherals (keyboards and mice), hard-copy devices (printers, fax machines, digital copiers, scanners), televisions (cathode ray tube [CRT] and flat-panel), and mobile devices (cell phones, personal digital assistants [PDAs], smartphones, and pagers). According to an EPA study published in 2008 on the disposal of electronic wastes, the total number of electronics sold in

Disposed electronic parts such as these circuit boards contain heavy metals and other toxic substances.

> ## What is eCycling?
>
> **e**Cycling is the reuse and recycling of electronic equipment. Donating used electronic equipment for use by others is the environmentally preferred alternative to discarding used electronics. If an electronic device cannot be reused, it should be recycled. The recycling rate for electronic devices is only 18 percent, with the remaining amount of disposed electronics being discarded in landfills.

1980 was approximately 18.6 million units. These included 17.6 million TVs (monochrome and color CRT units) and 4.5 million units of computers, including desktops, CRT monitors, keyboards, mice, and hard-copy peripherals, such as printers. There were no flat-panel TVs, portable computers, or cell phones. An updated study in 2011 estimates that 438 million units of electronic products were sold in 2009. In addition to new products that are sold, it is estimated that there are five million tons (4.5 million metric tons) of electronic products in storage (no longer in use) and 2.37 million tons (2.15 million metric tons) ready for end-of-life management (disposal and recycling) with a recycling rate of 25 percent. Although electronic waste accounts for only 1 to 2 percent of the total municipal solid waste, it is a matter of serious concern to ensure proper disposal of electronics.

What is the Sustainable Materials Management Electronics Challenge?

The Sustainable Materials Management (SMM) Electronics Challenge is an EPA initiative to encourage manufacturers and retailers to manage used electronics responsibly. The goals of the program are:

- To ensure responsible recycling through the use of third-party certified recyclers
- To increase accountability by publicly disseminating electronics collection and recycling data
- To encourage outstanding performance through awards and recognition

The EPA has challenged manufacturers and retailers to voluntarily commit to sending 100 percent of used electronics collected for reuse and recycling to third-party certified recyclers and increase the total amount of used electronics collected.

What is the environmental impact of disposing electronics in landfills?

The EPA believes that the disposal of electronics in properly managed municipal solid waste landfills does not threaten human health and the environment. However, recycling electronics will decrease the demand for additional mining of valuable resources and manufacturing new products. Recycling electronics recovers valuable materials, such as copper and engineered plastics, and as a result reduces greenhouse gas emissions and pollution and saves energy and resources by extracting fewer raw materials from the earth.

What are some benefits to recycling electronics?

- Recycling one million laptop computers saves the energy equivalent to the electricity used by 3,657 U.S. homes in one year.

- One metric ton of circuit boards can contain 40 to 800 times the concentration of gold ore mined in the United States.

- One metric ton of circuit boards can contain 30 to 40 times the concentration of copper ore mined in the United States.

- Recycling one million cell phones can recover 35,000 pounds (1,586 kilograms) of copper, 772 pounds (350 kilograms) of silver, 75 pounds (34 kilograms) of gold, and 33 pounds (15 kilograms) of palladium.

- The plastics recovered from cell phones may be recycled into other plastic products, such as outdoor furniture, replacement automotive parts, non-food containers, or new electronic devices.

SUSTAINABILITY

What is sustainability?

The most broad definition of sustainability is the practice of preserving the Earth's resources for future generations. Sustainability encompasses ecologic, economic, and social issues. It includes the use and care of resources, changing patterns of consumption, and control of population growth, which impact the total need for resources.

Is sustainability a new concept?

The concepts and philosophy of sustainability were a result of the environmental concerns of the 1960s and 1970s. In 1983, the United Nations (UN) established the World Commission on Environment and Development (WCED), also known as the Brundtland Commission, named after Gro Harlem Brundtland (1939–), the chair of the commission. In 1987, the WCED issued *Our Common Future* (also known as the Brundtland Report), a report that defined sustainable development as "development that meets the needs of the present without compromising the ability of future generations to meet their own needs." In 2015 the UN adopted the "2030 Agenda for Sustainable Development" to transform the world based on seventeen Sustainable Development Goals.

UN Sustainable Development Goals

1. End poverty in all its forms everywhere
2. End hunger, achieve food security and improved nutrition, and promote *sustainable* agriculture
3. Ensure healthy lives and promote well-being for all at all ages
4. Ensure inclusive and quality education for all and promote lifelong learning
5. Achieve gender equality and empower all women and girls
6. Ensure access to water and sanitation for all
7. Ensure access to affordable, reliable, *sustainable,* and modern energy for all
8. Promote inclusive and *sustainable* economic growth, employment, and decent work for all
9. Build resilient infrastructure, promote *sustainable* industrialization, and foster innovation
10. Reduce inequality within and among countries
11. Make cities inclusive, safe, resilient, and *sustainable*
12. Ensure *sustainable* consumption and production patterns
13. Take urgent action to combat climate change and its impacts
14. Conserve and *sustainably* use the oceans, seas, and marine resources
15. *Sustainably* manage forests, combat desertification, halt and reverse land degradation, and halt biodiversity loss
16. Promote just, peaceful, and inclusive societies
17. Revitalize the global partnership for *sustainable* development

Note how many times the word "sustainable" is used and the interest in preserving natural resources. The emphasis on equality and social justice also fosters an improved environment because people who are living in more prosperous circumstances are more able to conserve the environment, as well.

How do sustainable agriculture practices protect the environment?

The goals of sustainable agriculture practices are to preserve resources and protect human health. These goals are achieved through crop rotation, integrated pest management, soil and water tillage practices, and disease prevention in livestock through health maintenance instead of the use of antibiotics. Crop rotation involves planting different plant species on a parcel of land each year and letting the land lie fallow in certain years. The benefits of crop rotation include soil fertility without chemical fertilizers, controlling erosion, preventing insect infestation without the use of chemical pesticides, and discouraging weeds.

What is a "green product"?

Green products are environmentally safe products that contain no chlorofluorocarbons, are degradable (can decompose), and are made from recycled materials. "Deep-green" products are those from small suppliers who build their identities around their claimed environmental virtues. "Greened-up" products come from the industry giants and are environmentally improved versions of established brands.

Is it possible to buy green electronics?

Electronic manufacturers are designing products that are more environmentally friendly. Consumers should look for products that:

- Contain fewer toxic constituents
- Use recycled materials in the new product
- Are designed for easy upgrading and disassembly
- Are energy efficient
- Use minimal packaging
- Have leasing or takeback options for reuse or recycling
- Meet performance criteria that show they are environmentally preferable

BIOTECHNOLOGY AND GENETIC ENGINEERING

INTRODUCTION AND HISTORICAL BACKGROUND

What is biotechnology?

Biotechnology is the use of a living organism to produce a specific product. It includes any technology associated with the manipulation of living systems for industrial purposes. In its broadest sense, biotechnology includes the fields of chemical, pharmaceutical, and environmental technology as well as engineering and agriculture.

What is genetic engineering?

Genetic engineering, also popularly known as molecular cloning or gene cloning, is the artificial recombination of nucleic acid molecules in a test tube; their insertion into a virus, bacterial plasmid, or other vector system; and the subsequent incorporation of the chimeric molecules into a host organism in which they are capable of continued propagation. The construction of such molecules has also been termed gene manipulation because it usually involves the production of novel genetic combinations by biochemical means.

Genetic engineering techniques include cell fusion and the use of recombinant DNA or gene splicing. In cell fusion, the tough, outer membranes of sperm and egg cells are removed by enzymes, then the fragile cells are mixed and combined with the aid of chemicals or viruses. The result may be the creation of a new life form from two species (a chimera). Recombinant DNA techniques transfer a specific genetic activity from one organism to the next through the use of bacterial plasmids (small circular pieces of DNA lying outside the main bacterial chromosome) and enzymes, such as restriction endonucleases (which cut the DNA strands); reverse transcriptase (which makes a DNA strand from an RNA strand); DNA ligase (which joins DNA strands together); and Taq polymerase (which can make a double-strand DNA molecule from a single-strand "primer" molecule). The recombinant DNA process begins with the isolation and fragmentation of suitable DNA strands. After these fragments are combined with vectors, they are carried into bacterial cells, where the DNA fragments are "spliced" on to plasmid DNA that has been opened up. These hybrid plasmids are then mixed with host cells to form transformed cells. Since only some of the transformed cells will exhibit the desired characteristic or gene activity, the transformed cells are separated and grown individually in cultures. This methodology has been successful in producing large quantities of hormones (such as insulin) for the biotechnology industry. However, it is more difficult to transform animal and plant cells, yet the technique exists to make plants resistant to diseases and to make animals grow larger. Because genetic engineering interferes with the processes of heredity and can alter the genetic structure of our own species, there is much concern over the ethical ramifications of such power as well as the possible health and ecological consequences of the creation of these bacterial forms. Some applications of genetic engineering in various fields are:

- Agriculture: Crops having larger yields, disease- and drought-resistant; bacterial sprays to prevent crop damage from freezing temperatures; and livestock improvement through changes in animal traits.

- Industry: Use of bacteria to convert old newspaper and wood chips into sugar; oil- and toxin-absorbing bacteria for oil spill or toxic waste clean-ups; and yeasts to accelerate wine fermentation.

- Medicine: Alteration of human genes to eliminate disease (experimental stage); faster and more economical production of vital human substances to alleviate deficiency and disease symptoms (but not to cure them); substances include insulin, interferon (cancer therapy), vitamins, human growth hormone ADA, antibodies, vaccines, and antibiotics.

- Research: Modification of gene structure in medical research, especially cancer research.

- Food processing: Rennin (enzyme) in cheese aging.

When was the term "genetic engineering" first used?

The term "genetic engineering" was first used in 1941 by Danish microbiologist A. Jost in a lecture on sexual reproduction in yeast at the Technical Institute in Lwow, Poland.

What are some of the major achievements of biotechnology and genetic engineering?

Date	Discovery
1968	Stanley Cohen (1922–) uses plasmids to transfer antibiotic resistance to bacterial cells; receives the National Medal of Science in 1988 and National Medal of Technology in 1989
1970	Herbert Boyer (1936–) discovers that certain bacteria can "restrict" some bacteriophages by producing enzymes (restriction enzymes); isolates EcoRI; receives the National Medal of Technology in 1989 and the National Medal of Science in 1990
1972	Paul Berg (1926–) splices together DNA from the SV 40 virus and *E. coli*, making recombinant DNA; shares 1980 Nobel Prize with Walter Gilbert (1932–) and Fred Sanger (1918–2013)
1974	Stanley Cohen (1922–), Annie Chang, and Herbert Boyer (1936–) splice frog DNA into *E. coli*, producing the first recombinant organism
1975	DNA sequencing developed by Walter Gilbert (1932–), Allan Maxam (1942–), and Fred Sanger (1918–2013)
1978	Human insulin cloned in *E. coli* by biotech company Genentech
1986	Kary Mullis (1944–) develops the polymerase chain reaction (PCR), in which DNA polymerase can copy a DNA segment many times in a short period of time; receives the Nobel Prize in Chemistry in 1993
1989	Human Genome Project (HGP) begins
1990	Researchers at National Institutes of Health (NIH) use gene therapy to treat a human patient
1994	Introduction of first transgenic food, the Flavr Savr tomato
1996	A sheep named Dolly is cloned by Ian Wilmut (1944–)
1997	First human artificial chromosome is created
2000	Completion of the first working draft (90 percent complete) of the HGP
2003	Glofish, the first genetically modified pets, are marketed and sold in the United States
2003	Celera and NIH complete sequencing of the human genome
2006	FDA approves a vaccine against HPV (human papillomavirus)
2008	The first DNA molecule made almost entirely of artificial parts is made in Japan

What were some of the earliest uses of biotechnology?

 The earliest examples of biotechnology involved using yeast and bacteria to make bread, vinegar, and alcoholic beverages, such as beer and wine. The British Museum has a clay tablet dating back to 3100–3000 B.C.E. depicting the allocation of beer, which was used as rations for workers in ancient Mesopotamia.

Date	Discovery
2010	FDA approves an osteoporosis treatment that is one of the first medicines based on genomic studies
2010	J. Craig Venter Institute creates the first synthetic cell
2013	The Human Brain Project (HBP) begins as a collaborative project between research institutes in the United States, Europe, and Japan to simulate the human brain as completely as possible within ten years

What are some examples of genetic engineering in animals and microbes?

One of the earliest applications of biotechnology was the genetic engineering of a growth hormone (GH) produced naturally in the bovine pituitary. BG can increase milk production in lactating cows. Using biotechnology, scientists bioengineered the gene that controls bovine GH production into *E. coli*, grew the bacteria in fermentation chambers, and thus produced large quantities of bovine GH. The bioengineered bovine GH, when injected into lactating cows, resulted in an increase of up to 20 percent in national milk production. Using bovine GH, farmers are able to stabilize milk production in their herds, avoiding fluctuations in production levels. A similar regimen was adapted using the pig equivalent of growth hormone (porcine GH). Injected in pigs, porcine GH reduced back fat and increased muscle (meat) gain.

The first transgenic animal available as a food source on a large scale was the salmon, which reached U.S. food markets in 2001, following rigid evaluations of consumer and environmental safety. These salmon have the capability of growing from egg to market size (6 to 10 lbs.) in eighteen months, as compared to conventional fish breeding, which takes up to thirty-six months to bring a fish to market size. The use of transgenic salmon can help reduce overfishing of wild salmon stocks.

What was the first commercial use of genetic engineering?

Commercial recombinant DNA technology was first used to produce human insulin in bacteria. In 1982 genetically engineered insulin was approved by the FDA for use by diabetics. Insulin is normally produced by the pancreas, and for more than fifty years the pancreas of slaughtered animals such as swine or sheep was used as an insulin source. To provide a reliable source of human insulin, researchers harvested the insulin gene from cellular DNA. Researchers made a copy of DNA carrying this insulin gene and spliced it into a bacterium. When the bacterium was cultured, the microbe split from one cell into two cells, and both cells got a copy of the insulin gene. Those two microbes grew, then divided into four, those four into eight, the eight into sixteen, and so forth. With each cell division, the two new cells each had a copy of the gene for human insulin. And because the cells had a copy of the genetic "recipe card" for insulin, they could make the insulin protein. Using genetic engineering to produce insulin was both cheaper and safer for patients, as some patients were allergic to insulin from other animals.

What was the Human Genome Project?

The Human Genome Project was a major effort coordinated by the National Institutes of Health (NIH), the Department of Energy (DOE), and the International Human Genome Sequencing Consortium with researchers from the United Kingdom, France, Germany, Japan, and China. The project began in 1990 and was originally planned to last fifteen years, but effective resource and technological advances resulted in completion of the first working draft in June 2000 and the completed version in April 2003. The HGP identified approximately all of the 20,500 genes in human DNA and determined the sequences of 99.9 percent of the three billion chemical base pairs that make up human DNA. The goal was not only to pinpoint these genes but also to decode the biochemical information down to the four basic chemicals of all genes (nitrogen bases): A (adenine), C (cytosine), G (guanine), and T (thymine). Since these letters are linked in pairs of sequences in the double helix of DNA, this means that three billion pairs are involved in this process. The sequencing of the human genome has been hailed as the most groundbreaking scientific event of the twentieth century. It will, in equal measure, offer insight into our collective history and our individual identities and open up untold possibilities for the diagnosis, treatment, and prevention of disease. Not since James Watson (1928–) and Francis Crick's (1916–2004) discovery of the structure of DNA has a scientific investigation been greeted with such fanfare.

How much information does the human genome contain?

The amount of genetic information in the human genome is equivalent to the amount of information in two hundred telephone books of one thousand pages each! If the twenty-three chromosomes in one human cell were removed, unwound, and placed end to end, the DNA would stretch more than five feet (1.5 meters) and be fifty-trillionths of an inch wide.

What is a transposon?

A transposon is a nucleotide sequence in a gene that literally "jumps" around from one chromosome to another. Transposons can be problematic, as they may disrupt the nor-

---

When was the first genome sequenced?

The first genome was sequenced in 1976 by Walter Gilbert (1932–) and Fred Sanger (1918–2013). It was RNA bacteriophage MS2 with four genes. The following year they sequenced the genome of bacteriophage 174 with ten genes. In 1982, Sanger published the sequence of the first relatively large genome of bacteriophage. Bacteriophage has 48,502 bases of genomic DNA.

mal function of an important gene by their random insertion into the middle of that gene. However, transposons are found in almost all organisms, and their presence in a genome indicates that genetic information is not fixed within the genome. By studying results of the Human Genome Project, scientists predict that about 45 percent of all human genes may be derived from transposons.

Is size of a genome an indication of the complexity of an organism?

Comparing an organism's genome size and the total number of protein-coding genes indicates there is little correlation between the size of the genome and the organism's complexity.

Genome Size of Various Organisms

Organism	Estimated total size of genome (base pairs)	Estimated number of protein-coding genes
Saccharomyces cerevisiae (yeast)	12 million	6,000
Trichomonas vaginalis (single-celled parasite)	160 million	60,000
Plasmodium falciparum (malaria parasite)	23 million	5,000
Caenorhabditis elegans (nematode)	95.5 million	18,000
Drosophilia melanogaster (fruit fly)	170 million	14,000
Arabidopsis thaliana (mustard plant)	125 million	25,500
Oryza sativa (rice)	470 million	51,000
Gallus gallus (chicken)	1 billion	20,000–23,000
Canis familiaris (domestic dog)	2.4 billion	19,000
Mus musculus (laboratory mouse)	2.5 billion	30,000
Homo sapiens (human)	2.9 billion	20,500

What is TIGR?

TIGR is the Institute for Genomic Research, a nonprofit private research institute founded in 1992 by Craig Venter (1946–) that did research in the structural, functional, and comparative analysis of genomes and gene products using DNA sequencing along with computing and computational tools. In 1995, TIGR sequenced the first free living organism, *Haemophilus influenza*. Since then, TIGR has sequenced the fruit fly, mouse, rat, and more than fifty microbial genomes. In 2006, TIGR merged with the Center for the Advancement of Genomics (TCAG), the J. Craig Venter Science Foundation, the Joint Technology Center, and the Institute for Biological Energy Alternatives (IBEA) to form the J. Craig Venter Institute (JCVI).

What is proteomics?

Proteomics is the study of proteins encoded by a genome. This field extends the Human Genome Project and is a far more complex study than finding where genes are located on chromosomes. Proteins are dynamic molecules that can change according to the needs of a cell, and complete understanding of cell metabolism requires that scientists understand all of the proteins involved as well as their genes.

Who was awarded two Nobel Prizes for DNA sequencing?

Frederick Sanger (1918–2013) won the Nobel Prize in Chemistry in 1958 for his work on the structure of proteins (especially insulin) and shared the Nobel Prize in Chemistry in 1980, along with Walter Gilbert (1932–), for contributions in determining the nitrogen base sequences of DNA. This method was later referred to as the Sanger sequencing method for reading DNA.

Who was the first individual to find the gene for breast cancer?

Mary-Claire King (1946–) determined that in 5 to 10 percent of those women with breast cancer, the cancer is the result of a mutation of a gene on chromosome 17, the BRCA1 (Breast Cancer 1). She announced her findings at the American Society of Human Genetics annual meeting in 1990. The BRCA1 gene is a tumor suppressor gene and is also linked to ovarian cancer. A research team led by Mark Skolnick (1946–) succeeded in pinpointing the exact location on chromosome 17 in September 1994.

Can human genes be patented?

In 2013, the U.S. Supreme Court ruled in the case *Association for Molecular Pathology v. Myriad Genetics* that human genes and isolated DNA molecules may not be patented. The Court based its decision on the fact that a gene is a naturally occurring DNA segment and not patentable solely because it has been isolated. The decision was based on whether isolated genes are "products of nature," which may not be patented, or "human-made inventions," which are patentable. Although Myriad Genetics has isolated the BRCA1 and BRCA2 genes and had developed genetic testing for these genes, they could not patent the discovery. Synthetically created DNA, known as cDNA, may be patented because it does not occur naturally. In an earlier decision in the case *Diamond v. Chakrabarty* in 1980, the Supreme Court had ruled that genes could be patented if they were isolated from other genetic material. Following the 1980 decision, there was unprecedented growth and research in the biotechnology industry on the human genome and genomes of bacterial and viral pathogens, resulting in many new medicines, vaccines, and genetic tests.

What is a chimera?

The chimera from Greek mythology is a fire-breathing monster with a lion's head, a goat's body, and a serpent's tail. The chimera of biotechnology is an animal formed from two different species or strains—that is, a mixture of cells from two very early embryos. Most chimeras used in research are made from different mouse strains. Chimeras cannot reproduce.

What is bioinformatics?

Bioinformatics combines biology, computer science, and information technology to organize, catalog, and store large quantities of DNA sequence data. It also includes the computational tools to retrieve and analyze the data. GenBank® is the NIH genetic sequence database. It is an annotated collection of all publicly available DNA sequences. GenBank®, the DNA DataBank of Japan (DDBJ), and the European Molecular Biology Laboratory (EMBL) form the International Nucleotide Sequence Database Collaboration. The three organizations exchange data on a daily basis.

What is bioethics?

Bioethics is the study of the ethical aspects of biotechnology. Specific concerns include: 1) access to personal genetic information; 2) privacy and confidentiality of genetic information; 3) the effects of personal genetic information on the individual and how society perceives that individual; 4) use of genetic information in reproductive issues; 5) philosophical implications regarding how genes can affect behavior and enhancement of genes; and 6) commercialization issues concerning patents on genes and gene products.

What is bioterrorism?

Bioterrorism is the use of biological substances or toxins with the goal of causing harm to humans. Biotechnology can be used to manufacture biological weapons such as large amounts of anthrax spores. However, biotechnology can also be used positively to identify bioweapons. A new, faster method of PCR, called continuous flow PCR, uses a biochip and requires only nanoliter amounts of DNA to detect a bioweapon.

METHODS AND TECHNIQUES

What is cell culture?

Cell culture is the cultivation of cells (outside the body) from a multicellular organism. This technique is very important to biotechnology processes because most research programs depend on the ability to grow cells outside the parent animal. Cells grown in culture usually require very special conditions (e.g., specific pH, temperature, nutrients, and growth factors). Cells can be grown in a variety of containers, ranging from a simple petri dish to large-scale cultures in roller bottles, which are bottles that are rolled gently to keep culture medium flowing over the cells.

What is DNA sequence analysis?

DNA sequence analysis, or simply DNA sequencing, is part of a subfield of genomics known as structural genomics. DNA sequence analysis results in a complete description of a DNA molecule in terms of its order of nitrogen bases (A [adenine], T [thymine],

G [guanine], C [cytosine]). All sequencing projects are based on a similar technology: DNA fragmentation with restriction endonucleases, cloning of the DNA fragments, and overlapping analyses of the fragments to eventually provide a map.

What is the Sanger technique?

The chain termination method, also called the Sanger technique or method, was developed by Fred Sanger (1918–2013) in the 1970s. The basics of this method involve using dideoxynucleotides instead of the deoxynucleotides normally found in DNA. Dideoxynucleotides are nucleotides missing a hydroxyl group at the site where one nu-

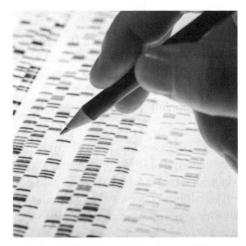

Analyzing DNA involves defining the complete arrangement of nitrogen bases in the molecule.

cleotide usually joins to another nucleotide. Essentially, a set of DNA fragments that match the chain to be sequenced is created. Each fragment is one nucleotide longer than the previous one. Once the final nucleotide in each fragment is identified, the sequence of the whole chain may be determined. If the hydroxyl group is missing, chain elongation stops. Gel electrophoresis is used to separate the fragments according to their length. The base sequence can be constructed by knowing which terminator base is associated with which fragment on the gel.

How does automated sequencing differ from manual sequencing?

The process can be done manually or by automation. A competent technician can determine the sequence of as many as 5,000 bases in a week, which were done manually when the Sanger technique was developed. The desire to sequence more fragments in shorter periods of time led to the development of automated sequencing. The first automated sequencer was built in 1985 as a collaborative effort between Leroy Hood (1938–) at California Institute of Technology and Michael Hunkapiller (1950–) of Applied Biosystems, Inc. (ABI). Hood attached a different color (blue, green, yellow, and red) fluorescent dye to the template used in the sequencing reaction. Each different color matched a different terminator base. The ABI Model 370 had an argon ion laser to excite the dyes, flat gel between two plates capable of sixteen-lane electrophoreses, and a Hewlett-Packard Vectra computer with 604 megabytes of memory for data analysis. Each lane could read over 350 bases, enabling a technician to read more sequence in a day than could be read manually in a week. Hood was inducted into the National Inventors Hall of Fame in 2007 for the development of the DNA sequencer. In 2011, he received the National Medal of Science. Newer models can read over 400,000 bases in a day. Therefore, the entire human genome could be read five times over within four months using 300 Model 3700 Automated Sequencers.

What is recombinant DNA?

Recombinant DNA is hybrid DNA that has been created from more than one source. An example is the splicing of human DNA into bacterial DNA so that a human gene product, such as insulin, is produced by a bacterial cell.

What is polymerase chain reaction?

Polymerase chain reaction (PCR) is a laboratory technique that amplifies or copies any piece of DNA very quickly without using cells. The DNA is incubated in a test tube with a special kind of DNA polymerase, a supply of nucleotides, and short pieces of synthetic, single-strand DNA that serve as primers for DNA synthesis. With automation, PCR can make billions of copies of a particular segment of DNA in a few hours. Each cycle of the PCR procedure takes only about five minutes. At the end of the cycle, the DNA segment—even one with hundreds of base pairs—has been doubled. A PCR machine repeats the cycle over and over. PCR is much faster than the days it takes to clone a piece of DNA by making a recombinant plasmid and letting it replicate within bacteria.

PCR was developed by biochemist Kary Mullis (1944–) in 1983 at Cetus Corporation, a California biotechnology firm. In 1993 Mullis, along with Michael Smith (1932–), won the Nobel Prize in Chemistry for the development of PCR. Mullis was inducted into the National Inventors Hall of Fame in 1998.

What is DNA amplification?

DNA amplification is a method by which a small piece of DNA is copied thousands of times using PCR. DNA amplification is used in genetic engineering to make multiple copies of a gene.

What is transformation?

Transformation is the process by which a cell or an organism incorporates foreign DNA. Transformation usually occurs between a plasmid and a bacterium. The transformed cell or organism then produces the protein encoded by the foreign DNA. In order to determine whether cells have been transformed, the foreign DNA usually contains a marker, such as the gene for penicillin resistance. Only cells with the resistance gene will be able to grow in a culture medium containing penicillin.

What is a bioreactor?

A bioreactor is a large vessel in which a biological reaction or transformation occurs. Bioreactors are used in bioprocessing technology to carry out large-scale mammalian cell culture and microbial fermentation.

Examples of Products and Organisms Made in Bioreactors

Product	Organisms
Single-celled protein	*Candida utilis* (yeast)
Penicillin (antibiotic)	*Penicillium chrysogenum* (fungus)

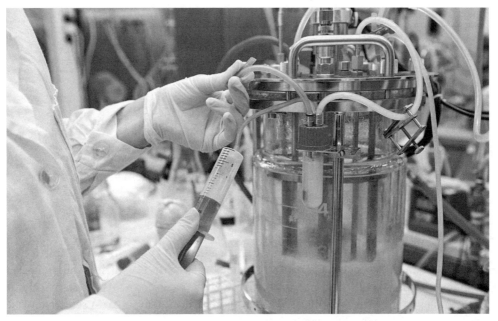

Biological transformations and reactions occur safely in a bioreactor to make cell cultures or ferment microbes.

Product	Organisms
Alpha amylase (enzyme)	*Bacillus amyloliquefaciens* (bacteria)
Riboflavin (vitamin)	*Eremothecium ashbyii* (bacteria)
Poliomyelitis vaccine	Monkey kidney or human cells
Insulin (hormone)	Recombinant *Escherichia coli* (bacteria)

What is a vector?

A vector is an agent used to carry genes into another organism. Specific examples of natural vectors include plasmids or viruses. In human gene therapy, vector viruses must be able to withstand the challenge of the patient's immune system. Once the vector manages to invade the immune system, it must be able to penetrate the cell membrane and, finally, must be able to combine its genome into that of the host cell. Vectors are also crucial to plant and animal genetic engineering.

What viruses are commonly used as vectors in human gene therapy?

Common Vectors in Gene Therapy

Type of Virus	RNA or DNA?	Example	Target Cells
Retroviruses	RNA	Human immunodeficiency virus (HIV)	any cell
Adenoviruses	Double-strand DNA	Viruses responsible for common respiratory, intestinal, and eye infections	lung

379

Type of Virus	RNA or DNA?	Example	Target Cells
Adeno-associated	Single-strand DNA	Can insert genetic material at specific site on chromosome 19	varies
Herpes simplex	Double-strand DNA	Infects neurons; causes cold sores	neurons

What is transduction?

Transduction is the process by which a vector (usually a bacteriophage) carries DNA from one bacterium to another bacterium. It can be used experimentally to map bacterial genes.

What is a restriction endonuclease?

A restriction endonuclease is an enzyme that cleaves DNA at specific sites. Restriction enzymes are made by bacteria as a means of bacterial warfare against invading bacteriophages (viruses that infect bacteria). These enzymes are now used extensively in biotechnology to cleave DNA into shorter fragments for analysis or to selectively cut plasmids so that foreign DNA can be inserted.

What is a gene library?

A gene library is a collection of cloned DNA, usually from a specific organism. Just as a conventional library stores information in books and computer files, a gene library stores genetic information either for an entire genome, a single chromosome, or specific genes in a cell. For example, one can find the gene library of a specific disease such as cystic fibrosis, the chromosome where most cystic fibrosis mutations occur, or the entire genome of those individuals affected by the disease.

How is a gene library made?

To create a gene library, scientists extract the DNA of a specific organism and use restriction endonucleases to cut it. The scientists then insert the resulting fragments into vectors and make multiple copies of the fragments. The number of clones needed for a genomic library depends on the size of the genome and the size of the DNA fragments. Specific clones in the library are located using a DNA probe.

What is a gene probe?

A gene probe is a specific segment of single-strand DNA that is complementary to a desired gene. For example, if the gene of interest contains the sequence AATGGCACA, then the probe will contain the complementary sequence TTACCGTGT. When added to the appropriate solution, the probe will match and then bind to the gene of interest. To facilitate locating the probe, scientists usually label it with a radioisotope or a fluorescent dye so that it can be visualized and identified.

How are genes physically found in a specific genome?

Finding one gene out of a possible 20,000 to 25,000 genes in the human genome is a difficult task. However, the process is made easier if the protein product of the gene is known. As an example, if a researcher is looking to find the gene for mouse hemoglobin, he or she would isolate the hemoglobin from mouse blood and determine the amino acid sequence. The amino acid sequence could then be used as a template to generate the nucleotide sequence. Working backward again, a complementary DNA probe to the sequence would be used to identify DNA molecules with the same sequence from the entire mouse genomic library.

However, if the protein product is not known, the task is more difficult. An example of this would be that of finding the susceptibility gene for late-onset Alzheimer's disease. DNA samples would be collected from members of the family of a patient with late onset Alzheimer's disease. The DNA would be cut with restriction endonucleases, and restriction fragment length polymorphisms (RFLPs) would be compared among the family. If certain RFLPs are only found when the disease gene is present, then it is assumed that the distinctive fragments are markers for the gene. Geneticists then sequence the DNA in the same area of the chromosome where the marker was found, looking for potential gene candidates.

What is an artificial chromosome?

An artificial chromosome is a new type of vector that allows cloning of larger pieces of DNA. It consists of a telomere at each end, a centromere sequence, and specific sites at which foreign DNA can be inserted. Once the DNA fragment is spliced in, the engineered chromosome is reinserted into a yeast cell. The yeast then reproduces the chromosome as if it were part of the normal yeast genome. As a result, a colony of yeast cells would then all contain a specific fragment of DNA.

What are the most frequently used artificial chromosomes?

Artificial chromosomes include YACs (yeast artificial chromosomes), BACs (bacterial artificial chromosomes), and MACs (mammalian artificial chromosomes).

What is antisense technology?

Antisense technology involves targeting a protein at the RNA level as a means of regulating gene expression. The technology involves segments of DNA that are complementary

What is the ENCODE project?

The ENCODE (ENCyclopedia Of DNA Elements) project refers to a long-term project initiated by the National Human Genome Research Institute (NHGRI) in 2003 to identify and locate all functional elements within the human genome.

to specific RNA targets. When the modified DNA binds to the RNA, the RNA can no longer produce a protein. Antisense technology is a new technique in cancer therapy. Human trials are ongoing on small-cell lung cancer and leukemia. Since cancer is a disease that can be characterized by excessive production of a particular protein, antisense technology works by interfering with the RNA that produces the protein.

What is a gene gun?

The gene gun was developed by John Sanford (1950–), a Cornell University plant breeder, with Edward Wolf and Nelson Allen, engineers at Cornell's Nanofabrication Facility. The goal was to overcome the difficulty in penetrating a plant's thick cell

A gene gun like this one is a device to inject genes through thick plant walls to incorporate them into the target plant.

walls to introduce new genes to incorporate specific genetic traits. In order to transfer genes into plants, gold or tungsten microspheres (1 micrometer in diameter) are coated with DNA from a specific gene. The microspheres are then accelerated toward target cells (contained in a petri dish) at high speed. Once inside the target cells, the DNA on the outside of the microsphere is released and can be incorporated into the plant's genome. Sanford, Wolf, and Allen filed for a patent in November 1984 and were granted U.S. Patent #4,945,050, "Method for Transporting Substances into Living Cells and Tissues and Apparatus Therefor," in July 1990.

What is a microarray?

A microarray is a technique in which PCR-amplified DNA fragments are placed on a thin glass or silicon plate by cross-linking the DNA to the glass or silicon. Fluorescent dye-labeled mRNA or complementary DNA is then hybridized to the sample. When hybridization occurs, a specific fluorescent color is produced. For example, if two samples, one labeled with a red dye and one with a green dye, are both hybridized to the same DNA sequence on the microarray, a yellow color is produced. The amount of color produced also allows scientists to detect the level of gene expression.

What is a gene chip?

A gene chip is part of the process of microarray profiling; it is also known as a biochip or a DNA chip. It is about the size of a postage stamp and is based on a glass wafer, holding as many as 400,000 tiny cells. Each tiny cell can hold DNA from a different human gene and can perform thousands of biological reactions in a few seconds. These chips can be used by pharmaceutical companies to discover what genes are involved in

When was the first artificial chromosome created?

The first artificial yeast chromosome (YAC) was developed by Jack William Szostak (1952–) in collaboration with Andrew Murray (1942–) in 1983.

various disease processes. They can also be used to type single-nucleotide polymorphisms (SNPs), which are base pair differences that are found approximately every 500 to 1,000 base pairs in DNA. There are more than 3 million SNPs in the human genome. They are very important in DNA typing because they represent about 98 percent of all DNA polymorphisms.

What is the ELISA test?

ELISA stands for enzyme-linked immunosorbent assay. It is a test used to determine if an individual has generated antibodies for a certain pathogen. In the test, inactivated antigens (for the pathogen) are coated onto a plate. In addition to antibodies from the patient serum, there is a second antihuman antibody linked to an enzyme. If the patient's antibodies bind to both the second antibody and the antigen, a color change occurs.

What is flow cytometry?

Flow cytometry is a biotechnology method in which single cells or subcellular fractions flowing past a light source are scanned based on measurement of visible or fluorescent light emission. The cells can be sorted into different fractions, depending on fluorescence emitted by each droplet. Flow cytometry can be used to identify particular cell types such as malignant cells, or immune cells such as T or B cells.

What is high throughput screening?

High throughput screening is a large-scale automated search of a genome in order to determine how a cell (or larger biological system) uses its genetic information to create a specific product or total organism. The term "platform" is used to describe all components needed for high throughput screening.

What is PAGE?

PAGE (PolyAcrylamide Gel Electrophoresis) is a type of separation method in which polyacrylamide (a polymer of acrylamide) is used instead of agarose in the gel bed of the electrophoresis chamber. Polyacrylamide is preferable to agarose because it forms a tighter gel bed that is then capable of separating smaller molecules, especially those with molecular weights that are very similar. PAGE is used in DNA sequencing, DNA fingerprinting, and for protein separation.

How did DNA typing solve an eighteenth-century French mystery?

Since the 1795 death of a ten-year-old boy in the tower of a French prison, historians have wondered whether this boy was the dauphin, the sole surviving son of King Louis XVI (1754–1793) and Marie Antoinette (1755–1793), who had each been executed by this time during the French Revolution. It was thought that the boy might have been a stand-in for the true heir; however, the dead boy's heart was saved after his death, and in December 1999, two tissue samples were taken from this heart tissue. DNA fingerprinting from the samples matched DNA extracted from the locks of the dauphin's hair kept by Marie Antoinette, from two of the queen's sisters, and from present-day descendants. DNA evidence thus supports that it was indeed the dauphin—Louis XVII (1785–1795)—who died in prison.

What methods of analysis are used for DNA profiling?

Both RFLP (restriction fragment length polymorphism) and PCR (polymerase chain reaction) techniques can be used for DNA profiling. The success of either method depends on identifying where the DNA of two individuals varies the most and how this variation can be used to discriminate between two different DNA samples. Since RFLP uses markers in regions that are highly variable, it is unlikely that the DNA of two unrelated individuals will be identical. However, this method requires at least 20 nanograms of purified, intact DNA. PCR-based DNA fingerprinting is a rapid, less expensive method and requires only a very small amount of DNA (as little as fifty white blood cells).

APPLICATIONS

What is bioprospecting?

Bioprospecting, also called biodiversity prospecting, is an important application of biotechnology. It involves the search for possible new plant or microbial strains, particularly from extreme environments, such as rain forests, deserts, hot springs, and coral reefs. These organisms are then used to develop new phytopharmaceuticals. There is some controversy as to who owns the resources of these countries: the countries in which the resources reside or the company that turns them into valuable products.

Can genetic engineering be used to save endangered species?

As endangered species disappear from natural habitats and are only found in zoos, researchers are looking for ways to conserve these species. Using cryopreservation, the Zoological Society of San Diego has created a "frozen zoo" that stores viable cell lines from more than 3,200 individual mammals, birds, and reptiles, representing 355 species and subspecies. Researchers maintain that there should still be continued efforts to preserve species in their natural habitats, but by preserving and studying animal DNA in the laboratory, scientists can learn genetic aspects crucial to the species' survival.

How can biotechnology be used to manufacture vaccines?

Vaccine development is risky using conventional methods because vaccines must be manufactured inside living organisms, and the diseases themselves are extremely dangerous and infectious. Using genetic engineering, specific pathogen proteins

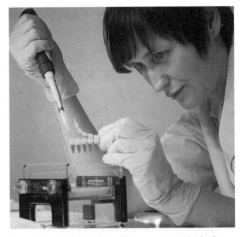

Gene therapy repairs or replaces genes in the body that cause disease. Often, this is done with the help of viruses as a means of delivering the genes inside cells.

that trigger antibody production are isolated and inserted into a bacterial or fungal vector. The organisms are then cultured to produce large quantities of the protein.

What is gene therapy?

Gene therapy is an experimental technique that uses genes to treat or prevent disease. One approach involves replacement of an "abnormal" disease-causing gene with a "normal" gene. The normal gene is delivered to target cells using a vector, which is usually a virus that has been genetically engineered to carry human DNA. The virus genome is altered to remove disease-causing genes and insert therapeutic genes. Target cells are infected with the virus. The virus then integrates its genetically altered material containing the human gene into the target cell, where it should produce a functional protein product. Another approach is to inactivate the mutated gene that is not functioning properly.

What are the risks and challenges of gene therapy?

There are technical challenges and risks associated with gene therapy. First, there is the challenge of being able to introduce new genes into the body and keep them working. The body may react to the newly introduced viruses with the new genetic material as intruders, resulting in an immune system attack. The immune system response may cause inflammation and possibly organ failure. In 1999, eighteen-year-old Jesse Gelsinger (1981–1999) died of complications due to a severe immune response to the viral carrier used to deliver the therapeutic genes. All gene therapy clinical trials in the United States were stopped for a period of time after Gelsinger's death. Another risk of gene therapy is that the viral carrier may infect healthy cells, in addition to the targeted cells, causing illness or disease. Infection by the virus is possible if the virus reverts to its original ability to cause disease. Finally, in some cases, the new genes were inserted in the wrong spot in the DNA, leading to the formation of tumors. Finally, there is the challenge of whether gene therapy is commercially viable.

What are some genetic disorders treated by gene therapy?

Clinical trials using gene therapy have been tested for a number of genetic disorders, including muscular dystrophy, cystic fibrosis, severe combined immunodeficiency (SCID), inherited degenerative blindness, hemophilia, and hemoglobin defects, such as thalassemia. Most clinical trials consist of small groups of individuals.

Who was the first person to receive gene therapy?

The first gene therapy clinical trial was in 1990 at the National Institutes of Health Clinical Center. On September 14, 1990, Dr. W. French Anderson (1936–), Dr. Michael Blaese (1939–), and Dr. Kenneth Culver treated a four-year-old girl who had adenosine deaminase (ADA) deficiency. ADA causes severe combined immunodeficiency (SCID), which leaves the patient defenseless against all infections, including bacterial, viral, and fungal. In the procedure, white blood cells were removed and normal genes for making adenosine deaminase were inserted into them. The corrected genes were then reinserted into the young girl. As of 2014, she is living a normal life. Gene therapy continues to be an experimental therapy and is only available through a clinical trial for treatment of diseases that have no other cures.

How can genes be used to detect single gene disorders?

Genetic testing can be used to determine those at risk for a particular inherited condition. There are more than two hundred single gene disorders that can be diagnosed in prenatal individuals using recombinant DNA techniques. Also, since some genetic disorders appear later in life, children and adults can be tested for genetic disorders before becoming symptomatic. If the locus of the disease-causing gene is known, gene markers can be used to determine which family members are at risk. An example of an adult-onset genetic disorder is polycystic kidney disease, which occurs between the ages of thirty-five and fifty. These cysts produced by the disease will eventually destroy the kidneys. Prior knowledge of the condition allows both patient and doctor to closely monitor any changes in the kidneys.

How can genes be used to detect multigene disorders?

DNA chips and microarray technology can be used as a diagnostic tool when searching for multiple mutations that can be part of a multigene disease, such as some types of cancers. DNA chip technology is now being used to screen for mutations in the p53 gene. Approximately 60 percent of all cancers are linked to mutations of the p53 gene. Eventually, these methods will be used to collectively generate an individual genetic profile of all mutations associated with known genetic disorders.

What are the types of genetic testing?

There are more than one thousand genetic tests available and in use to identify changes in chromosomes, genes, or proteins. The three methods used for genetic testing are:

1. Molecular genetic tests, which study single genes or short lengths of DNA to identify variation or mutations that lead to a genetic disorder.

2. Chromosomal genetic tests, which analyze whole chromosomes or long lengths of DNA to see if there are large genetic changes, such as an extra copy of a chromosome, that cause a genetic condition.

3. Biochemical genetic tests that study the amount or activity level of proteins. Abnormalities in either can indicate changes to the DNA that result in a genetic disorder.

Genetic testing may be conducted at any point from testing for carriers to use as a diagnostic tool. These are some of the types of genetic testing:

- *Carrier testing* is used to identify individuals who carry one copy of a gene mutation that, when present in two copies, causes a genetic disorder. When both parties are tested, it can be used to provide information about the couple's risk of having a child with a genetic disorder.

- *Preimplantation testing* is a specialized technique used to detect genetic changes in embryos that were created using assisted reproductive techniques. It allows selection of embryos without genetic changes to be chosen for implantation in the woman's uterus.

- *Prenatal testing* is used to detect changes in a fetus's genes or chromosomes before birth. Although prenatal testing cannot identify all possible inherited disorders and birth defects, it is helpful when there is an increased risk that the baby will have a genetic or chromosomal disorder.

- *Newborn screening* is used shortly after birth to identify genetic disorders that can be treated early in life. All states in the United States require testing for phenylketonuria and congenital hypothyroidism. Phenylketonuria is a genetic disorder that causes mental retardation if left untreated. Congenital hypothyroidism is a disorder of the thyroid gland.

- *Diagnostic testing* may be performed before birth or at a later point in a person's life to identify and confirm or rule out a specific condition based on physical signs and symptoms.

- *Forensic testing* uses DNA sequences to identify an individual for legal purposes. This type of testing can identify crime or catastrophe victims, rule out or implicate a crime suspect, or establish biological relationships between individuals, such as paternity testing.

What is gene doping?

Gene doping is an extension of gene therapy. Whereas gene therapy introduces new DNA to a person to fight a disease or genetic condition, gene doping introduces new DNA to enhance an athlete's performance. An example of the misuse of gene therapy for athletic performance is the gene for erythropoietin (EPO). The hormone EPO controls the production of red blood cells. A higher red blood cell count will give an athlete more oxygen-carrying red blood cells, thereby enhancing his or her performance. The hormone was abused by cyclist Lance Armstrong (1971–). Introducing the EPO gene to the body would result in the individual producing more red blood cells. There is concern that the misuse of recombinant human EPO may lead to serious illness, such as heart disease or stroke, or even death. While tests exist to detect the introduction of EPO, genetic testing is not yet sophisticated enough to

Bicyclist Lance Armstrong was stripped of his Tour de France medals after finally confessing to increasing his red blood cells with doping.

determine the introduction of the EPO gene. Both the International Olympic Committee and the World Anti-Doping Agency have banned gene doping.

What is terminator gene technology?

This is a method of biotechnology in which crops that are bioengineered for a specific desirable trait (such as drought resistance) would contain a lethal gene that would cause any seeds produced by the plant to be nonviable. The lethal gene could be activated by spraying with a solution sold by the same company that originally marketed the bioengineered plant. Thus, the plants would still provide seeds with nutritional value, but these seeds could not be used to produce new plants. This technique would allow companies to control the product's genes so that they would not spread into the general plant population. Seeds would have to be purchased by the grower for each season.

What is a VNTR?

VNTR refers to variable number of tandem repeats, which is a kind of molecular marker that contains sequences of DNA that have end-to-end repeats of different short DNA sequences. When the repeating unit is two to four base pairs long, it is called an STR (short tandem repeat), and when the repeating unit is two to thirty base pairs long, it is called a VNTR. An example of a VNTR is GGATGGATGGATGGATGGAT, which is four tandem repeats of the sequence GGAT. In addition, VNTRs are highly variable regions, and thus, there may be many alleles (gene variations) at many loci (physical location of a gene on a chromosome). VNTRs were discovered by Alec Jeffreys (1950–) and colleagues as they were working on the basics of DNA fingerprinting.

What is the biological basis for DNA fingerprinting?

DNA fingerprinting, also known as DNA typing or DNA profiling, is based on the unique genetic differences that exist between individuals. DNA fingerprinting is based on a comparison of VNTRs. Most DNA sequences are identical, but out of one hundred base pairs (of DNA), two people will generally differ by one base pair. Since there are three billion base pairs in human DNA, one individual's DNA will differ from another's by three million base pairs. To examine an individual's DNA fingerprint, the DNA sample is cut with a restriction endonuclease, and the fragments are separated by gel electrophoresis. The fragments are then transferred to a nylon membrane, where they are incubated in a solution containing a radioactive DNA probe that is complementary to specific polymorphic sequences.

When was DNA fingerprinting developed?

Sir Alec Jeffreys (1950–), a geneticist, developed DNA fingerprinting in the early 1980s when he was studying inherited genetic variations between people. He was one of the first scientists to describe small DNA changes, referred to as single-nucleotide polymorphisms (SNPs). From SNPs, he began to look at tandem repeat DNA sequences, in which a short sequence of DNA was consecutively repeated many times. Jeffreys was inducted into the National Inventors Hall of Fame in 2005 for his work on genetic fingerprinting.

What types of samples can be used for DNA fingerprinting?

Any body fluid or tissue that contains enough protein for analysis can be used for DNA fingerprinting, including hair follicles, skin, ear wax, bone, urine, feces, semen, or blood. DNA evidence may also be gathered from dandruff; from saliva on cigarette butts, chewing gum, or envelopes; and from skin cells on eyeglasses.

What is the accuracy of DNA fingerprinting?

The accuracy of DNA fingerprinting depends on the number of VNTR or STR loci that are used. At present, the FBI uses thirteen STR loci in its profile, with the expected frequency of this profile to be less than one in ten billion. As the number of loci analyzed increases, the probability of a random match becomes smaller.

Can an innocent person be convicted based on DNA analysis?

Current methods of DNA analysis are very sensitive, as only a few cells are needed for DNA extraction. However, it is possible for an innocent person's DNA to be found at any crime scene, either from accidental deposition or by direct deposition by a third party. Also, a partial DNA profile from a crime scene could match that of an innocent person whose DNA is already in a DNA data bank. In addition, close relatives of suspected criminals could also be partial matches to a DNA profile.

What factors might influence the interpretation of DNA fingerprinting?

Ruling out procedural sources of error (such as contamination from another sample or improper sample preparation), DNA obtained from siblings will share both alleles at a particular locus, so the DNA profile may not be conclusive. DNA from identical twins will have identical DNA profiles. The frequency of identical twins in a population is one in every 250 births. In all cases, the probability calculations that are used in DNA profiling are based on a large population; therefore, if a population is small or reflects interbreeding, probabilities must be adjusted.

What is the Innocence Project?

The Innocence Project is a public law clinic affiliated with the Benjamin Cardozo Law School at Yeshiva University in New York City that uses biotechnology, specifically DNA evidence, to reopen cases of people who have been wrongly convicted of crimes.

What are applications of DNA fingerprinting?

DNA fingerprinting is used to determine paternity; in forensic crime analysis; in population genetics to analyze variation within populations or ethnic groups; in conservation biology to study the genetic variability of endangered species; to test for the presence of specific pathogens in food sources; to detect genetically modified organisms either within plants or food products; in evolutionary biology to compare DNA extracts from fossils to modern-day counterparts; and in the identification of victims of a disaster. DNA fingerprinting has also gained importance in archaeology. For example, it is used to study the DNA of the ancient pharaohs to determine how they were related.

What was the first use of DNA fingerprinting in a criminal investigation?

The first case of DNA fingerprinting occurred in 1986 in England. Between 1983 and 1986, two schoolgirls were murdered in a village near Leicestershire, England. The only clue left at the respective crime scenes was semen, and a suspect had already been arrested. However, police asked a scientist to help them positively identify whether the DNA profiles from both crime scenes matched that of the suspect. Using VNTR analysis, the scientist showed that although the DNA profiles from both crime scenes matched, they did not match that of the suspect. The suspect was released, and police then obtained a blood sample from every adult male in the area of the village. A local bakery worker, who had paid another coworker to provide his DNA sample, was eventually arrested and forced to provide a blood sample. His DNA sample matched the samples from the crime scene, and the worker confessed to the murders.

What is the largest forensic DNA investigation in U.S. history?

The identification of the remains of the victims from the September 11, 2001, terrorist attacks in New York City has comprised the largest and most difficult DNA identification to date. After 1.6 million tons of debris were removed from the site of the attacks on the World Trade Center, only 239 intact bodies of the 2,752 people killed were found, along with more than 20,000 pieces of human remains. One of the challenges of DNA identification in the World Trade Center tragedy was the condition of the DNA samples from the victims. Many of the fragments had been burned in the fires and intense heat or decomposed due to the long duration from the time of the disaster until discovery. Since most DNA profiling technology relied on larger fragments, techniques were developed that utilized "mini" short tandem repeats (STRs) and single-nucleotide polymorphisms (SNPs).

In order to match DNA profiles to the bodies, personal items such as razor blades, combs, and toothbrushes were collected from the victims' homes. When possible, cheek swabs were taken from the victims' family members for comparison with remains. Some remains have still not been identified since the DNA technology is not sophisticated enough for identification based solely on muscle, skin, and hair samples. These samples are being preserved with the hopes that one day the technology will allow identification. There are still victims who have not been linked to any remains and remains that are not connected to any of the identified victims. In one case, more than 300 tissue fragments were linked to a single victim.

How did DNA analysis determine the fate of the Romanov family during the Bolshevik Revolution in 1917?

In July 1918, the Bolshevik revolutionaries claimed they had executed Tsar Nicholas Romanov II (1868–1918) and his family. There was no information about where the family was buried, and there were rumors that some family members had survived the execution. In 1991, nine skeletons were found in a shallow grave twenty miles from Ekaterinburg, Russia. The British Forensic Science Service and Russian authorities began an extensive forensic examination, including DNA testing. First, sex testing using a gene that differs in the X and Y chromosomes was used to determine the gender of each of the skeletal remains. The DNA of living relatives (including Prince Philip, Duke of Edinburgh [1921–]) of the Romanov family was compared to the DNA of the remains.

After Russia's Czar Nicholas II and his family were executed during the Russian Revolution, the skeletal remains had to be identified using DNA analysis.

391

Short tandem repeat (STR) tests were used to determine the relationships between the remains. The remains of the tsar, his wife, and three of their children were identified as part of the group of the skeletal remains.

How many DNA profiles are contained in NDIS (the National DNA Index System)?

The National DNA Index System (NDIS) is one part of CODIS—the national level. It contains DNA profiles contributed by federal, state, and local participating forensic laboratories. As of June 2015, there were 11,822,927 offender profiles (including convicted offender, detainee, and legal profiles), 2,028,734 arrestee profiles, and 638,162 forensic profiles in the NDIS. The CODIS primary metric, the "Investigation Aided," tracks the number of criminal investigations where CODIS has added value to the investigative process. CODIS has produced over 288,298 hits, assisting in more than 274,648 investigations. CODIS hits are either offender hits or forensic hits. An offender hit is when the identity of a potential suspect is generated. A forensic hit occurs when the DNA profiles obtained from two or more crime scenes are linked but the source of the profiles remains unknown.

What is germline therapy?

Germline therapy involves gene alteration within the gametes (eggs or sperm) of an organism. Alterations in gamete genes transfer the gene to all cells in the next generation. As a result, all members of future generations will be affected.

What is xenotransplantation?

Xenotransplantation is the transplantation of tissue or organs from one species to another. Xenotransplantation of whole organs may be attempted as a last possible effort while waiting for a human organ to become available since the potential for rejection of the organ is very great. Using only some tissue or a component of an organ, such as pig heart valves for human valve replacement, has been much more successful.

When were the first experiments with xenotransplantation?

Records of xenotransplantation date back to the seventeenth century. Most early experiments in xenotransplantation were not successful. In one case, bone from a dog was used to repair the skull of an injured Russian aristocrat. Although the procedure was considered a success, it angered the church. In the early twentieth century, French surgeon Alexis Carrel (1873–1944) experimented with transplanting veins. Other scientists attempted to transplant animal kidneys into humans, but each attempt failed.

Modern research and experimentation in xenotransplantation date to the 1960s. Dr. Keith Reemtsma (1925–2000) is recognized as the first to perform animal-to-human transplants in 1963–1964. He transplanted chimpanzee kidneys into human patients. Most of the patients lived only eight to sixty-three days, but one was able to survive for nine months before succumbing to infection.

What are the dangers involved with transgenic animal transplants?

Recognizing transplanted tissue or an organ as a foreign body and subsequently attempting to reject it is the usual immune response. This is true for both same species transplantation and xenotransplantation. A further risk of xenotransplantation is the risk of transplanting animal viruses along with the transfer cells or organs. Since the patient is already immunosuppressed, the patient could die from the viral infection, or the virus could be spread to the general population.

What is a knockout mouse?

A knockout mouse is a mouse that has specific genes mutated ("knocked out") so that the lack of production of the gene product can be studied in an animal model. Knockout mice are frequently used in pharmaceutical studies to test the potential for a particular human enzyme as a therapy target. Mice have identical or nearly identical copies of human proteins, making them useful in studies that model effects in humans.

What is biopharming?

Biopharming is a relatively new technology in which transgenic plants are used to "grow" pharmaceuticals. In this technique, scientists bioengineer medically important proteins into a corn plant, thus spurring the corn plant to produce large amounts of a particular medicinal protein. This is less expensive than using a microbial fermentation chamber.

What are some examples of genetic engineering in plants?

Genetically engineered plants include transgenic crop plants that are resistant to herbicides used in weed control. These transgenic crops carry genes for resistance to herbicides such that all plants in a field are killed with the exception of the modified plant. Transgenic soybeans, corn, cotton, canola, papaya, rice, and tomatoes are used by many farmers in the United States. Plants resistant to predatory insects have also been genetically engineered.

What is GMO?

The acronym GMO stands for "genetically modified organisms." The DNA and genetic structure of these organisms have been altered by incorporating a single gene or mul-

As awareness grows about GMOs, so do the number of protests around the country. The debate as to how safe or dangerous these foods actually are, however, is still a subject for disagreement among scientists.

tiple genes from another organism or species. Historically, almost all domesticated animals and crop plants have been genetically modified over thousands of years by human selection and cross-breeding. One of the desired advantages of GMO crops is that they are bred to be pest-resistant, thus allowing farmers to reduce the use of pesticides, including herbicides, insecticides, fungicides, rodenticides, and fumigants. Recent studies have shown that the use of pesticides decreased by 2 percent or 31 million pounds (14 million kilograms) during the first six years (1996–2001) of commercial use of GMO crops. During the years 1999 to 2011, the use of herbicides has grown, however, because of certain weeds becoming resistant to the genetically engineered herbicide-tolerant plants. The use of insecticides decreased by 10 to 12 million pounds (4.5 to 5.4 million kilograms) annually, approximately 28 percent, between 1996 and 2011. Additional agricultural research is needed to assess the risk of exposure to pesticides and the safety of pesticide-resistant crops while developing newer strains of genetically engineered crops.

Who regulates GMOs?

In the United States, GMOs are regulated by the Food and Drug Administration (FDA), Environmental Protection Agency (EPA), and U.S. Department of Agriculture (USDA).

How widespread is the use of GMO crops?

GMO crops were first planted commercially in 1996. In 2013, 433 million acres (175.2 million hectares) of GMO crops were grown in 27 countries globally. The United States, Brazil, Argentina, India, and Canada were the largest producers of GMO (biotech) crops. The United States produces 173.2 million acres (70.1 million hectares) of GMO crops, which represents 40 percent of the global acreage of biotech crops. The most commonly grown biotech crops are maize, cotton, soybeans, and canola.

What is Frankenfood?

Frankenfood is a term coined by environmental and health activist groups to denote any food that has been genetically modified (GM) or that contains genetically modified organisms (GMO). Opposition to GM food is based on concerns that the gene pool of "natural" plants could be altered permanently if exposed to pollen from genetically altered plants. There is also fear that people and animals that consume GM food might have allergic reactions to altered protein or could develop health problems later.

What is a Flavr Savr® tomato?

The Flavr Savr® tomato was produced in response to consumer complaints that tomatoes were either too rotten to eat when they arrived at the store or too green. Growers had found that they could treat green tomatoes in the warehouse with ethylene, a gas that causes the tomato skin to turn red. However, the tomato itself stayed hard. In the late 1980s, researchers at Calgene (a small biotech company) discovered that the enzyme polygalactouronase (PG) controlled rotting in tomatoes. The scientists reversed the DNA sequence of PG; the effect was that tomatoes turned red on the vine, yet the skin of the tomatoes remained tough enough to withstand the mechanical pickers. However, before the Flavr Savr® tomato was introduced to the market, Calgene disclosed to the public how the tomato was bioengineered. This caused a public protest that led to a worldwide movement against genetically modified organisms (GMO).

What is Golden Rice?

In 1999 Ingo Potrykus (1933–), then of the Institute for Plant Sciences at the Swiss Federal Institute of Technology, and Peter Beyer (1952–) of the University of Freiburg in Germany invented Golden Rice. Golden Rice was created using genes from maize and a common soil organism to create a rice variety that contains beta carotene. The importance of this is that in the human body, beta carotene is converted to vitamin A, a crucial vitamin missing from the diet of millions of poor people, whose diets consist of large quantities of rice, around the world. Lack of vitamin A causes the death of an estimated one million of Asia's poorest children due to weakened immune systems; vitamin A de-

Is Golden Rice available to consumers?

The first field trial of Golden Rice was planted in Louisiana and harvested in September 2004. A new version with even more beta carotene was developed in 2005. Multilocation field trials were started in the Philippines in 2012 and are ongoing. Other field trials are scheduled in Bangladesh and Indonesia.

ficiency is also linked to blindness. The rice was called "golden rice" because it turned yellow like a daffodil.

What is a biopesticide?

A biopesticide is a chemical derived from an organism that interferes with the metabolism of another species. An example is the Bt toxin *Bacillus thuringiensis*, which interferes with the absorption of food in insects but does not harm mammals.

What is Starlink corn?

Starlink is a bioengineered corn variety that was genetically modified to include a gene from the bacterium *Bacillus thuringiensis* (Bt), which produces a protein (called an endotoxin) that kills some types of insects. Bt endotoxin has been registered as a biopesticide in the United States since 1961, and the Bt endotoxin has been used by organic farmers for biological pest control. The endotoxins only become activated in the guts of susceptible insects. Because of the significant losses to corn crops caused by the European corn borer, scientists targeted the corn plant itself as a candidate for insertion of the Bt gene.

What other crops contain the Bt gene?

Bt cotton is another bioengineered crop besides corn; cotton is highly susceptible to pests and, thus, cotton farming typically involves the use of a high level of pesticide. A positive effect of Bt cotton is the reduced use of such pesticides and less deposition of these chemicals into the environment.

What was the controversy surrounding Bt corn and the monarch butterfly?

In 1999 a study was released, based on controlled laboratory feeding experiments, that showed that corn pollen from Bt-altered plants would kill monarch butterflies. In the study, three-day-old monarch butterfly larvae were fed milkweed leaves dusted with Bt corn pollen. The larvae ate less, grew slower, and had a higher mortality rate than those fed milkweed with no corn pollen or milkweed coated with non-Bt corn pollen. However, the laboratory study did not provide information on the number of Bt pollen grains that were consumed by the monarch larvae in order to observe the lethal effects. Also, no information was provided on the effects on older, larger larvae, which would be expected to have a higher tolerance to Bt toxicity. Headlines such as "Attack of the Killer Corn" and "Nature at Risk" triggered regulatory action on the part of the European Union to ban the importation and use of Bt corn varieties in Europe. Extensive field and laboratory studies were conducted in 1999 and 2000 in both the United States and Canada to evaluate the impact of Bt corn on the monarch butterfly. At the end of the two-year study, it was concluded that Bt corn does not pose a risk to monarch populations. The results of the research were published in the *Proceedings of the National Academy of Sciences* on October 9, 2001. A week later, on October 16, 2001, the Environmental Protection Agency extended the registration of Bt corn.

What is green fluorescent protein?

Green fluorescent protein is a protein found in a luminescent jellyfish (*Aquorea victoria*) that lives in the cold waters of the northern Pacific. Bioluminescence is the production of light by living organisms. These jellyfish contain two proteins: a bioluminescent protein called aequorin that emits blue light and an accessory green fluorescent protein (GFP). However, what we actually see when the jellyfish fluoresces is the conversion of the blue light emitted by aequorin to a green light—a metabolic reaction facilitated by the GFP. Since GFP is simply a protein, it is often used both as a marker for gene transfer and for localization of proteins. There are a variety of green fluorescent proteins that can glow different colors.

What is the GloFish®?

The fluorescent red zebrafish (*Danio rerio*) called GloFish® are produced by bioengineering the fish with the gene for red fluorescent protein from sea anemones and coral. The fish were originally bred by scientists at the National University of Singapore to detect water pollution. Research continues on developing zebrafish that will selectively fluoresce when exposed to estrogen or heavy metals in the water. In the United States, the FDA, which has jurisdiction over genetically modified organisms, maintains that the fish should pose no threat to native fish populations, as they are tropical aquarium fish used exclusively for that purpose. Concerns about the GloFish® include the ethics of the use of genetic engineering for commercial purposes, the potential use of ornamental fish (such as the GloFish®) as a food fish, or the chance that the Glofish® might proliferate without safety monitoring. GloFish® are not intended for human consumption.

What is cloning?

A clone is a group of cells derived from the original cell by fission (one cell dividing into two cells) or by mitosis (cell nucleus division with each chromosome splitting into two). Cloning perpetuates an existing organism's genetic makeup. Gardeners have been making clones of plants for centuries by taking cuttings of plants to make genetically identical copies. For plants that refuse to grow from cuttings, or for the animal world, modern scientific techniques have greatly extended the range of cloning. The technique for plants starts with taking a cutting of a plant that best satisfies the criteria for reproductive success, beauty, or some other standard. Since all of the plant's cells contain the genetic information from which the entire plant can be reconstructed, the cutting can be taken from any part of the plant. Placed in a culture medium having nutritious chemicals and a growth hormone, the cells in the cutting divide, doubling in size every six weeks until the mass of cells produces small, white, globular points called embryoids. These embryoids develop roots, or shoots, and begin to look like tiny plants. Transplanted into compost, these plants grow into exact copies of the parent plant. The whole process takes eighteen months. This process, called tissue culture, has been used to make clones of oil palm, asparagus, pineapples, strawberries, brussels sprouts, cauliflower, bananas, carnations, ferns, and others. Besides making highly productive copies of the best plant available, this method controls viral diseases that are passed through normal seed generations.

Can human beings be cloned?

In theory, yes. There are, however, many technical obstacles to human cloning, as well as moral, ethical, philosophical, religious, and economic issues to be resolved before a human being could be cloned. At the present time, most scientists would agree that cloning a human being is unsafe under current conditions.

How could a human be cloned?

Nuclear transplantation or somatic cell nuclear transfer is used to move the cell nucleus and its genetic material from one cell to another. Somatic cell nuclear transfer may be used to make tissue that is genetically compatible with that of the recipient and could be used in the treatment of a specific disease. Or, if the material is moved to an egg cell lacking its own nucleus, the transfer could result in the formation of a clone embryo.

Has human cloning been successful?

Despite several attempts, human cloning has not been successful. One of the technical difficulties in human cloning is that the spindle proteins are located very close to the chromosomes in human eggs. Removing the egg's nucleus to make room for the donor nucleus also removes the spindle proteins, interfering with cell division.

What was the first animal to be successfully cloned?

In 1962 British molecular biologist John B. Gurdon (1933–) cloned a frog. He transplanted the nucleus of an intestinal cell from a tadpole into a frog's egg that had had its nucleus removed. The egg developed into an adult frog that had the tadpole's genome in all of its cells and was therefore a clone of the tadpole. Gurdon shared the 2012 Nobel Prize in Physiology or Medicine with Shinya Yamanaka (1962–) for the discovery that mature cells can be reprogrammed to become pluripotent. Pluripotent cells are able to differentiate to all the cell types that make up the body.

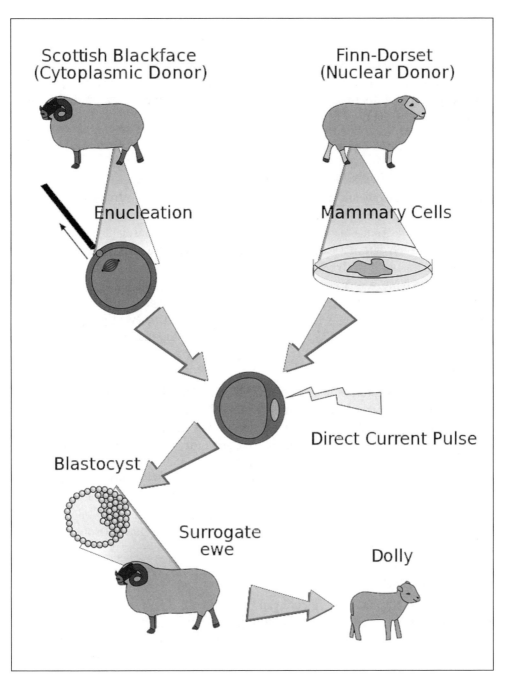

The process that created the genetic clone Dolly the sheep.

What was the first mammal to be successfully cloned?

The first mammal cloned from adult cells was Dolly, a ewe born in July 1996. Dolly was born in a research facility in Scotland. Ian Wilmut (1944–) led the team of biologists that removed a nucleus from a mammary cell of an adult ewe and transplanted it into an enucleated egg extracted from a second ewe. Electrical pulses were administered to fuse the nucleus with its new host. When the egg began to divide and develop into an embryo, it was transplanted into a surrogate mother ewe. Dolly was the genetic twin of the ewe that donated the mammary cell nucleus. On April 13, 1998, Dolly gave birth to Bonnie—the product of a normal mating with a Welsh mountain ram. This event is important as it demonstrates that Dolly was a healthy, fertile sheep, able to produce healthy offspring.

What is bioprinting?

Bioprinting is an adaptation of 3-D printing. Building on the technology of 3-D printing, the pioneers of bioprinting, Gabor Forgács (1949–) and Thomas Boland, recognized the potential of creating human tissues layer by layer using 3-D printing. Bioprinting consists of a multistep process. Cells are first grown in a growth medium to create a bioink made of cell aggregates. The cells are then loaded into cartridges very similar to inkjet printer cartridges with a long nozzle for printing. The bioprinter "prints" (deposits) a pattern of cells layer by layer, dictated by computer software and interspersed with a hydrogel substrate. The hydrogel serves as a support for the cells while they grow. It is then removed and the tissue may be used for research. The goal is to one day be able to print cells, tissues, and even organs for transplantation.

What is *Sugababe*?

Sugababe is a replica of artist Vincent van Gogh's (1853–1890) severed ear. It was created by artist Diemut Strebe by collecting genetic material from Lieuwe van Gogh (1992–), the great-great-grandson of the artist's brother, Theo van Gogh (1857–1891). The cells were grown and printed using a 3-D bioprinter. They were then shaped to resemble Vincent van Gogh's ear.

What are some applications of biometrics?

The term "biometrics" is derived from the Greek words *bio*, meaning "life," and *metrics*, meaning "to measure." It uses physical characteristics, such as fingerprints, hand geometry, eye structure, retina scans, iris features, palm veins, facial features, voice recognition, and gait (walking) patterns of an individual for identification and recognition purposes. A fingerprint scan to unlock a computer screen, instead of a traditional password, is one use of biometrics. Other situations that may use biometric tools are to limit physical access to buildings, gain access to bank ATM machines, identification in criminal cases, and passport control.

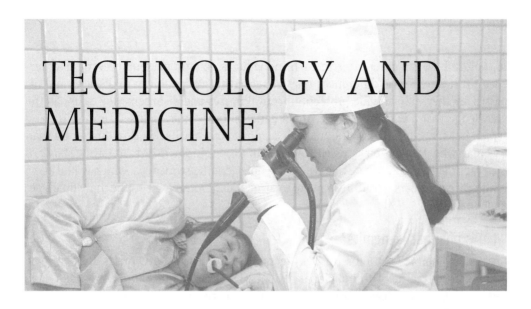

TECHNOLOGY AND MEDICINE

INTRODUCTION

What is the role of technology in medicine?

Technology plays a significant role in the advancement of medicine. The development of diagnostic tools allows health care professionals to identify diseases and conditions. Earlier diagnosis often leads to better, more successful outcomes. Technology continues to improve the tools and devices used to manage chronic conditions, such as diabetes and heart disease. New medications and other treatments have been developed to treat diseases, such as cancer. Advances in surgical procedures, such as robotic surgery, have made some surgical procedures less invasive and traumatic for the patient. Intensive care units and the continuous monitoring of patients with advanced machinery have improved the survival rate of some of the sickest patients.

What is telehealth?

Telehealth, also called e-health or m-health (mobile health), is the term used to describe using digital information and electronic communication devices and technologies to manage health and well-being. Telehealth includes electronic health records, e-visits and consultations with health care providers, remote monitoring of vital signs and other health conditions, such as glucose levels for diabetics, online support groups, and online health management tools and apps.

What is personalized medicine?

Personalized medicine, also known as genomic medicine, focuses on an individual's genetic makeup to screen, detect, and treat various health conditions. The completion of the Human Genome Project in April 2003 allowed medical researchers the opportunity to explore how the differences in each individual's genetic makeup present as differences in health. These differences may increase the susceptibility to diseases and illnesses, or they may provide protection from illnesses and disorders. For example, the knowledge that a woman has the BRCA gene mutation may be used to monitor and screen her more frequently and make decisions about her health care differently than a woman who does not have the BRCA gene mutation. In the future, it may be standard medical practice for everyone to have their whole genome sequenced, which would allow health care providers to routinely make treatment decisions based on the individual's genetic makeup.

What is pharmacogenomics?

Pharmacogenomics is the use of DNA technology to develop new drugs and optimize current drug treatment to individual patients. For example, the interaction of a drug with a specific protein can be studied and then compared to a cell in which a genetic mutation has inactivated that protein. Its potential is to tailor drug therapy to an individual's genome, a tailoring that could reduce adverse drug reactions and increase the efficacy of drug treatment. Cancer treatments are being developed based on individual genetic markers and the type of cancer cells. Different patients with the same disease may receive different treatments based on their individual genetic markers. Other benefits of pharmacogenomics include better vaccines, advanced screening for diseases, and more powerful drugs.

Gleevec (approved by the FDA in 2001) is an example of a drug developed through pharmacogenomics. It was developed by Novartis to treat a rare type of genetically caused chronic myeloid leukemia. In this type of leukemia, pieces of two different chromosomes break off and reattach on the opposite chromosome, causing a chromosome translocation. This abnormality causes a gene for a blood cell enzyme to continually manufacture the enzyme, resulting in high levels of white blood cells in the bone marrow and blood. Gleevec was specifically engineered to inhibit the enzyme created by the translocation mutation and to thus block the rapid growth of white blood cells.

What is tissue engineering?

Tissue engineering is used to create semisynthetic tissues that are used to replace or support the function of defective or injured body parts. It is a broad field, encompassing cell biology, biomaterial engineering, microscopic engineering, robotics, and bioreactors, where tissues are grown and nurtured. Tissue engineering can improve current medical therapies by designing replacements that mimic natural tissue function. Commercially produced skin is already in use for treating patients with burns and diabetic ulcers.

IMAGING TECHNIQUES

How do physicians and other health care providers explore the inside of the body?

Until the end of the nineteenth century, there were no non-invasive techniques to explore the internal organs of the body. Medical practitioners relied on descriptions of symptoms as the basis of their diagnosis. X-rays, discovered at the very end of the nineteenth century, provided the earliest technique to explore the internal organs and tissues of the body. During the twentieth century, significant advances were made in the field of medical imaging to explore the internal organs.

What are X-rays?

X-rays are electromagnetic radiation with short wavelengths (10^{-3} nanometers) and a great amount of energy. They were discovered in 1898 by William Conrad Roentgen (1845–1923). X-rays are frequently used in medicine because they are able to pass through opaque, dense structures such as bone and form an image on a photographic plate. They are especially helpful in assessing damage to bones, identifying certain tumors, and examining the chest—heart and lungs—and abdomen.

What are the disadvantages of X-rays as a diagnostic tool?

A major disadvantage of X-rays as a diagnostic tool is that they provide little information about the soft tissues. Since they only show a flat, two-dimensional picture, they cannot distinguish between the various layers of an organ, some of which may be healthy and others diseased.

What are CAT or CT scans?

CAT or CT scans (computer-assisted tomography or simply computerized tomography) are specialized X-rays that produce cross-sectional images of the body. An X-ray-emitting device moves around the body region being examined. At the same time, an X-ray-detecting device moves in the opposite direction on the other side of the body. As these two devices move, an X-ray beam passes through the body from hundreds of different angles. Since tissues and organs absorb X-rays differently, the intensity of X-rays reaching the detector varies from position to position. A computer records the measurements made by

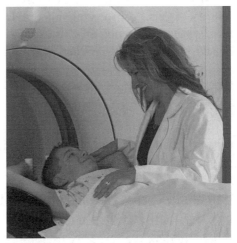

CT scans still use X-rays, but they do so in a way that produces 3-D images that are more useful than regular X-ray machines in diagnosing what is going on inside the body.

the X-ray detector and combines them mathematically. The result is a sectional image of the body that is viewed on a screen.

How are CT scans used in the study of the human body?

CT scans are used to study many parts of the body, including the chest, belly, and pelvis, extremities (arms and legs), and internal organs, such as pancreas, liver, gallbladder, and kidneys. CT scans of the head and brain may detect an abnormal mass or growth, stroke, area of bleeding, or blood vessel abnormality. Patients complaining of pain may have a CT scan to determine the source of the pain. Sometimes a CT scan will be used to further investigate an abnormality found on a regular X-ray.

What is an advantage of positron emission tomography (PET imaging) over CT scans and X-rays?

Unlike traditional X-rays and CT scans, which reveal information about the structure of internal organs, positron emission tomography (PET imaging) is an excellent technique for observing metabolic processes. Developed during the 1970s, PET imaging uses radioactive isotopes to detect biochemical activity in a specific body part.

What is the procedure for a PET scan?

A patient is injected with a radioisotope, which travels through the body and is transported to the organ and tissue to be studied. As the radioisotopes are absorbed by the cells, high-energy gamma rays are produced. A computer collects and analyzes the gamma-ray emission, producing an image of the organ's activity.

How are PET scans used to detect and treat cancer?

PET scans of the whole body may detect cancers. While the PET scans do not provide cancer therapy, they are very useful in examining the effects of cancer therapies and treatments on a tumor. Since it is possible to observe biochemical activities of cells and tumors using PET scans, biochemical changes to tumors following treatment may be observed.

Is it possible to study the blood flow to the heart or brain?

PET scans provide information about blood flow to the heart muscle and brain. They may help evaluate signs of coronary heart disease and reasons for decreased function in certain areas of the heart. PET scans of the brain may detect tumors or other neurological disorders, including certain behavioral health disorders. Studies of the brain using PET scans have identified parts of the brain that are affected by epilepsy and seizures, Alzheimer's disease, Parkinson's disease, and stroke. In addition, they have been used to identify specific regions of the healthy brain that are active during certain tasks.

What is nuclear magnetic resonance?

Nuclear magnetic resonance (NMR) is a process in which the nuclei of certain atoms absorb energy from an external magnetic field. Scientists use NMR spectroscopy to identify unknown compounds, check for impurities, and study the shapes of molecules. They use the knowledge that different atoms will absorb electromagnetic energy at slightly different frequencies.

What is nuclear magnetic resonance imaging?

Magnetic resonance imaging (MRI), sometimes called nuclear magnetic resonance imaging (NMRI), is a noninvasive, nonionizing diagnostic technique. It is useful in detecting small tumors, blocked blood vessels, or damaged vertebral disks. Because it does not involve the use of radiation, it can often be used where X-rays are dangerous. Large magnets beam energy through the body, causing hydrogen atoms in the body to resonate. This produces energy in the form of tiny electrical signals. A computer detects these signals, which vary in different parts of the body and according to whether an organ is healthy or not. The variation enables a picture to be produced on a screen and interpreted by a medical specialist.

What distinguishes MRI from computerized X-ray scanners is that most X-ray studies cannot distinguish between a living body and a cadaver, while MRI "sees" the difference between life and death in great detail. More specifically, it can discriminate between healthy and diseased tissues with more sensitivity than conventional radiographic instruments like X-rays or CAT scans. CAT (computerized axial tomography) scanners have been around since 1973 and are actually glorified X-ray machines. They offer three-dimensional viewing but are limited because the object imaged must remain still.

Who proposed using magnetic resonance imaging for diagnostic purposes?

The concept of using MRI to detect tumors in patients was proposed by Raymond Damadian (1936–) in a 1972 patent application. Damadian was inducted into the National Inventors Hall of Fame in 1989 for his pioneering work to develop MRI. The fundamental MRI imaging concept used in all present-day MRI instruments was proposed by Paul Lauterbur (1929–2007) in an article in *Nature* in 1973. Lauterbur and Peter Mansfield (1933–) were awarded the Nobel Prize in Physiology or Medicine in 2003 for their discoveries concerning magnetic resonance imaging. Lauterbur and Mansfield were inducted into the National Inventors Hall of Fame in 2007 for their work on magnetic resonance imaging. The main advantages of MRI are that it not only gives superior images of soft tissues (like organs), but it can also measure dynamic physiological changes in a noninvasive manner (without penetrating the body in any way). A disadvantage of MRI is that it cannot be used for every patient. For example, patients with implants, pacemakers, or cerebral aneurysm clips made of metal cannot be examined using MRI because the machine's magnet could potentially move these objects within the body, causing damage.

What is ultrasound?

Ultrasound, also called sonography, is another type of 3-D computerized imaging. Using brief pulses of ultrahigh-frequency acoustic waves (lasting 0.01 second), it can produce a sonar map of the imaged object. The technique is similar to the echolocation used by bats, whales, and dolphins. By measuring the echo waves, it is possible to determine the size, shape, location, and consistency (whether it is solid, fluid-filled, or both) of an object.

Why is ultrasound used frequently in obstetrics?

Ultrasound is a very safe, noninvasive imaging technique. Unlike X-rays, sonography does not use ionizing radiation to produce an image. It gives a clear picture of soft tissues, which do not show up in X-rays. Ultrasound causes no health problems (for the mother or unborn fetus) and may be repeated as often as necessary.

Which imaging technique is used to examine breast tissue and diagnose breast diseases?

Mammography is the specific imaging technique used to examine breast tissue and diagnose breast diseases. A small dose of radiation is passed through the breast tissue. Mammography has become a very important tool in the early diagnosis of breast cancer. Small tumors may be visible on a mammogram years before they may be felt physically by a woman or her health care provider.

MEDICAL DEVICES AND TOOLS

What technological advances were important to the development of the endoscope?

One of the greatest challenges to physicians was the ability to see inside the human body without performing surgery. The obstacles to seeing inside the body included that most body parts are not straight, e.g., the many curves and twists in the intestine, and the inside of the body cavity is dark. One of the earliest systems used to explore the inner cavity of the body was devised by German physician Philipp Bozzini (1773–1809). He demonstrated it to the Academy of Medicine of Vienna in 1806. The device, called the Lichtleiter, consisted of a tube, a mirror, and a candle, which directed the light into the body and was redirected to the eyepiece. Needless to say, it was not a very comfortable experience for the patient since the tubing was rigid and there was the risk of burns from the candle. The invention of the incandescent light by Thomas Edison (1847–1931) in 1878 helped solve the problem of using an open candle for the light source.

Physicians continued to make modifications and improvements to the light, tube, and mirror system, but it was not until development of fiber optics in the early 1950s that modern endoscopes were developed. British physicist Harold Hopkins (1918–1994) designed an endoscope using a bundle of flexible glass fibers to transmit light. His idea,which he called a "fiberscope," was published as a letter in *Nature* in January 1954. The basic design was improved by Basil Hirschowitz (1925–2013) with his colleagues Larry Curtiss and C. Wilbur Peters (1918–1989), creating the first fully flexible fiber-optic endoscope. The scope consisted of a bundle of thin glass fibers bundled together. Curtiss had suggested covering the thin fibers with an additional layer of glass with a lower refractive index. The light would travel down the glass fiber, bounce off the glass fiber that surrounded it, and continue down the fiber. The scope was demonstrated in 1957, and the ability to see inside the body and around corners and bends in the body became a reality.

The newest advances in endoscopy include instruments with tiny digital cameras that display images on screen. Endoscopy ultrasonography uses ultrasound to visualize internal tissues. There are also capsule endoscopes that are swallowed by the patient. The swallowed capsules transmit images wirelessly.

What are some newer technologies associated with thermometers?

The mercury-filled glass thermometer was one of the most common thermometers to record body temperature until the end of the twentieth century. Concerns with mercury being a hazardous substance led to the use of alcohol or another non-toxic compound in thermometers. Digital thermometers have replaced glass thermometers as the most commonly used thermometer. They are inexpensive, easy to use—orally, rectally, or axillary (under the armpit)—and accurate.

The ear or tympanic thermometer was invented by Theodor H. Benzinger (1905–

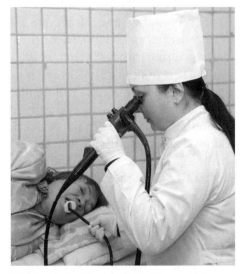

Endoscopes use thin glass optical fibers which can then go inside patients without incisions so that medical professionals can see inside the body.

409

1999) in 1964. It measures the infrared heat waves released in the eardrum. They are quick and accurate for home use. Other infrared thermometers are used by placing the thermometer on the side of the forehead. They measure the infrared heat waves released by the temporal artery.

Some of the newest technological advances in thermometers are no-contact thermometers. They are usually held from one to six inches away from a person's forehead, depending on the manufacturer. These thermometers measure infrared energy emanating from the body. The thermometer then translates the energy reading to a temperature reading of the person. Tests have indicated the readings are accurate as a diagnostic tool for fever. The major advantages of no-contact thermometers are that they are non-invasive and do not require physical contact with a potentially ill individual, reducing the risk of cross-infection. They have been used at airports for screening for SARS and Ebola.

What devices have been developed to protect health care workers from needlesticks and injuries from other sharp objects?

Health care works are at risk of exposure to blood-borne pathogens, such as human immunodeficiency virus (HIV), Hepatitis B virus (HBV), Hepatitis C virus (HCV), and other disease-causing pathogens from needlesticks. The recognition of the danger to health care workers from needlesticks and other sharp objects, such as scalpels, led to the design of newer, safer devices with built-in safety features. Options for safer devices include needleless devices, devices with integrated safety features, and devices with safety features as an accessory to the device. In the latter case, implementation of the safety feature is dependent upon the health care worker activating the safety feature. Devices that have a safety feature as an integral part of the design are most effective in protecting health care workers. Examples of safety features are:

- self-sheathing needle shields or blade shields where a shield is moved forward over the used needle and then locked in place

- retractable needles or sharp objects where the needle automatically retracts into the syringe or device

- self-blunting technology where a blunt tip is moved forward through the needle past the sharp point

When was the first successful pacemaker invented?

Pacemakers are devices used to control certain abnormal heart rhythms. They send electrical impulses to the patient's heart. John Hopps (1919–1998), a Canadian electrical engineer, invented an external pacemaker in 1950. However, the pacemaker was so large and cumbersome, it was painful to the patient. Two years later, in 1952, Paul Zoll (1911–1999) in Boston, in collaboration with the Electrodyne Company, developed the first successful pacemaker. The device was worn externally on the patient's belt. It relied on an electrical wall socket to stimulate the patient's heart through two metal electrodes attached to the patient's chest.

Wilson Greatbatch (1919–2011), working with Earl Bakken (1924–), developed an internal pacemaker. It was first implanted in a patient by surgeons William Chardack (1915–2006) and Andrew Gage (1922–) in 1960. Greatbatch was inducted into the National Inventors Hall of Fame in 1986 for the invention of the pacemaker. He received the National Medal of Technology in 1990. It is estimated that the invention of the implantable pacemaker has saved over two million lives.

What are some newer technologies being investigated for pacemakers?

Most conventional pacemakers are the size of a silver dollar and consist of two parts: 1) a battery-powered pulse generator that produces the stimulation and 2) insulated wires, called "leads," that connect the generator to the heart and deliver the stimulation as needed. These devices are implanted during a surgical procedure requiring an incision in the chest.

The newest pacemakers are smaller than a penny, wireless, and can be implanted into a patient without major surgery. Wireless pacemakers contain a sensor electrode to detect information about the heart rate, software to process the information, a generator to provide the stimulation to the heart, and a battery to run the device. It is expected that the battery will last eight to ten years. Since it is implanted directly into the wall of the heart, there is no need for "leads," which in the past have on occasion malfunctioned, dislodged, or become infected. Instead of requiring surgery to be implanted, they are small enough to be inserted using a catheter that is inserted through a vein in the leg to the heart. Tiny, wireless pacemakers were approved for use in Europe in 2013 and are in clinical trials for U.S. FDA approval. Although the products currently being tested are for only the right ventricle of the heart, research is continuing to develop wireless pacemakers for dual chambers—upper and lower—of the heart since the majority of patients require pacing in both chambers.

When was an external defibrillator first used?

Defibrillation is a technique used to restore normal heart rhythms following sudden cardiac arrest. One of the earliest defibrillators used alternating current (AC) to treat a fourteen-year-old boy during surgery in 1947. The defibrillator consisted of two silver paddles that were placed directly on the heart. Paul Zoll (1911–1999) produced the first external defibrillator in 1956. By the early 1960s, research had shown that direct current (DC) was safer and better for defibrillation. The first successful use of defibrillation before arriving at the hospital was in 1966 by ambulance-transported physicians in Belfast, Northern Ireland. Emergency medical technicians (EMTs) first performed defibrillation without the presence of physicians in 1972 in Portland, Oregon. During the following decades, first responders and police officers were trained in the use of defibrillators.

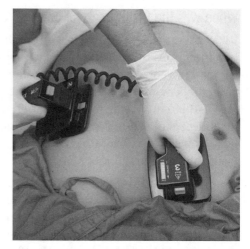
Defibrillators employ a DC current to stimulate a heart back into beating after a cardiac arrest.

How did changes to defibrillators allow them to become available widely in public areas?

Automated external defibrillators (AEDs) became available in the 1990s. The device consists of a battery and pad electrodes. Unlike earlier defibrillators, they are small, portable devices. AEDs are able to automatically analyze the heart rhythms of victims in cardiac arrest to determine whether delivering a shock is recommended. The American Red Cross began including AED training as a part of cardiopulmonary resuscitation (CPR) training in 1999. The newest models of AEDs give full instructions to the user so they can be used by even untrained bystanders. Their use has greatly increased the survival rate of individuals who suffer from sudden cardiac arrest.

Is there a defibrillator that can be worn on a regular basis to prevent sudden cardiac arrest?

LifeVest® is a wearable defibrillator. It was developed by M. Stephen Heilman (1933–) and approved for use by the FDA in 2001. The LifeVest® is a garment with sensing electrodes that continuously monitors the patient's heart rate. If life-threatening abnormal heart rhythms are detected, the device will first alert a patient that it will deliver a shock. A conscious patient has the option to delay the shock. In the event the patient is unconscious, the LifeVest® automatically releases a conductive gel and then delivers a shock. A monitor that is worn around the waist collects ECG data from the sensing electrodes. The data can be sent to the patient's physician or other health care professional via modem. LifeVest® is a technology that saves patients lives.

Which machine was essential for the development of open-heart surgery?

The heart-lung machine was a major technical advance for the development of open-heart surgery. Dr. John H. Gibbon Jr. (1903–1973) performed the first open-heart surgery with the use of the heart-lung machine that he designed in 1953. The patient, an eighteen-year-old girl, was totally dependent on the heart-lung machine for twenty-six minutes while Gibbon closed a septal defect between the upper chambers of the heart. The functions of circulation and respiration are performed by the heart-lung machine and bypass the heart and lungs during surgeries of the heart.

How are lasers used in medicine?

The term "laser" is an acronym for light amplification by stimulated emission of radiation. Lasers produce intense, monochromatic light in waves that are synchronized, i.e., moving in exactly the same direction at the same time. The single-wavelength light allows different types of tissue to be targeted because different colors absorb particular wavelengths. For example, hemoglobin, which gives red blood cells their color, absorbs the green light of an argon laser, which is used to stop bleeding.

The three types of lasers used most frequently in medical treatments are carbon dioxide lasers, neodymium:yttrium-aluminum-garnet (Nd:YAG) lasers, and argon lasers. The carbon dioxide laser is mostly used as a surgical tool. It converts light energy to heat, which is strong enough to minimize bleeding while it vaporizes the tissue. The Nd:YAG laser is able to penetrate tissue more deeply than other lasers. It makes blood clot quickly, allowing surgeons to reach parts of the body that would otherwise require open, invasive surgery. The argon laser provides limited penetration, which is required for eye surgery.

How is technology changing drug delivery systems?

Drugs and medications have traditionally been administered orally, by injection, inhalation (e.g., inhalers for asthma), or absorption through the skin (e.g., topical creams and ointments for rashes). Biomedical engineers are researching new methods to deliver drugs and medications. Challenges for drug delivery research include finding better delivery systems targeted to maximize the efficacy of the drug with fewer side effects. One newer method to deliver drugs is the use of microneedle arrays. Dozens of microneedles, each thinner than a strand of hair, are either coated or filled with the medication. The needles penetrate the skin, but they are so small that they do not reach the nerves in the skin, so the medication is delivered without the pain associated with injections.

What are transdermal patches?

Transdermal patches are medicated adhesive pads that are placed on the skin and release drugs gradually for up to a week. The first transdermal patches were approved for use by the FDA in 1979 to deliver scopolamine to treat motion sickness. Some substances, including drugs, are introduced through the skin by means of an adhesive patch that includes a small reservoir containing the drug. The drug passes from the reservoir through a permeable membrane at a known rate. It then diffuses into the epidermis and enters the blood vessels of the dermis. Transdermal patches are used to control motion sickness, prevent chest pain associated with heart disease, control blood pressure, and deliver nicotine to help individuals stop smoking.

How does iontophoresis differ from a standard transdermal patch?

A transdermal patch is a passive method to deliver a drug. Iontophoresis is considered an active method to deliver medications. It uses an electrical current to carry the medication through the sweat glands and hair follicles to tissues below the skin. The medication is placed on a patch. A low-level, weak electrical current carries the drug through the skin to the underlying tissue.

How do insulin pumps help manage diabetes?

Insulin pumps are small devices that continuously deliver insulin. There are two basic types of insulin pumps—durable pumps and patch pumps. Durable pumps include tubing that is inserted into the body. The insulin is delivered through the tubing to the body. Patch pumps do not have tubing. The insulin is held in a reservoir on the inside of the patch. A small needle places a cannula under the skin. The needle then retracts back, so it is not in the body during delivery of insulin. Insulin pumps can deliver very small, precise amounts of insulin and rapidly adjust for meals, allowing better management of diabetes.

Are pumps available to deliver drugs other than insulin?

Pumps that deliver medication to control chronic pain have been approved by the FDA since 1991. Drug pumps (intrathecal drug delivery systems) deliver pain medication into the intrathecal space that surrounds the spinal cord. It reduces chronic pain since the medication is released directly to the receptors near the spine, interrupting the pain signals before they reach the brain. Another advantage of drug pumps is that many patients have fewer side effects from medications.

How is patient-controlled analgesia (PCA) administered?

Patient-controlled analgesia is a drug delivery system that dispenses a preset intravenous (IV) dose of a narcotic analgesic for reductions of pain whenever a patient pushes a switch on an electric cord. The device consists of a computerized pump with a chamber containing a syringe holding up to 60 milliliters of a drug. The patient administers a dose

---

What health information is available to users of wearable technology?

Wearable technology is available in many different forms for health and fitness monitoring. Smart watches, wristbands, belts, goggles, glasses, and other gadgets collect data. Many of these devices count steps like pedometers, calculate calories burned during exercise, and monitor heart rate and sleep. Other gadgets include devices that are embedded in socks to track the effectiveness of a person's stride by measuring foot-landing techniques and weight distribution while walking or running.

of narcotic when the need for pain relief arises. A lockout interval device automatically inactivates the system if the patient tries to increase the amount of narcotic within a pre-set time period.

What is the future of wearable technology in health care?

Researchers and innovators are developing more wearable devices to monitor and potentially manage health conditions. Wearables may one day monitor pain and then dispense electric pulses to control the pain level. They also have the potential to manage conditions such as autism, attention deficit disorder, and epilepsy. One of the challenges in the development of wearable technology for health care is maintaining confidentiality of health information while utilizing wireless technologies.

Have continuous glucose monitors replaced traditional fingerstick glucose tests?

Continuous glucose monitoring (CGM) devices do not replace traditional fingerstick glucose tests. Continuous glucose monitoring devices measure the glucose level in the interstitial fluid (fluid and tissue that surround the cells) once every one to five minutes, depending on the model of the device. Since they do not measure the glucose level in the blood, the values obtained using CGM devices and a fingerstick will differ. Traditional fingersticks are still required for accurate real-time monitoring. Decisions regarding treatment option should be based on readings from fingersticks and not CGM values.

A CGM device consists of a sensor, transmitter, and receiver. The sensor is a thin wire, approximately the diameter of two human hairs, that is inserted into the skin of the abdomen or upper arm. An adhesive pad keeps it in place. The disposable sensors are replaced every three to five days, depending on the device. The transmitter sends information to the receiver wirelessly via radio waves. Data from readings can be stored and analyzed to find trends in glucose levels based on time of day, exercise, other medications, and food intake. Most monitors have alarm settings to alert users when their blood glucose levels are approaching levels that are either too high or too low, when glucose levels are rising or falling too rapidly, or that predict when a value will be out of the target range.

Can the results of continuous glucose monitoring be shared on mobile devices?

The FDA approved the first system of mobile medical apps for continuous glucose monitoring in 2015, allowing people to automatically and securely share data from a CGM with other people in real-time using an Apple mobile device, such as an iPhone. This newest innovation is especially helpful for parents to monitor their children's glucose levels remotely.

ASSISTIVE TECHNOLOGIES AND PROSTHETIC DEVICES

What are the major categories of disabilities?

Disabilities are divided into four major categories: 1) cognitive disabilities, which include learning disorders and other intellectual challenges, 2) hearing loss or impaired hearing, 3) vision loss, low vision, or colorblindness, and 4) physical disability, including paralysis and difficulties with walking or other movement. Some of the assistive technologies that are designed to help these disabilities include computers and iPads for learning disabilities, hearing aids, computer programs with screen-enlargement programs for vision impairment, walkers, wheelchairs, and scooters for people with physical disabilities.

What are assistive technologies?

Assistive technologies are the tools, equipment, products, and devices that help individuals with disabilities function successfully in their everyday lives at school, work, home, and in the community. Disabilities may be the result of injuries, diseases, or disorders. While some disabilities are permanent, others, such as the recuperation period after hip or knee replacement surgery, are short-term. Other terms used to describe assistive technologies are adaptive devices, accessible technology, independent living aids, mobility aids, prosthetics, medical rehabilitative aids and devices, and rehabilitative engineering and technology.

What types of assistive devices are available for individuals with hearing loss?

Hearing aids are basic devices that make sounds louder, so a person with hearing loss can listen, communicate, and participate more fully in daily activities. Hearing aids have a microphone, amplifier, and speaker. The sound is received through a microphone, which converts the sound waves to electrical signals and sends them to an amplifier. The amplifier increases the power of the signals and then sends them to the ear through a speaker. Hearing aids may be analog or digital. Analog aids convert sound waves into electrical signals, which are then amplified. Digital aids convert sound waves into numerical codes before amplifying them. Digital aids can be programmed to amplify some frequencies more than others and can be programmed to focus on sounds coming from a specific direction.

What is a cochlear implant?

Individuals with sensorineural deafness, damage to the tiny hair cells in the part of the inner ear called the cochlea, sometimes choose to have a cochlear implant. A cochlear implant does not create or restore normal hearing but allows a deaf person to understand speech. A cochlear implant is a small, complex, electronic device surgically implanted under the skin behind the ear. There are both external and internal parts of the device. The external parts include: 1) a microphone, which picks up sound from the environment; 2) a speech processor, which selects and arranges this sound; and 3) a transmitter, which receives signals from the speech processor, converts them into electrical impulses or signals, and sends them to a receiver. The internal part is a receiver, which collects the impulses from the stimulator and sends them to electrodes that have been surgically inserted into the inner ear (cochlea). The first cochlear implant was approved for use by the Food and Drug Administration (FDA) in 1984 for adults and in 1990 for children. Since 2000, cochlear implants have been FDA-approved for use in children beginning at 12 months of age. As of

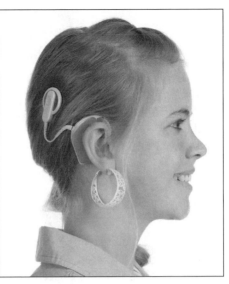

Some hearing-loss or legally deaf patients whose deafness is a result of damage to the cochlea can have their hearing improved with a cochlear implant.

the end of 2012, approximately 324,200 people worldwide have received cochlear implants. In the United States, approximately 58,000 adults and 38,000 children have received cochlear implants.

What are assistive listening devices?

Assistive listening devices (ALDs) amplify the sound a person wants to hear. They are especially helpful in situations where there is a lot of background noise. Some types of ALDs are designed for use in large facilities, such as theaters, airports, and classrooms, while others are designed for use in small settings or one-on-one conversations. They may be used with or without hearing aids or cochlear implants. Personal amplifiers are useful when watching television, traveling in a car, or being outdoors. Similar in size to a cell phone, these devices increase sound levels and decrease background noise for the listener. The amplified sound is picked up by the receiver the listener is wearing as a headset or earbuds.

What ALDs are best suited for large facilities?

The ALDs for larger facilities are hearing loop (or induction loop) systems, frequency-modulated (FM) systems, and infrared systems. A hearing loop system has four parts: a sound source, e.g., a public address system, microphone, or home TV or telephone, 2) an amplifier, 3) a thin wire that encircles a room or branches out beneath carpeting, and 4) a receiver worn in the ear or as a headset. Sound travels through the loop and creates an electromagnetic field that is picked up directly by a hearing loop receiver. Since the sound is picked up directly by the receiver, it is much clearer, without as much background noise.

Frequency-modulated (FM) systems use radio signals to transmit amplified sounds. The speaker uses a microphone connected to a transmitter, and the listener wears the receiver, which is tuned to a specific frequency or radio channel.

Infrared systems use infrared light to transmit sound. A transmitter converts sound into a light signal, which is beamed to a receiver worn by the listener. The receiver decodes the infrared light signal back to sound. An advantage of infrared systems is that the signal cannot pass through walls, so it can be used in situations where confidentiality should be maintained, such as in courtrooms.

What are some newer technologies that assist individuals with low vision?

Magnifying glasses and other enlargers, such as closed-circuit TV (CCTV) magnifiers (which are expensive), have been standard assistive devices for individuals with low vision. Electronic readers (e-readers) use a newer technology; they are more affordable than CCTV and may be a better option for some people with low vision. E-readers have adjustable font sizes and contrast settings to enlarge and improve the brightness of text on the screen. They also have a feature that converts text to speech, so the written text may be read aloud to the user.

Smartphones and tablets offer a variety of apps to assist individuals with low vision. Apple products offer EyeNote, which is a free app that scans and identifies the denomination of U.S. paper money by reading aloud or emitting an ascending number of beeps or pulsed vibrations for each bill. There is also an app that measures visual function with a set of vision tests. Significant changes in vision are recorded and sent wirelessly to the person's health care provider. Voice recognition systems allow users to use voice commands to dictate texts for emails without typing or to access emails or information on their digital calendars.

How have advances in technology changed prosthetic devices?

Prosthetic devices are artificial gadgets or apparatuses that replace parts of the body that are missing due to injury or disease. Prosthetic devices for fingers, hands, arms, toes, feet, or legs are among the most common types of artificial devices. The earliest prosthetic devices were made of iron or wood and had minimal functionality. Modern prosthetic devices are designed to mimic their natural counterparts in both appearance and function. They are constructed of newer, lightweight materials, such as plastics, metal alloys, and carbon-fiber composites. The fields of nanotechnology, robotics, and tissue engineering all play a significant role in the development of modern prosthetic devices.

Advances in materials and design have made it possible for even athletes to compete with prosthetic legs.

Can the brain be used to control prosthetics?

Researchers and scientists have developed brain-computer interfaces (BCIs), also known as brain-machine interfaces (BMIs), that allow communication between brain activity and an external prosthetic device. The human brain controls movement by sending electrical signals to the muscles. In an amputee, the signals stop at the point of the amputation. The technology of BCIs transfers the electrical signals to the prosthetic device. Tiny sensors are implanted in the brain, which decode signals from the neurons into motor commands that are sent to the prosthetic device. The prosthetic device then performs the task, e.g., picking up an object.

MEDICAL AND
SURGICAL PROCEDURES

When was the first robotic surgery performed?

The first robotic surgery was performed in 1985 using the PUMA 560, a robotic arm invented by Dr. Yik San Kwoh (1946–). The procedure was a neurosurgical biopsy on a fifty-two-year-old man suspected of having a brain tumor. Three years later the first robot-assisted laparoscopic surgery was performed. The Food and Drug Administration approved the da Vinci Surgical System for laparoscopic procedures in 2000. The da Vinci system consisted of two components—a viewing and control console and a surgical arm unit.

What are some advantages of robotic surgery?

Robotic surgery is similar to laparoscopic surgery because it is also minimally invasive, requiring much smaller incisions than traditional surgery. Once the robotic arm is in

place, it is easier for the surgeon to manipulate the surgical tools than by using
traditional endoscopic techniques. Examples of procedures that may be done ro-
botically are:

- Gallbladder removal
- Hip replacement
- Hysterectomy
- Tubal ligation
- Kidney transplant or kidney removal
- Coronary artery bypass
- Mitral valve replacement
- Prostate removal

How does telesurgery differ from robotic surgery?

Telesurgery combines high-speed communication networks with robotic surgery
systems, allowing physicians to perform
surgery in locations physically removed
from the patient. The first transatlantic

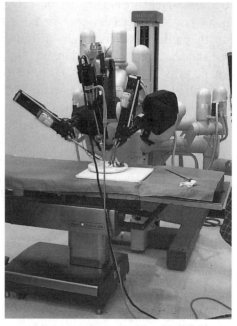

A laproscopic surgery robot can be used for performing delicate surgery that is minimally invasive and lessens the chance of exposure to germs through human contact.

surgery was performed by French surgeon Jacques Marescaux (1948–) on September 7, 2001. Dr. Marescaux was in New York, and the patient was 4,300 miles (7,000 kilometers) away in Strasbourg, France. The surgery, known as "Operation Lindbergh," was re-
moval of the gallbladder. Fiber-optic communications allowed Marescaux's hand
movements to be conveyed to the robotic instruments in France. The instruments in
France included an endoscopic camera, and the images from the camera were transmitted back to Marescaux. There was a delay of only about 150 milliseconds in each direction, short enough so it did not hinder the procedure. The ability to use telesurgery
for patients in remote areas or in military settings has the potential to increase acces-
sibility of specialized procedures.

What is electrical muscle stimulation?

Neuromuscular electrical stimulation (NMES) is a technique used by occupational and
physical therapists to reeducate and retrain muscles. It is used when the muscle is not
contracting properly as a result of a stroke or orthopedic surgery. Electrodes are placed
on the muscle, and an electrical impulse is sent to the muscle. The muscle contracts in
response to the stimulus similar to muscle contraction during exercise or other movement. The goal of using electrical stimulation is to reeducate the muscle to regain normal muscle function, so it contracts on its own.

When were dialysis machines first invented?

Dialysis is an important technology to prevent renal failure when the kidneys do not function or function improperly. The earliest dialysis machines date back to the 1940s. Willem Kolff (1911–2009) is considered the father of dialysis. He constructed the first machine for dialysis using sausage skins and various spare parts in 1943. He performed the first dialysis treatment on a patient in 1945. It was not until the 1960s when Belding Scribner (1921–2003) used a plastic tube coated with Teflon as a connection device or "shunt" between an artery and a vein that dialysis could be performed repeatedly without damaging the blood vessels. Dialysis performs the functions of the kidney by removing wastes from the body.

Willem Kolff, the father of dialysis

Are there newer technologies than dialysis to treat kidney failure?

A wearable artificial kidney was invented by Dr. Victor Gura and developed by Blood Purification Technologies Inc. It was approved for clinical trials in 2014. The battery-powered device connects to a catheter to filter toxins and excess fluid from the blood. The goal is to improve the quality of life for individuals who suffer from renal disease.

What is assisted reproductive technology (ART)?

The Centers for Disease Control and Prevention (CDC) defines assisted reproductive technology as including all fertility treatments in which both the sperm and eggs are handled during the treatment. Most ART procedures involve surgically removing eggs from a woman's ovaries, combining them with sperm in the laboratory, and returning them to the woman's body or donating them to another woman. This definition excludes treatments in which only the sperm are handled, e.g., artificial insemination or procedures in which a woman takes medicine to stimulate egg production without retrieving the eggs.

How do the various ART methods differ?

One of the most successful and effective ART methods is in vitro fertilization (IVF). It may be used when the woman's fallopian tubes are blocked or when the man produces too few sperm. The woman takes a drug that stimulates the ovaries to produce multiple eggs. Once mature, the eggs are retrieved and put in a dish in a laboratory setting with the man's sperm for fertilization. The sperm are usually able to fertilize the eggs on their own. After three to five days, healthy embryos are implanted in the woman's uterus.

Zygote intrafallopian transfer (ZIFT), also called tubal embryo transfer, is similar to IVF. Fertilization occurs in the laboratory, then the very young embryo is transferred to the fallopian tube instead of the uterus.

Gamete intrafallopian transfer (GIFT) involves transferring eggs and sperm into the woman's fallopian tube. Fertilization occurs in the woman's body. It is not as common a procedure as either IVF or ZIFT.

Couples who experience serious problems with the man's sperm or who have been unsuccessful with IVF may try intracytoplasmic sperm injection (ICSI). In ICSI, a single sperm is injected into a mature egg, then the embryo is transferred to the uterus or fallopian tube.

What technologies were important to the development of fetal surgery?

Ultrasound imaging techniques play a major role in identifying conditions that warrant fetal surgery. The use of endoscopes and minimally invasive surgery also helped change the procedures for fetal surgery. Fetal surgery, also called prenatal surgery, is performed to correct problems, such as heart defects, urinary blockages, bowel obstructions, and airway malfunctions, that are best treated prior to birth to ensure a healthier baby at birth.

When was the first fetal surgery performed?

Dr. Michael R. Harrison (1943–) performed the first open fetal surgery in 1981. He inserted a catheter to allow urine to drain in a fetus whose bladder was dangerously enlarged due to a blockage in the urinary tract. The surgery was a success, and the baby survived.

Open fetal surgery is the most invasive type of fetal surgery. The mother is anesthetized, and an incision is made in her lower abdomen, similar to a Cesarean section, to expose the uterus. Once the surgeon reaches the uterus, the uterus is opened with a special instrument that prevents bleeding. The fetus is partially or completely removed from the womb, surgery to correct the problem is performed, the fetus is returned to the uterus, and the incision is closed.

Are there less invasive forms of fetal surgery?

Fetoscopic surgery, often called Fetendo fetal surgery, is a minimally invasive type of fetal surgery. It was developed in the 1990s and named "Fetendo" because the surgeons felt the technique was similar to playing video games, such as Nintendo. Surgeons use very small endoscopes to see the fetus with very small surgical instruments to manipulate and repair problems in the fetus. The procedure is performed either through the mother's skin or through small openings (incisions) in the mother's abdomen. It is often the method used to treat twin-twin syndrome, a condition of the placenta that causes one twin to grow at the expense of the other twin due to abnormal blood connections in the shared placenta.

Fetal image-guided surgery is even less invasive than Fetendo. Doctors use ultrasound images to manipulate the fetus. It generally does not require an incision in the mother's abdomen nor does it use endoscopy as a means to visualize the fetus. Manipulations may include placing a catheter in the bladder, abdomen, or chest while using the sonogram views of the fetus.

Who was the first to use a laser treatment for a medical procedure?

Dr. Charles J. Campbell (1926–2007) and Charles J. Koester (1929–) used a ruby laser to destroy a retinal tumor in 1961 at Columbia University Presbyterian Hospital in New York City.

Today, lasers are used by many medical specialists, including cardiologists, dentists, dermatologists, gastroenterologists, gynecologists, neurosurgeons, oncologists, ophthalmologists, orthopedists, otolaryngologists, pulmonologists, and urologists. Some procedures that use lasers include the removal of tumors and treatment of canker sores, non-cancerous enlargement of the prostate gland, ovarian cysts, varicose veins, glaucoma, nearsightedness, astigmatism, and other vision problems.

Is it possible to remove tattoos with lasers?

Leon Goldman (1906–1997), a dermatologist, was an innovative pioneer in the use of lasers. In 1961, he was the first physician to treat melanoma with a laser. He then adapted the technique to use lasers to remove tattoos. Tattoos, along with birthmarks and other skin pigmentations, can be removed since lasers break the pigment colors into tiny pieces that are absorbed by the body. Lasers with different wavelengths of focused light have different colors, allowing different colors of a tattoo to be removed.

What is LASIK?

LASIK—laser-assisted in situ keratomileusis—is eye surgery that permanently changes the shape of the cornea. The cornea is the clear covering on the front of the eye. A difference between the shape of the cornea and the length of the eye can cause blurry vision. LASIK corrects the shape of the cornea, making it thinner and restoring clear vision. The procedure takes about ten to fifteen minutes for each eye. Eyedrops are given to numb the surface of the eye. A flap of corneal tissue is created. The flap is then peeled back, so the laser can reshape the corneal tissue underneath. Once the cornea is reshaped, the surgeon replaces the flap. In most cases, patients will have improved vision and will no longer need to wear glasses or contact lenses.

Further Reading

BOOKS

Ackerman, Diane. *The Human Age: The World Shaped by Us*. New York: W.W. Norton & Company, 2014.

Barnham, Keith. *The Burning Answer: The Solar Revolution*. New York: Pegasus Books, 2015.

Berlin, Leslie. *The Man behind the Microchip: Robert Noyce and the Invention of Silicon Valley*. New York: Oxford University Press, 2005.

Black, Jeremy. *The Power of Knowledge: How Information and Technology Made the Modern World*. New Haven, CT: Yale University Press, 2014.

Brain, Marshall. *The Engineering Book: From the Catapult to the Curiosity Rover: 250 Milestones in the History of Engineering*. New York: Sterling, 2015.

Bridgman, Roger Francis. *1,000 Inventions & Discoveries*. New York: DK Publishing, 2004.

Brockman, John. *The Greatest Inventions of the Past 2,000 Years*. New York: Simon & Schuster, 2000.

Browne, John. *Seven Elements That Changed the World: An Adventure of Ingenuity and Discovery*. New York: Pegasus Books, 2014.

Bruno, Leonard C. *On the Move: A Chronology of Advances in Transportation*. Detroit: Gale Research, 1993.

———. *The Tradition of Technology: Landmarks of Western Technology in the Collections of the Library of Congress*. Washington, DC: Library of Congress, 1995.

Bryson, Bill. *A Short History of Nearly Everything*. New York: Broadway Books, 2003.

Bunch, Bryan. *The History of Science and Technology: A Browser's Guide to the Great Discoveries, Inventions, and the People Who Made Them, from the Dawn of Time to Today*. Boston: Houghton Mifflin Company, 2004.

Cantor, Doug, ed. *The Big Book of Hacks*. San Francisco: Weldon Owen, 2012.

Carlisle, Rodney P. *Scientific American Inventions and Discoveries*. Hoboken, NJ: Wiley, 2004.

Chaline, Eric. *Fifty Minerals that Changed the Course of History*. Buffalo, NY: Firefly Books, 2012.

Chamayou, Gregoire. *A Theory of the Drone*. Translated by Janet Lloyd. New York: The New Press, 2015.

Ching, Francis D. K. *Building Construction Illustrated*, 5th ed. Hoboken, NJ: Wiley, 2014.

Chiras, Daniel D. *The Homeowner's Guide to Renewable Energy: Achieving Energy Independence through Solar, Wind, Biomass, and Hydropower*, revised and updated ed. Gabriola, BC: New Society Publishers, 2011.

Cianci, Philip J. *HDTV and the Transition to Digital Broadcasting: Understanding New Television Technologies*. Boston: Focal Press, 2007.

Clegg, Brian. *Final Frontier: The Pioneering Science and Technology of Exploring the Universe*. New York: St. Martin's Press, 2014.

Collin, Robin Morris, and Robert William Collin. *Encyclopedia of Sustainability*. Santa Barbara, CA: Greenwood Press, 2010.

Cotterell, Arthur. *Chariot: From Chariot to Tank, the Astounding Rise and Fall of the World's First War Machine*. Woodstock, NY: The Overlook Press, 2004.

Coulson, Michael. *The History of Mining: The Events, Technology and People Involved in the Industry That Forged the Modern World*. Petersfield, UK: Harriman House, 2012.

Coupérie-Eiffel, Philippe. *Eiffel by Eiffel*. Edited by Dominique Bouvard. Translated by Joseph Laredo. Zürich, Switzerland: Olms, 2014.

Crosby, Alfred W. *Throwing Fire: Projectile Technology through History*. New York: Cambridge University Press, 2002.

Dear, I.C.B., and Peter Kemp, eds. *The Oxford Companion to Ships and the Sea*. 2nd ed. New York: Oxford University Press, 2005.

DuBravac, Shawn. *Digital Destiny: How the New Age of Data Will Transform the Way We Work, Live and Communicate*. Washington, DC: Regnery Publishing, 2015.

Evans, Harold. *They Made America: From Steam Engine to the Search Engine*. Boston: Little, Brown, 2004.

Francis, Raymond L. *The Illustrated Almanac of Science, Technology, and Invention: Day by Day Facts, Figures and the Fanciful*. New York: Plenum Trade, 1997.

Frenzel, Louis E., Jr. *Electronics Explained: The New Systems Approach to Learning Electronics*. Amsterdam; Boston: Newnes, 2010.

Graf, Bernhard. *Bridges that Changed the World*. New York: Prestel, 2002.

Greider, Katharine. *National Geographic Science of Everything*. Washington, DC: National Geographic, 2013.

Grossman, Elizabeth. *High Tech Trash: Digital Devices, Hidden Toxics, and Human Health*. Washington, DC: Island Press, 2006.

Harrison, Ian. *The Book of Inventions*. Washington, DC: National Geographic Society, 2004.

Hart-Davis, Adam, ed. *Engineers*. New York: DK Publishing, 2012.

Heilbron, J. L., ed. *The Oxford Companion to the History of Modern Science*. New York: Oxford University Press, 2003.

Ikenson, Ben. *Patents: Ingenious Inventions: How They Work and How They Came to Be*. New York: Black Dog & Leventhal Publishers, 2004.

Johnson, Steven. *How We Got to Now: Six Innovations That Made the Modern World*. New York: Riverhead Books, 2014.

Kelly, John E., III. *Smart Machines: IBM's Watson and the Era of Cognitive Computing*. New York: Columbia Business School Publishing, 2013.

Kelly, Kevin. *What Technology Wants*. New York: Viking, 2010.

Langmead, Donald, and Christine Garnaut. *Encyclopedia of Architectural and Engineering Feats*. Santa Barbara, CA: ABC-CLIO, 2001.

Langone, John. *How Things Work: Everyday Technology Explained*, new ed. Washington, DC: National Geographic Society, 2006.

Lee, W. David, with Jeffrey Drazen, Phillip A. Sharp, and Robert S. Langer. *From X-rays to DNA: How Engineering Drives Biology*. Cambridge, MA: The MIT Press, 2014.

Lepik, Andres. *Skyscrapers*, new ed. London: Prestel, 2008.

LeVine, Steve. *The Powerhouse: Inside the Invention of a Battery to Save the World*. New York: Viking, 2015.

Macaulay, David. *The New Way Things Work*. Boston: Houghton Mifflin, 1998.

Macy, Christine. *Dams*. New York: W.W. Norton & Co., 2010.

Madrigal, Alexis. *Powering the Dream: The History and Promise of Green Technology*. Cambridge, MA: Da Capo Press, 2011.

Magoun, Alexander B. *Television: The Life Story of a Technology*. Westport, CT: Greenwood Press, 2007.

Malone, Michael S. *The Intel Trinity: How Robert Noyce, Gordon Moore, and Andy Grove Built the World's Most Important Company*. New York: Harper Business, 2014.l

McCoy, John F., ed. *Space Sciences*, 2nd ed. Detroit: Macmillan Reference USA, 2012.

McCullough, David G. *The Great Bridge: The Epic Story of the Building of the Brooklyn Bridge*. New York: Simon & Schuster, 2012.

Moss, Frank. *The Sorcerers and Their Apprentices: How the Digital Magicians of the MIT Media Lab Are Creating the Innovative Technologies that Will Transform Our Lives*. New York: Crown Business, 2011.

Naughton, John. *From Gutenberg to Zuckerberg: Disruptive Innovation in the Age of the Internet*. New York: Quercus, 2014.

Nocks, Lisa. *The Robot: The Life Story of a Technology*. Westport, CT: Greenwood Press, 2007.

Ochoa, George, and Melinda Corey. *The Wilson Chronology of Science and Technology*. New York: H.W. Wilson, 1997.

Park, Chris, and Michael Allaby. *A Dictionary of Environment and Conservation*. 2nd ed. Oxford: Oxford University Press, 2013.

Perlin, John. *Let It Shine: The 6,000-Year Story of Solar Energy*. Novato, CA: New World Library, 2013.

Pernick, Ron. *Clean Tech Nation: How the U.S. Can Lead in the New Global Economy*. New York: Harper Business, 2012.

Petroski, Henry. *Engineers of Dreams: Great Bridge Builders and the Spanning of America*. New York: Knopf, 1995.

———. *Remaking the World: Adventures in Engineering*. New York: Alfred A. Knopf, 1997.

Plowden, David. *Bridges: The Spans of North America*, new ed. New York: Norton, 2002.

Restivo, Sal, and Peter H. Denton, eds. *Battleground: Science and Technology*. Westport, CT: Greenwood Press, 2008.

Schaeffer, John. *Real Goods Solar Living Sourcebook: Your Complete Guide to Living Beyond the Grid with Renewable Energy Technologies and Sustainable Living*. Gabriola Island, BC: New Society Publishers, 2014.

Schmidt, J. Eric, and Jared Cohen. *The New Digital Age: Reshaping the Future of People, Nations and Business*. New York: Alfred A. Knopf, 2013.

Stryker, Cole. *Hacking the Future: Privacy, Identify, and Anonymity on the Web*. New York: Overlook Duckworth, 2012.

Thackray, Arnold. *Moore's Law: The Life of Gordon Moore, Silicon Valley's Quiet Revolutionary*. New York: Basic Books, 2015.

Tillemann, Levi. *The Great Race: The Global Quest for the Car of the Future*. New York: Simon & Schuster, 2015.

Van Dulken, Stephen. *Inventing the 20th Century: 100 Inventions That Shaped the World: From the Airplane to the Zipper*. Washington Square, NY: New York University Press, 2000.

Webb, Richard C. *Tele-visionaries: The People behind the Invention of Television*. Hoboken, NJ: Wiley-Interscience, 2005.

Wells, Matthew. *Skyscrapers: Structure and Design*. New Haven, CT: Yale University Press, 2005.

Wengenmayr, Roland, and Thomas Bührke. *Renewable Energy: Sustainable Concepts for the Energy Change*, 2nd ed. Weinheim: Wiley-VCH, 2013.

Woyke, Elizabeth. *The Smartphone: Anatomy of an Industry*. New York: The New Press, 2014.

Youngblood, Norman. *The Development of Mine Warfare: A Most Murderous and Barbarous Conduct*. Westport, CT: Praeger Security International, 2006.

Zehner, Ozzie. *Green Illusions: The Dirty Secrets of Clean Energy and the Future of Environmentalism*. Lincoln: University of Nebraska Press, 2012.

JOURNALS AND PERIODICALS

Air and Space Smithsonian. This is probably the best overall journal for individuals interested in aviation, space, and related topics.

Automotive News. The leading source of news and information on all aspects of the automotive industry worldwide.

AutoWeek. A weekly publication geared to the general car enthusiast, which covers news, car reviews, motorsport competitions, and industry trends.

Aviation Week and Space Technology. This is the premier trade journal for both general and specialist readers that covers aviation, aerospace, aeronautics, and numerous other industry topics.

Bell Labs Technical Journal. This is the in-house scientific journal for scientists of Bell Labs/Alcatel-Lucent and is published by John Wiley & Sons. This publication has been the source of numerous contributions in areas as diverse as telephony and acoustics to wireless and data communications.

Civil Engineering. The flagship journal of the American Society of Civil Engineers which covers articles on engineering science and technology, history, projects, trends, current issues, public policy, law, and ethics.

CQ: Amateur Radio. This publication is focused on the practical aspects of this hobby and geared to active operators.

Electronic Design. Covers current topics in electronic design including automation, test and measurement, and communications. There are both short, practical articles and longer ones, providing advice and solutions.

ENR. The major trade publication for engineers in the construction industry. Covers news, information, and lists on all construction topics, including business management and the environment.

IEEE Spectrum. This periodical is the flagship publication of the IEEE (Institute of Electrical and Electronics Engineers) and presents major articles and information on developments in technology, engineering, and science.

Iron and Steel Technology. This technical journal contains news and articles for metallurgical, engineering, operating and maintenance personal of the iron and steel and allied industries. It is the official publication of the Association for Iron & Steel Technology.

Issues in Science and Technology. Discusses topics in public policy for readers at all levels, as it relates to science, engineering, and medicine.

Make: Technology on Your Time. This magazine is for do-it-yourself "technophiles" and includes challenging projects and applications of online how-to videos.

Manufacturing Engineering. The monthly trade journal published by the Society of Manufacturing Engineers that includes news, trends, product reviews, current issues, case histories and other topics.

Mechanical Engineering. The monthly journal of the American Society of Mechanical Engineers with informative articles and news on all aspects of mechanical engineering.

Modern Steel Construction. This periodical is the official publication of the American Institute of Steel Construction and is devoted exclusively to the design, fabrication, and construction of structural steel buildings and bridges.

MSDN Magazine. This publication provides comprehensive coverage of Microsoft technologies and connects with specialists for practical solution to real-world problems such as building applications for the desktop or mobile devices.

Nature. The most prestigious and most frequently cited weekly science journal that publishes major and significant advances and breakthroughs in all fields of science.

New Scientist. This British-based non-peer-reviewed international science publication contains current developments, news, reviews, and commentary on science and technology, as well as speculative articles, ranging from the technical to the philosophical.

The Oil and Gas Journal. The weekly trade publication that provides news, reviews, referenced papers, and statistics on all aspects of the oil and gas industry.

Popular Mechanics. This magazine will be of interest to the "do it yourself" individual in such areas as home improvement, automotive, outdoors, and technology.

Popular Science. This magazine covers the latest scientific and technological breakthroughs and developments for the non-specialist science reader, including inventions and consumer products and services.

Science. This is the flagship publication of the American Association for the Advancement of Science (AAAS) and includes original research articles, reports, analyses of current research and science policy, news, book reviews, and additional selections.

Scientific American. One of the most popular and well-known scientific publications that contains in-depth, authoritative, accessible, and timely articles that are of interest to both specialist and non-specialist readers.

Sierra. This magazine is published by the Sierra Club and includes articles of interest to individuals who care deeply about nature and enjoy exploring the outdoors while protecting the planet.

Smithsonian. A very popular and highly regarded magazine that covers history, science, nature, and art through a wonderful combination of writing and photography.

Technology Review. An important journal that covers emerging technologies and analyzes their significance for leaders in the field. It includes feature articles, essays, book and product reviews, and trends.

Wired. Covers the Internet and other digital topics from pop culture and media to business and technology. Includes feature articles, and columns that cover news and trends in fields as diverse as technology and entertainment.

WEBSITES

American Iron and Steel Institute (AISI). (http://www.steel.org/). The AISI site is the online resource for steel and covers topics as diverse as making steel and public policy priorities.

American Petroleum Institute (API). (http://www.api.org/). The site of the API provides information as diverse as meetings and training courses to standards, statistics and other publications.

American Physical Society (APS). (http://www.aps.org/). The site of the APS, which is an organization that advances the knowledge of physics through publications, meetings and events, programs, and related activities.

ASM International. (http://www.asminternational.org/). The site of American Society for Metals international, which is the world's largest association of metals-centric materials, engineers, and scientists.

Battelle Memorial Institute. (http://www.battelle.org/). The site of the Battelle Memorial Institute, which is a private, nonprofit, applied science and technology laboratory that partners with companies to develop new products and technologies such as the photocopier, product bar codes, and major atomic energy activities.

Biotechnology Industry Organization (BIO). (http://www.bio.org/) . The site of BIO, which is the world's largest biotechnology organization and is involved in research and development to heal, fuel, and feed the world.

Copper Development Association. (http://www.copper.org/). The Copper Development Association is the market development, engineering and information services arm of the copper industry.

Energy Information Administration (EIA). (http://www.eia..gov/). The EIA provides official energy statistics from the U.S. government.

Energy Star. (http://www.energystar.gov/). Energy Star, a U.S. Environmental Protection Agency (EPA) voluntary program, helps individuals and businesses save money and protect the environment through energy efficiency.

Food and Agricultural Organization (FAO) of the United Nations. (http://www.fao.org/). The site of the FAO of the United Nations has numerous information resources and links on agriculture, fisheries, forestry, hunger, sustainable development, and related topics.

Institute of Electrical and Electronics Engineers (IEEE). (http://www.ieee.org/). The site for IEEE, the world's largest professional association for the advancement of technology.

International Atomic Energy Agency (IAEA). (http://www.iaea.org/). The IAEA works for safe, secure, and peaceful uses of nuclear energy.

International Energy Agency (IEA). (http://www.iea.org/). The IEA is an organization of twenty-nine countries that works to ensure access to reliable, affordable, and clean energy.

Johns Hopkins University Applied Physics Laboratory (APL). (http://www.jhuapl.edu/). The APL conducts research, development, and analysis in numerous areas of technology and engineering.

Maker Faire. (http://www.makerfaire.com/). This is the site for the Maker Faire, which is a gathering of technology enthusiasts, educations, crafters, engineers, hobbyists, artists, authors, students, and exhibitors who attend these meetings to show what they have made and to share what they have learned.

National Academy of Engineering (NAE). (http://www.nae.edu/). The NAE provides engineering leadership in service and independent advice to the federal government on matters involving engineering and technology.

National Academy of Sciences (NAS). (http://www.nas.edu/). The NAS is an organization of the leading scientific researchers in the U.S. and provides objective science-based advice to the federal government on critical issues affecting the nation on matters related to science and technology.

National Aeronautics and Space Administration (NASA). (http://www.nasa.gov/). The site for NASA provides the latest news, images, and videos from America's space agency, pioneering the future in space exploration and scientific discovery.

National Institute of Standards and Technology (NIST). (http://www.nist.gov/). The site for NIST, which promotes innovation and industrial competitiveness by advancing measurement science, standards and technology through ongoing research and development.

National Inventors Hall of Fame. (http://www.invent.org/). Documents the inventions and life-changing achievements of selected U.S. patent holders and preserves their legacies for future generations.

National Nanotechnology Initiative. (http://www.nano.gov/). This is the site of the U.S. National Nanotechnology Initiative that supports nanotechnology research and development in academic, government, and industry laboratories in the United States.

National Resources Defense Council (NRDC). (http://www.nrdc.org/). The NRDC works to protect wildlife and wild places and to ensure a healthy environment for all life on earth.

Organovo. (http://www.organovo.com/). Organovo is a medical laboratory and research company. The site for this company focuses on designing and developing a range of tissue and disease models for medical research and therapeutic applications.

Rubber Manufacturers Association (RMA). (http://www.rma.org/). The site for the RMA, which represents tire manufacturers that produce tires in the U.S. with emphasis on tire safety and tire recycling.

Safer Car. (http://www.safercar.gov/). This site is dedicated to educating the public on how to purchase safe vehicles, and it offers information on vehicle recalls, car seats, and related topics.

Satellite Industry Association (SIA). (http://www.sia.org/). This trade association represents the leading global satellite operations, service and launch providers, and manufacturers and ground equipment suppliers. This site provides information on membership, events, news, resources, policy, and related topics.

Science—How Stuff Works. (http://science.howstuffworks.com/). This site has explanations and illustrations related to earth sciences, life sciences, and other wonders of the physical world.

Smart Grid. (http://www.smartgrid.gov/). This site is concerned with the implementation of newer technologies in the U.S. to meet the needs of the twenty-first century economy.

Smithsonian Institution. (http://www.si.edu/). This is the official website of the Smithsonian Institution, the world's largest museum and research facility.

U.S. Department of Energy (DOE). (http://www.energy.gov/). The site of the U.S. DOE, which provides significant information on all aspects of energy science and technology.

U.S. Environmental Protection Agency (EPA). (http://www.epa.gov/). The site of the U.S. EPA, which is the leading government body for protecting and improving the environment.

U.S. Fish and Wildlife Service (FWS). (http://www.fws.gov/). The FWS works to conserve, protect, and enhance fish, wildlife plants and their natural habitats.

U.S. Geological Survey (USGS). (http://www.usgs.gov/). The USGS provides information to describe and understand the earth; minimize loss of life and property from natural disasters; and, manage water, energy, mineral, and biological resources.

U.S. Patent and Trademark Office (USPTO). (http://www.uspto.gov/). The USPTO is the federal agency that grants U.S. patents and registers trademarks. The site has extensive information on getting started, applying, and maintaining both patents and trademarks and extensive learning and resource materials.

World Nuclear Association (WNA). (http://www.world-nuclear.org/). The WNA is the international organization that promotes nuclear energy and supports the companies that comprise the global nuclear industry. The site for this organization has an extensive information library of over 180 frequently updated pages on nuclear power and nuclear technologies.

Index

Note: (ill.) indicates photos and illustrations.

A

academic-generated research reports, 17–18
Acela Express, 259
acid rain, 332–33, 333–34
acid-free paper, 188
acidic paper, 49–50
active solar energy systems, 289–90
Advanced Mobile Phone System (AMPS), 77
aflatoxin, 130
Afsluitdijk, 211
Age of Scientific Revolution, 6
Agricola, Georgius, 16, 139
Aiken, Howard, 28, 28 (ill.)
air bags, 235–37, 236 (ill.)
air conditioner, 232, 310
air pollution
    acid rain, 332–33, 333–34
    Air Quality Index, 330–31
    asbestos, 336
    catalytic converter, 330
    Clean Air Act, 332
    definition, 328
    formaldehyde contamination, 336
    health effects of, 329, 329 (ill.)
    indoor, 334–35
    motor vehicle exhaust, 330
    nitrogen oxide emissions, 333
    primary air pollutants vs. secondary air pollutants, 329
    radon, 337
    scrubbers, 334, 334 (ill.)
    smog, 331 (ill.), 331–32
    sources of, 328
    sulfur dioxide emissions, 333
    volatile organic compounds (VOCs), 336–37
Air Quality Index, 330–31
Airbus A380, 251
aircraft
    Airbus A380, 251
    amphibian plane, 254

atomic bomb, 257
avionics, 252
B-17 Flying Fortress, 256
balloon, 251–52, 252 (ill.)
bird shot test, 253
black box, 252–53
blow outs, 251
Boeing 747-8 Intercontinental, 251
drones, 258, 258 (ill.)
fastest, 250, 250 (ill.)
Flying Tigers, 257
helicopter, 254–55
highest flying, 250–51
Hindenburg, 247–48, 248 (ill.)
hovercraft, 248
largest, 251
lift, 247
Mach number, 249
MiG designation, 257
nonstop, unrefueled, round-the-world flight, 249
nonstop transatlantic flight, 248–49
parachute jump, 252
seaplane, 254
Sopwith Camel, 256, 256 (ill.)
Spruce Goose (airplane), 253–54, 254 (ill.)
stealth technology, 257–58
supersonic flight, 249
wind tunnel, 253
Wright brothers, 247
airplanes. *See* aircraft
air-supported domes, 204–5
Akasaki, Isamu, 21, 22, 314
Akashi-Kaikyo bridge, 212, 212 (ill.)
alarm clock, 117 (ill.), 117–18
Alberts, Bruce, 23
Alcock, John W., 248
alerting devices, 418
algorithm, 29
alkali metals, 156
alkaline Earth metals, 156
Allen, Nelson, 382

alloy, 163–64
alluvial mining, 140
all-weather tires, 230–31
alphabets, 50–53
alternative fuel vehicle, 227
alternative fuels, 272–73
aluminum, 158–60, 160 (ill.), 359, 359 (ill.), 360
aluminum foil, 161
A.M., 114
AM radio stations, 62–63
Amano, Hiroshi, 21, 314
amber, 173
ambergris, 174
American Association for the Advancement of Science, 14
American Philosophical Society, 14
American Standard Code for Information Exchange (ASCII), 53
Ammann, Othmar H., 214
ammonia, 184
Ampère, André-Marie, 100
amphibian plane, 254
analog TV, 68–69
analytical engine, 27, 41
Ancient World, 6
Anderson, W. French, 386
Android operating system, 37
animal cloning, 398–400
animals, 372
Antarctic ozone hole, 342
Antheil, George, 64
anthrax, 129
antilock braking systems, 233
antisense technology, 381–82
Antonov An-225, 251
Appliance Labeling Rule, 312
application software, 35–36
April, 109
aqua regia, 183
arch bridge, 211
Arctic National Wildlife Refuge oil, 269
ard, 119
area, 97–98

433

Armstrong, Edwin Howard, 62
Armstrong, Lance, 388, 388 (ill.)
Armstrong, Neil, 8
Arrhenius, Svante, 337
Arthur M. Bueche Award, 22
artificial chromosome, 381, 383
artificial kidney, 421
asbestos, 151 (ill.), 151–52, 336
ASCII code, 53
assembly language, 42–43
assisted reproductive technology, 421–22
assistive listening devices, 417–18
assistive technologies, 416
Atomic and Electronic Age, 7
atomic bomb, 134–35, 135 (ill.), 257
atomic time, 101
August, 109
automated external defibrillators, 412
automated sequencing, 377
automatic transmission, 231
automobiles
    air bags, 235–37, 236 (ill.)
    air conditioner, 232
    all-weather tires, 230–31
    alternative fuel vehicle, 227
    antilock braking systems, 233
    automatic transmission, 231
    body number plate, 232
    braking distance, 233–34
    colors, 237–38
    cost to operate, 232–33
    early, 226
    electric, 227–28
    engine, 232
    fatal accidents, 234–35
    gasoline vs. ethanol and gasohol, 227
    horsepower, 226–27
    invention of, 225
    laser speed guns, 238
    license plates, 231–32
    Michelin tire, 229–30
    nuclear-powered, 227
    parking meter, 239
    police radar, 238
    registration, 232
    rumble seat, 231, 231 (ill.)
    safety innovations, 233
    safety recalls, 234
    seatbelts, 235
    side-impact air bags, 236–37
    skidding distances, 234
    Smart car, 229, 229 (ill.)
    snow tires, 230–31
    speed traps, 238–39
    taxicab, 239–40
    three-point safety belt, 235
    tires, 229–31
    tubeless tires, 230
    vehicle identification number (VIN), 232
    Visual Average Speed Computer and Recorder (VASCAR), 238
avionics, 252
avoirdupois measurements, 98–99
awards
    National Academy of Engineering, 21–22
    National Medal of Science, 23–24
    National Medal of Technology and Innovation, 22 (ill.), 22–23
    Nobel Prize, 20, 21
    Turing Award, 24–25
Axelrod, Robert, 23
axle, 120, 120 (ill.)

B

B-17 Flying Fortress, 256
Ba Na cable car, 261–62
Babbage, Charles, 27, 41
Babylonian calendar, 104
Bache, Alexander Dallas, 14
Bachman, Charles W., 23
Bacillus thuringiensis (Bt), 396
BackRub, 84
Backus, John, 43
Bacon, Francis Thomas, 291–92
Baekeland, Leo Hendrik, 181
Bain, Alexander, 75 (ill.), 76
Baird, John Logie, 67
bakelite, 181
Bakewell, Frederick, 76
Bakken, Earl, 411
Bakken Formation, 268
Ballard, Robert, 245–46
balloon, 251–52
Baltimore fire, 20
bar code, 56–57
Barbier, Charles, 51
Barnes, John Arundel, 88
Bartholdi, Frédéric-Auguste, 220, 221
basic oxygen furnace, 169–70
battery electric vehicles, 228
Bauer, George, 16
bauxite ore, 159–60
Bayer, Carl Josef, 160
Bayer process, 160–61
bazooka, 132
B.C.E., 111
Beaufort, Francis, 100
Becket, Frederick, 170
Bell, Alexander Graham, 74, 74 (ill.), 142
Bell, Lawrence, 249
benchmark, 100
Benedictus, Edouard, 177
Benz, Karl, 225
Benzinger, Theodor H., 409
Berenbaum, May, 23
Berg, Paul, 371
Berkeley Pit open-pit mine, 142, 354
Bernard M. Gordon Prize, 22
Berners-Lee, Tim, 80, 81
Bernoulli, Daniel, 247
Bernoulli's principle, 247
Bessemer, Henry, 168, 169 (ill.)
Bessemer process, 168–69
Beyer, Peter, 395
biblical measurements, 91
Big Bertha, 134, 134 (ill.)
billiard balls, 180 (ill.)
binary format, 29
Bingham Canyon Mine, 141 (ill.), 141–42
biochemical cycle, 322
biochemical genetic tests, 387
biodegradable plastics, 359–60
bioethics, 376
bioinformatics, 376
biological warfare, 129–30
biomass energy, 292–93
biometrics, 400–401
biopesticide, 396
biopharming, 393
bioprinting, 400
bioprospecting, 384
bioreactors, 357, 378–79, 379 (ill.)
bioremediation, 327
biotechnology and genetic engineering
    achievements, 371–72
    animal cloning, 398–400
    animals, 372
    antisense technology, 381–82
    artificial chromosome, 381, 383
    automated sequencing vs. manual sequencing, 377
    bioethics, 376
    bioinformatics, 376
    biometrics, 400–401
    biopesticide, 396
    biopharming, 393
    bioprinting, 400
    bioprospecting, 384
    bioreactors, 378–79, 379 (ill.)
    bioterrorism, 376
    breast cancer, 375
    Bt corn and monarch butterfly, 396
    Bt gene, 396
    cell culture, 376
    chimera, 375
    cloning, 397–400
    CODIS, 392
    commercial recombinant DNA technology, 372

crop plants, 393
definitions, 369–70
DNA amplification, 378
DNA analysis, 389–92
DNA fingerprinting, 389–90
DNA profiling, 384
DNA sequence analysis, 376–77, 377 (ill.)
DNA sequencing, 375
DNA typing, 384
Dolly (sheep), 400
earliest uses of, 371
ELISA test, 383
ENCODE project, 381
endangered species, 384–85
Flavr Savr tomato, 395
flow cytometry, 383
Frankenfood, 395
gene chip, 382–83
gene discovery, 381
gene doping, 388
gene gun, 382, 382 (ill.)
gene library, 380
gene probe, 380
gene therapy, 385 (ill.), 385–86
genes and patenting, 375
genetic disorders, 386
genetic testing, 387
genome, 373, 374
germline therapy, 392
GloFish, 397
GMO (genetically modified organisms), 393–95, 394 (ill.)
Golden Rice, 395–96
green fluorescent protein, 397
high throughput screening, 383
human cloning, 398
human gene therapy, 379–80
Human Genome Project, 373
Innocence Project, 390
Institute for Genomic Research, 374
knockout mouse, 393
microarray, 382
microbes, 372
multigene disorders, 386
NDIS (National DNA Index System), 392
PAGE (PolyAcrylamide Gel Electrophoresis), 383
polymerase chain reaction, 378
proteomics, 374–75
recombinant DNA, 378
restriction endonuclease, 380
restriction fragment length polymorphisms, 381
Romanov family, 391 (ill.), 391–92
Sanger technique, 377
September 11, 2001, terrorist attack DNA investigation, 391
single gene disorders, 386
Starlink corn, 396
Sugababe, 400
terminator gene technology, 388
transduction, 380
transformation, 378
transgenic animal transplants, 393
transposon, 373–74
vaccines, 385
vector, 379–80
viruses, 379–80
VNTR (variable number of tandem repeats), 388–89
xenotransplantation, 392–93
bioterrorism, 376
bird shot test, 253
Birt, Mildred, 67 (ill.)
Birtwhistle, W. K., 191
bit, 29
bituminous coal, 145
black box, 252–53
Blackberry, 86, 86 (ill.)
Blackbird, 250–51
Blackout Bomb, 136
Blackwell, David, 23
Blaese, Michael, 386
blogs, 87
blue light-emitting diodes, 314
boats and ships
by and large, 241
dead reckoning, 241
figurehead, 241
gopher wood, 241–42
hospital ship, 244
Liberty ship, 244
loadline, 244
mark twain, 242
Noah's ark, 241–42
nuclear-powered vessels, 246–47
passenger ship, 246
Queen Mary, 246
starboard, 241
steam-powered, 242–43, 243 (ill.)
Titanic, 245 (ill.), 245–46
tonnage, 243–44
tugboats, 244
Bockscar (aircraft), 257
body number plate, 232
Boeing 747-8 Intercontinental, 251
Boeing assembly plant, 200
Bohlin, Nils, 235
Boland, Thomas, 400
bone china, 179
booting, 37
Borchers, W., 170
Borglum, Gutzon, 221
Borglum, Lincoln, 221
boron carbide, 165
Bosch, Carl, 184
botulism, 129
Bowerman, William, 13
Bowie, Jim, 128
Bowie, Rezin, 128
Bowie knife, 128
Boyer, Herbert, 371
Bozzini, Philipp, 409
B.P., 111
Bradford, William, 188
Braille, Louis, 51
Braille alphabet, 51–52
brain-computer interfaces, 419
braking distance, 233–34
brass, 165
Braun, Ferdinand, 66
Brearley, Harry, 170
breast cancer, 375
breast diseases, 408
breathalyzer, 126–27, 127 (ill.)
bridges
Brooklyn Bridge, 213
causeway, 215 (ill.), 215–16
covered, 212
floating, 214–15
Golden Gate Bridge, 216
kissing bridge, 211
longest spans, 212 (ill.), 212–13
Mississippi River, 216
Pennsylvania, 214
pontoon, 215
suspension, 214
types of, 211–12
Brin, Sergey, 84, 84 (ill.)
British Petroleum offshore oil spill, 348
broadband Internet service, 81
broadband technologies, 81–82
bronze, 164
Brooklyn Bridge, 213
Brown, Adolphe F., 126
Brown, Arthur W., 248
Brown, Louise, 422
Brown, Robert A., 22
Brown, Roy, 255
Brown, Samuel, 225
brownfields, 354
browsers, 82
Bruhn, Friedrich Wilhelm, 239
Brundtland, Gro Harlem, 366
Bryant, Gridley, 260
Bt corn, 396
Bt gene, 396
BTU, 305
Bueche, Arthur M., 22
bug, 45
buildings. See also structures
air-supported domes, 204–5
cable-supported domed roof, 204
chimney, 196

crown molding, 196
domes, 203–5
doorjamb, 196
evergreen tree, 201
flue, 196
geodesic dome, 203
green building, 193–94
green roofs, 195 (ill.), 195–96
International Code Council, 197
largest, 200
Leaning Tower of Pisa, 199 (ill.), 199–200
LEED (Leadership in Energy and Environmental Design), 194–95
malls, 199
megatall, 201
Mellon Arena, 203
O_2 Dome, 204 (ill.), 205
Pentagon, 200
pisé, 197–98
rammed earth, 197–98
retractable roof stadium, 203–4
R-value, 196–97
self-supporting structure, 202
shopping center, 199
skyscrapers, 200–202
STC (Sound Transmission Class), 197
supertall, 201
tallest, 201–2
World Trade Center, 202
yurts, 198, 198 (ill.)
Bulletin of the Atomic Scientists, 118
bulletproof glass, 177
Burj Khalifa, 201, 202 (ill.)
Burns, Bob, 132
Burton, William, 270
Bushnell, David, 131
butterfly, 396
by and large, 241
Byron, Augusta Ada, 41
Byron, Lord, 41
byte, 29

C

C computer language, 42–43
cable car, 261–62
cable modem, 81
cable-stayed bridge, 213
cable-supported domed roof, 204
cabooses, 260, 260 (ill.)
calcium carbide, 165
calendar year, 103
calendars, 104–6
California Geysers project, 297
caller ID, 75
calorie, 264–65
Campbell, Charles J., 423

Campbell-Swinton, Alan, 66
cancer, 406
cannel coal, 145
Cannon King, 132, 134
cantilever bridge, 212, 213
capacitive screens, 39
Cape Wind project, 296
carat, 154
carbide compounds, 164–65
carbon black, 185–86
carbon cycle, 323
carbon dioxide, 338–39, 340
carbon management, 340
carbon sink, 340
carbon-carbon composites (CCCs), 180
cardboard, 189
Cardullo, Mario W., 65
Carnot, Nicolas, 120
Carothers, Wallace H., 191
Carrel, Alexis, 393
carrier testing, 387
cars. *See* automobiles
Carson, Rachel, 317, 318, 319, 319 (ill.), 352
cashmere, 190
catalytic converter, 330
CAT/CT scans, 405 (ill.), 405–6
causeway, 215 (ill.), 215–16
CD-R, 79
CD-RW discs, 79
CDs, 35, 79–80
cell culture, 376
cell phones, 77–78
Celsius, Anders, 100
central processing unit (CPU), 32
century, 103
ceramics, 179
cesium beam, 114
Chang, Annie, 371
Chaotianmen Bridge, 213
Chapin, Daryl, 291
Chardack, William, 411
Charles Stark Draper Prize, 21–22
Charlotte Dundas (steamboat), 242
chartered railroad, 260
chat rooms, 88
Chaucer, Geoffrey, 18
chemical engineering, 1–2
chemical releases, 350
chemical warfare, 128–29
Chennault, Claire Lee, 257
Chernobyl accident, 285 (ill.), 285–86
chess, 47
chimera, 375
chimney, 196
China Syndrome, 284–85
China time zones, 112
Chinese calendar, 104
chlorofluorocarbons, 341–42

Chorin, Alexandre J., 24
chromated copper arsenite, 185
chromosomal genetic tests, 387
Churchill, Winston, 130
cinnabar, 152
civil engineering, 1–2, 205
Civil War, 131
Clark, Graeme M., 22
Clean Air Act, 332
clean coal technologies, 339
Clean Water Act of 1972, 344–45
Clemens, Samuel L., 242
Cleopatra, 155
clepsydras, 116
Clermont (steamboat), 242–43
client/server principle, 82
climate change
Antarctic ozone hole, 342
carbon dioxide, 338–39, 340
carbon management, 340
carbon sink, 340
chlorofluorocarbons, 341–42
clean coal technologies, 339
enhanced greenhouse effect, 339
greenhouse effect, 337–38
greenhouse gases, 337 (ill.), 338–39
hole in ozone layer, 342–43
Kyoto Protocol, 343
measurement of, 339–40
Montreal Protocol, 343
ozone, 341 (ill.), 341–43
physical indicators of, 343
clockwise, 116
cloning, 397–400
clostridium perfringens, 130
cloverleaf interchange, 209
coal. *See also* mining
advances in mining technology, 146
bituminous, 145
cannel, 145
China as largest producer of, 146, 146 (ill.)
damp, 145
fly ash, 146
formation of, 145
fossil fuel, 265, 266, 266 (ill.)
red dog, 147
reserves, 146–47
types of, 145
underground mining methods, 145
uses for, 146
coal gasification, 339
coast-to-coast highway, 208
COBOL (common business oriented language), 43
cochlear implant, 416–17, 417 (ill.)

codes
  ASCII, 53
  definition, 21
  International Standard Book
    Number (ISBN), 58–59
  Navajo Code Talkers, 54, 54 (ill.)
  palindromic square, 55–56
  Price Look-Up (PLU) codes, 57
  QR code, 57, 57 (ill.)
  Standard Book Numbering, 58
  UPC, 56–57
codes of practice, 19
CODIS, 392
cogeneration, 278
cognitive disabilities, 416
Cohen, Stanley, 371
coins, 167
color standarization scheme, 150
colorblindness, 416
colored glass, 176
Colt, Samuel, 131
Colt revolver, 131
coltan, 162
commercial recombinant DNA technology, 372
commercial sector, 265
communications satellite, 73
compact discs (CDs), 35, 78–79
compact fluorescent lamps, 313
compass, 119–20
composite materials, 182–83
computer virus, 45–46
computers
  Aiken, Howard, 28
  algorithm, 29
  Android operating system, 37
  assembly language, 42–43
  bit vs. byte, 29
  booting, 37
  bug, 45
  C computer language, 42–43
  central processing unit (CPU),
    32
  COBOL (common business ori-
    ented language), 43
  CPU clock, 32
  3-D printing, 38, 38 (ill.)
  data mining, 57–58
  disk operating system (DOS), 36
  "Do not fold, spindle, or muti-
    late," 42
  expert system, 37
  flash drives, 35
  floppy disks, 34
  FORTRAN (FORmula TRANsla-
    tor) computer language, 43
  games, 46 (ill.), 46–47
  glitch, 44–45
  graphical user interface (GUI),
    36–37
  hard drive, 33

  hardware vs. software, 30
  Hopper's Rule, 40
  impact of, 29–30
  Intel processors, 31
  invention of, 27–28
  Java, 44
  language, 41–44
  Linux operating system, 36
  Lisa (personal computer), 37
  machine language, 42–43
  MANIAC, 28
  Moore's Law, 31
  mouse, 33 (ill.), 33–34
  object-oriented computer lan-
    guage, 43
  open-source software, 44
  operating system, 36
  operating system software vs.
    application software, 35–36
  PASCAL computer language, 43
  phishing, 58
  pixel, 38–39
  portable, 38
  portable storage media, 34–35
  programmers, 41
  programming languages, 41–42
  punched cards, 40 (ill.), 40–41,
    42
  RAM vs. ROM, 32–33
  sensors, 39
  silicon chip, 30–31, 31–32
  storage space, 29
  10-codes, 54–55
  touch screen, 39–40
  The Turk, 47, 47 (ill.)
  Universal Serial Bus (USB), 34
  virus, 45–46
  Zip® disks, 34
concrete-arch bridge, 213
Conti, Prince Piero, 297
continuous casting steel, 170–71
continuous glucose monitoring, 415
contour mining, 140
conventional gas, 276
cooling degree day, 305
Cooper, Martin, 77
Coordinated Universal Time, 112
copper, 162–63, 163 (ill.), 164, 165
Coptic calendar, 104
copyrights, 12
cord of wood, 187, 293
Cormack, Allan M., 406
Corning Glass Works, 178
CorningWare, 179–80
corrosive materials, 349
corrugated cardboard, 189
COR-TEN steel, 171
corundum, 150
cosmetics, 155
covered bridges, 212
CPU clock, 32

Craford, M. George, 22
creosote, 185
Crick, Francis, 373
Cronkite, Walter, 73
crop plants, 393
crop rotation, 367–68
crown glass, 175
crown molding, 196
cruise missile, 136
Crutzen, Paul, 342
cryptanalysis, 53–54
cubic centimeters, 121
cubic zirconia, 155
cubic zirconium, 155
Cugnot, Nicolas-Joseph, 225
Culin, Curtis G., 255
culin device, 255
Cullinan, Thomas M., 155
Cullinan Diamond, 154–55
Culver, Kenneth, 386
Cunningham, Ward, 87–88
curbside recycling program, 358
Curie, Marie, 100, 157
Curie, Pierre, 157
Curtiss, Larry, 409
cybernetics, 125

D

da Vinci Surgical System, 419
Daimler, Gottlieb, 225, 240
Damadian, Raymond, 407
damp, 145
dams, 220
data mining, 57–58
Davy, Humphry, 144, 312, 312 (ill.)
dawn, 102
daybreak, 101
Daylight Savings Time, 111
days, 108
DDT, 352, 352 (ill.)
De re Metallica (Agricola), 16, 16
  (ill.), 139
dead reckoning, 241
deadweight tonnage, 243
December, 110
"Deep Blue" chess computer, 47
defibrillator, 411–12, 412 (ill.)
deforestation, 323 (ill.), 323–24
Deike, George H., 144
Delamare-Deboutteville, Edouard,
  225
Delano, Edward, 209
design patents, 9
Devol, George, 124
Dewees, William, 188
diabetes, 414
diagnostic testing, 387
dial-up Internet service, 81
dialysis machines, 421
diamonds, 152–55, 154 (ill.)

diapers, 188–89
Dickson, J. T., 191
difference engine, 27
Digges, Leonard, 101
Digital Subscriber Line (DSL), 81
digital technology, 77
digital TV, 68–69
digital video recording (DVR), 71–72
digital watch, 118
dikes, 211
dimensional standards, 19
dirty bomb, 137
disabilities, 416
disaster prevention, 20
discarded tires, 364
dishwasher, 363–64
disk operating system (DOS), 36
displacement tonnage, 243
distance to the horizon, 100
diverging diamond interchange, 209, 209 (ill.)
DNA amplification, 378
DNA analysis, 389–92
DNA chip technology, 386
DNA fingerprinting, 389–90
DNA profiling, 384
DNA sequence analysis, 376–77, 377 (ill.)
DNA sequencing, 375
DNA typing, 384
"Do not fold, spindle, or mutilate," 42
documentation standards, 19
Dolby, Ray M., 70
Dolby noise reduction system, 70
Dolly (sheep), 400
domes, 203–5
donkey engine, 122
doomsday clock, 118–19
doorjamb, 196
Doppler, Christian, 238
Dorsey, Jack, 89
Dougherty, Dale, 8
Drake, Edwin L., 266
Drake oil well, 266–67
Draper, Charles Stark, 21–22
Dresselhaus, Mildred, 13
drift mines, 143
driving speed, 306
drones, 258, 258 (ill.)
drug delivery systems, 413
dry ice, 184 (ill.), 184–85
dry measures, 96–97
Dubus-Bonnel, Ignace, 183
Dupuis, Russell, 22
Duryea, Charles, 226
Duryea Motor Wagon Company, 226, 226 (ill.)
dusk, 102
DVDs, 35, 79–80

dynamic time, 101
dynamite, 186

E

Eads, James B., 216
Early Bird, 73
Earth Day, 319
earth station, 68
Eckert, John Presper, Jr., 28
ecological footprint, 316
eCycling, 365
Edison, Thomas, 11, 13, 144, 312, 409
Edwards, Robert, 422
Egyptian calendar, 104
Egyptian measurements, 91
Ehrlich, Paul R., 315
Eiffel, Gustave, 219
Eiffel Tower, 219, 219 (ill.)
Eisenhower, Dwight D., 206
electric arc furnace, 169–70
electric automobiles, 227–28
Electrical Age, 6–7
electrical appliances, 310–12
electrical engineering, 1–2
electrical muscle stimulation, 420
electricity
    fuel sources, 301
    generation of one killowatt hour, 307
    generator, 301
    home use, 308, 309
    hydropower, 298–300
    movement of, 301, 302 (ill.)
    Niagara Power Plant, 299
    nuclear power plant, 279
    power grids, 302
    power outages, 302
    power plant, 301
    renewable and alternative energy sources, 290
    Smart Grid, 302–3
    smart meters, 303
    transmission lines, 302
    wind turbines, 294
electronic products and devices, 308, 364 (ill.), 364–66
electronic stability control systems, 233
elements, 156
elevators, 20
Elion, Gertrude Belle, 13
ELISA test, 383
Ellis, Charles, 216
Elsener, Karl, 128
email, 85–86
emerald, 151
Empire State Building, 202, 203
ENCODE project, 381
endangered species, 384–85

endoscope, 408–9, 409 (ill.), 410
energy. *See also* electricity; fossil fuels; nuclear energy; renewable and alternative energy sources
    Appliance Labeling Rule, 312
    blue light-emitting diodes, 314
    BTU, 305
    calorie, 264–65
    conservation, 309–14
    consumption, 305–9
    cooling degree day, 305
    definition, 263
    driving speed, 306
    economic sector, 265
    electrical appliances, 310–12
    electronic devices, 308
    Energy Star, 309–10
    energy-efficient light bulbs, 313–14
    fluorescent light, 314
    fuel equivalents to produce one quad, 304
    gas lighting, 313
    gas mileage, 306
    gasoline, 306
    heating degree day, 305
    heating values of fuels, 304
    incandescent light bulb, 312, 314
    insulation, 310
    lighting brightness, 312–13
    lowering thermostat, 310, 310 (ill.), 311
    measurements of, 264, 303–5
    power, 264
    raising setting on air conditioner, 310
    residential end-use, 308–9
    role of, 263
    source comparisons, 303–4
    sources, 264
    types of, 263
    underinflated tires, 306, 306 (ill.)
    visible-spectrum LED, 314
    weight per gallon of common fuels, 303
Energy Star, 309–10
energy-efficient light bulbs, 313–14
Engel, Joel S., 77
Engelbart, Douglas C., 33–34
Engelberger, Joseph, 124
engine, 232
engineering
    accidental discoveries, 3
    branches of, 1–2
    first recognized as profession, 2
    twentieth-century, 7–8
engineers, 1, 2
English Channel tunnel, 210
enhanced greenhouse effect, 339

Enigma cipher machine, 53–54
Enola Gay (aircraft), 257
Enterprise, USS, 246–47
environment. *See also* air pollution; climate change; hazardous waste; land use; recycling; solid waste; water pollution
    Carson, Rachel, 317, 318, 319, 319 (ill.)
    crop rotation, 367–68
    Earth Day, 319
    ecological footprint, 316
    environmental impact statements, 322
    Environmental Protection Agency, 319
    impact of technology on, 315
    legislation, 320–21
    problems, 316
    scientific method, 315–16
    sustainability, 366–68
    sustainable materials management, 322
    technological milestones, 316–18
environmental impact statements, 322
Environmental Protection Agency, 287, 319
enzyme-linked immunosorbent assay, 383
Ephemeris Time, 101
Ephemeris Time scale, 112, 114
Ereky, Karl, 370
erythropoietin, 388
escalators, 20
essential oils, 186–87
estuary, 327
ethanol, 227, 274
European Article Number (EAN), 56–57
eutrophication, 327–28
Evelyn, John, 316
evening, 102
evergreen tree, 201
expert system, 37
extended-range electric vehicles, 228
external defibrillator, 411–12
external hard drives, 34–35

F

Fabry, Charles, 317
Facebook, 89
Fahrenheit, Gabriel, 100
fall, 110
Faraday, Michael, 301
Farman, Joseph, 342
Farnsworth, Philo T., 67
Fat Man, 135
fathoms, 242

fax machine, 75–76
feathers, 99, 99 (ill.)
February, 109
female tanks, 131
Ferebee, Thomas W., 257
Ferris, George Washington Gale, 221
Ferris wheel, 221–22
ferrous metals, 167
fetal image-guided surgery, 423
fetal surgery, 422
Fetendo fetal surgery, 423
fetoscopic surgery, 423
Feynman, Richard, 3
fiber optic technology, 81
fiberglass, 183
fiber-optic cable, 76
Fieser, Louis, 128
figurehead, 241
File Transfer Protocol (ftp), 82
fine china, 179
fingerstick glucose tests, 415
fire hydrant nozzles, 125
fireworks, 186
fission, 278–79
flail, 128
Flanigen, Edith M., 23
flash drives, 35
flash memory sticks, 35
flat-panel screens, 69
Flavr Savr tomato, 395
float glass process, 175–76
floating bridges, 214–15
floppy disks, 34
floral clock, 116
flow cytometry, 383
Flowers, T. H., 28
flue, 196
flue gas desulfurization technology, 334
fluorescent light, 314
fly ash, 146
Flyer (airplane), 247
Flying Tigers, 257
FM radio stations, 62
Fogarty, Thomas J., 23
fool's gold, 166
Ford, Henry, 226, 228
Ford Nucleon, 227
forensic testing, 387
Forest, Lee de, 61
forest area, 323
Forgács, Gabor, 400
formaldehyde contamination, 336
FORTRAN (FORmula TRANslator) computer language, 43
Fortrel EcoSpun, 361
Fossett, Steve, 251–52, 252 (ill.)
fossil fuels
    alternative fuels, 272–73
    Arctic National Wildlife Refuge oil, 269

Bakken Formation, 268
    Canada as largest oil supplier to U.S., 267
    cleanest, 265
    coal, 265, 266, 266 (ill.)
    cogeneration, 278
    conventional gas, 276
    costs of, 273
    Drake oil well, 266–67
    ethanol, 274
    gasohol, 273–74
    gasoline, 270–72
    gasoline additives, 271
    hydraulic fracturing, 268 (ill.), 269, 277
    hydrocarbon cracking, 270
    leaded gasoline, 270–71
    lead-free gasoline, 270–71
    liquid natural gas, 275
    natural gas, 265, 274–76, 275 (ill.)
    North Dakota crude oil, 267, 268–69
    nuclear reactors, 278–79
    octane numbers, 271–72, 272 (ill.)
    offshore drilling, 269
    oil, 265, 266–67, 268–70
    oil fields, 266–67
    Pennsylvania crude oil, 269
    pipeline system, 275
    refining process, 269–70
    reformulated gasoline, 271
    renewable and alternative energy sources vs., 288
    reserves, 278
    shale, 267
    shale gas, 276–77
    synthetic fuels, 277
    Texas crude oil, 267
    unconventional gas, 276
Founders Award, 22
4G LTE, 78
four-stroke engine, 121
Fox River hydroelectric-generating plant, 299
fracking, 268 (ill.), 269, 277
Frankenfood, 395
frankincense, 174
Franklin, Benjamin, 14
Fregin, Douglas, 86
French republican calendar, 105–6
freshwater depletion and shortage, 345
Friday, 108
Fried, Henry, 116
Frigg, 108
Fritz J. Russ and Dolores H. Russ Prize, 22
Frontinus, 18
ftp, 82

fuel cell, 291–92
fuel-cell electric vehicles, 228
Fukushima nuclear accident, 286 (ill.), 286–87
Fuller, Buckminster, 203
Fuller, Calvin, 291
fuller's earth, 172
Fulton, Robert, 242–43
funicular railway, 261
fuzzy search, 83

G

Gabor, Dennis, 122
Gage, Andrew, 411
galena, 151
Galvani, Luigi, 165
gamete intrafallopian transfer, 422
gandy dancer, 261
garbage, 354
garbage incinerator, 355
Gardiner, Brian, 343
Garnerin, André-Jacques, 252
gas lighting, 313
gas mileage, 306
gasohol, 227, 273–74
gasoline, 227, 270–72, 306, 338
gasoline additives, 271
Gateway Arch, 216–17, 217 (ill.)
Gatling, Richard J., 132
Gatty, Harold, 249
Geiger, David, 205
Gelsinger, Jesse, 385
gene chip, 382–83
gene cloning, 369
gene doping, 388
gene gun, 382, 382 (ill.)
gene library, 380
gene probe, 380
gene therapy, 385 (ill.), 385–86
generator, 301
generic product name, 11–12
genetic disorders, 386
genetic engineering. See biotechnology and genetic engineering
genetic testing, 387
genetically modified organisms (GMO), 393–95, 394 (ill.)
genome, 373, 374
genomic medicine, 403–4
geodesic dome, 203
George III, King, 359
geosynchronous orbit, 73
geothermal energy, 297–98
geothermal power plant, 297–98, 298 (ill.)
German silver, 167
germline therapy, 392
Ghawar field, 267
Gibbon, John H., Jr., 412–13

Gilbert, Walter, 371, 373, 373 (ill.), 375
Ginsburg, Charles, 70
glass, 175–78
glass blocks, 177
glass windows, 177
glazes, 179
Gleevac, 404
glitch, 44–45
GloFish, 397
GMOs (genetically modified organisms), 393–95, 394 (ill.)
Gogh, Lieuwe van, 400
Gogh, Theo van, 400
Gogh, Vincent van, 400
gold, 99, 99 (ill.), 165–66
gold leaf, 166, 166 (ill.)
Golden Gate Bridge, 216
Golden Rice, 395–96
Goldman, Leon, 423
Goldwasser, Shafi, 25
Goodlin, Chalmers, 249
Google, 83–84
gopher wood, 241–42
Gordon, Bernard M., 22
Gore, Al, 80
Gosling, James, 44, 44 (ill.)
government-sponsored research reports, 17–18
Grand Coulee Dam, 300
grandfather clock, 116
graphical user interface (GUI), 36–37
Gray, Elisha, 74
gray water, 345
Great Pyramid at Giza, 216
Greatbatch, Wilson, 411
Greek measurements, 91
green building, 193–94
green electronics, 368
green fluorescent protein, 397
Green Globes, 193–94
green product, 368
green roofs, 195 (ill.), 195–96
greenhouse effect, 337–38
greenhouse gases, 337 (ill.), 338–39
Greenwich Mean Time, 112
Grégoire, Marc, 182
Gregorian calendar, 103–4, 105, 106, 108
Gregory XIII, Pope, 103, 107–8
grey literature, 16–17
grocery bags, 361
gross tonnage, 243
Grove, William, 291
Groves, Leslie R., 134
Gura, Victor, 421
Gurdon, John B., 398
Gurevich, Mikhail I., 257

H

Haber, Fritz, 184
hacker, 86–87
Hale, Arthur, 209
Hale, Gerald A., 239
Hall, Charles Martin, 161
Halley, Edmond, 15, 316
Hall-Héroult process, 160
Hallidie, Andrew S., 261
halogen incandescents, 313
Harari, Eli, 23
hard drive, 33
hardware, 30
Harley-Davidson, 240
Harrison, Michael R., 422
Harvard Mark I computer, 28
Haüy, Valentin, 51–52
Haven, Charles D., 177
Hayes, Denis, 319
Haynes, Elwood, 170
hazardous waste
 brownfields, 354
 chemical releases, 350
 classification of, 349
 DDT, 352, 352 (ill.)
 Persistent Organic Pollutants, 351–52
 polychlorinated biphenyls (PCBs), 352
 polyvinyl chloride (PVC) plastics, 352–53
 source reduction activities, 350
 Superfund Act, 353
 Superfund sites, 353 (ill.), 353–54
 Toxic Release Inventory, 349–50
 Toxic Substances Control Act, 353
HD Radio® Technology, 63
HDPE (high-density polyethylene), 360
hearing aids, 416
hearing loss, 416
heart-lung machine, 412–13
heating degree day, 305
heat-resistant glass, 178, 178 (ill.)
heavy truck, 239
Heilman, M. Stephen, 412
helicopter, 254–55
Henry I, King of England, 93, 93 (ill.)
Héroult, Paul Louis-Toussaint, 161
Hertz, Heinrich, 32
Hetrick, John W., 235
high technology centers, 4
high throughput screening, 383
high-density polyethylene (HDPE), 360
highway toll collection system, 66
Higinbotham, William, 46

Hindenburg, 247–48, 248 (ill.)
Hindu calendars, 105
Hirschowitz, Basil, 409
Hitchings, George, 13
Hitler, Adolf, 132
Hochmair, Erwin, 22
Hochmair-Desoyer, Ingeborg J., 22
Holdren, John P., 315
hole in ozone layer, 342–43
Hollerith, Herman, 40–41, 42
holography, 122
Holonyak, Nick, Jr., 22, 314
Home Box Office (HBO), 68
Home Insurance Company Building, 200–201
Homer M. Hadley Memorial Bridge, 214
Honda Civic GX Natural Gas Vehicle, 227
Hood, Leroy, 377
Hood Canal Bridge, 214
Hoover, Herbert, 16
Hoover, Lou Henry, 16
Hoover Dam, 220
Hopkins, Samuel, 10
Hopper, Grace Murray, 41, 43, 45
Hopper's Rule, 40
Hopps, John, 411
Horse Hollow Wind Energy Center, 295
horsepower, 226–27
Horton, Joseph W., 117
hospital ship, 244
Hounsfield, Godfrey N., 406
hovercraft, 248
Howard, Henry Taylor, 68
http, 82
Hughes, Howard, 253–54
Hull, Charles, 38
human cloning, 398
human gene therapy, 379–80
Human Genome Project, 373
Humvee, 255
Hunkapiller, Michael, 377
Hunter, Matthew A., 162
Hurst, George Samuel, 39
Hutchins, Levi, 117
Huygens, Christiaan, 116
Hyatt, John Wesley, 180–81
hybrid electric vehicles, 228
hydraulic fracturing, 268 (ill.), 269, 277
hydrocarbon cracking, 270
hydroelectric power, 300
hydrogen peroxide, 184
hydrologic cycle, 322
hydropower, 298–300
Hypertext Transfer Protocol (http), 82

I

IEDs (improvised explosive devices), 136–37, 137 (ill.)
ignitable materials, 349
Imhotep, 2, 2 (ill.)
impaired hearing, 416
Imperial Yard, 19
improvised explosive devices (IEDs), 136–37, 137 (ill.)
in vitro fertilization, 421–22
incandescent light bulb, 312, 314
indoor air pollution, 334–35
Industrial Revolution, 6, 6 (ill.)
industrial robots, 123–24
industrial sector, 265
industrial waste, 344 (ill.)
information superhighway, 80
Innocence Project, 390
Institute for Genomic Research, 374
Institute of Medicine, 14, 15
insulation, 310
insulin pumps, 414
Intel processors, 31
interconnections, 302
Intergovernmental Panel on Climate Change, 343
internal combustion engine, 120, 121 (ill.)
International Building Code, 21
International Code Council, 21, 197
International Date Line, 112
International Fixed calendar, 105
International Radiotelephone Spelling Alphabet, 50–51
International Residential Code, 21
International Standard Book Number (ISBN), 58–59
Internet
    Blackberry, 86, 86 (ill.)
    blogs, 87
    broadband service, 81
    broadband technologies, 81–82
    chat rooms, 88
    client/server principle, 82
    definition, 80
    dial-up service, 81
    email, 85–86
    Facebook, 89
    fuzzy search, 83
    Google, 83–84
    hacker, 86–87
    information superhighway, 80
    podcasts, 88
    public-key cryptography, 84–85
    search engine, 83
    SixDegrees, 88
    social networking, 88–89
    spam, 86
    top-level domain (TLD) suffixes, 83

Twitter, 89
Uniform Resource Locator (URL), 82–83
usage changes, 85
Web 2.0, 87
wikis, 87–88
wireless fidelity (WiFi), 81–82, 82 (ill.)
World Wide Web, 80–81
Interstate Highway System, 205–6
inventions, 13
iontophoresis, 414
IP address, 82–83
Iron Age, 168
iron ore deposits, 168
Ironbridge power plant, 293
ISBN system, 58–59
Itaipú hydroelectric power plant, 300
Ivanpah Solar Energy Generating System, 291

J

jackhammer, 125
Jacquard, Joseph Marie, 40
Jacquet-Droz, Henri-Louis, 124
Jacquet-Droz, Pierre, 124
January, 109
January 1 as first day of year, 103–4
Japanese calendar, 105
Jarvis, Richard, 78
Java, 44
Jefferson, Thomas, 13, 13 (ill.), 96, 221
Jeffreys, Alec, 389
Jenney, William Le Baron, 200
Jersey barrier, 210
Jewish calendar, 104
Jones, Albert, 189
Jost, A., 370
joule, 264–65
Joule, James Prescott, 100, 264
Julian calendar, 105, 108
Julian Day calendar, 105
Julian Day Count, 106
Julius Caesar, 103, 103 (ill.), 107–8
July, 109
June, 109

K

Kailath, Thomas, 24
Kaiser, Henry J., 244, 253–54
Kamen, Dean, 240
Kasparov, Garry, 47
Kay, Alan, 43
KDKA (Pittsburgh, Pennsylvania), 61–62
Keith, Carl D., 330

Kelly, William, 168, 169 (ill.)
Kempelen, Wolfgang von, 47
Kennedy, John F., 74
Kevlar, 191
Khwarizmi, Muhammad ibn Musa al-, 29
kidney, 421
Kilby, Jack, 30–31
kilogram, 94–95
kinetic energy, 263
King, Mary-Claire, 375
Kingdom Tower, 201
kissing bridge, 211
Klinman, Judith P., 24
Knight, J. P., 209
knives, 128
knockout mouse, 393
Koechlin, Maurice, 219
Koester, Charles J., 423
Kolff, Willem, 421, 421 (ill.)
Kroll, William J., 162
Krupp, Alfred (grandfather), 132
Krupp, Alfred (grandson), 132, 134
Krupp, Bertha, 134
Krupp, Friedrich, 132, 134
KVLY-TV transmitting tower, 202
Kwoh, Yik San, 419
Kwolek, Stephanie, 191
Kyoto Protocol, 343

L

Lacey V. Murrow-Lake Washington Bridge, 214
lake color, 328
Lake Pontchartrain Causeway, 215 (ill.), 215–16
Lamarr, Hedy, 64, 64 (ill.)
Lamport, Leslie, 25
land use
    biochemical cycle, 322
    bioremediation, 327
    carbon cycle, 323
    deforestation, 323 (ill.), 323–24
    estuary, 327
    eutrophication, 327–28
    forest area, 323
    hydrologic cycle, 322
    lake color, 328
    phytoremediation, 328
    rain forest, 324
    tropical forests, 324–25
    tropical rain forest, 324
    watershed, 327
    wetlands, 325–27, 326 (ill.)
landfill, 361
landfill gas, 357
landfill gas emissions projects, 357
landfills, 356 (ill.), 356–57
Langen, Eugen, 120
laparoscopic surgery, 419, 420 (ill.)

Larderell, Italy, power plant, 297
Las Vegas High Roller, 222, 222 (ill.)
laser, 423
laser speed guns, 238
lasers, 413
LASIK, 423–24
Laurer, George, 56–57
Lauterbur, Paul, 407–8
Lazaridis, Mike, 86
Lazarus, Emma, 221
LCD technology, 69–70
leaded gasoline, 270–71
Leadership in Energy and Environmental Design (LEED), 194–95
lead-free gasoline, 270–71
Leaning Tower of Pisa, 199 (ill.), 199–200
leap second, 112, 114
leap year, 108
Lebon, Philippe, 313
LEED (Leadership in Energy and Environmental Design), 194–95
Leibniz, Gottfried Wilhelm, 27
Leith, Emmett, 122
Lemoine, M. Louis, 125
Lenin (ship), 247
Lenoir, Jean-Joseph-Étienne, 120
Lenoire, J. J. Étienne, 225
Leonsis, Ted, 89
Leopold, Aldo, 317
lever, 120, 120 (ill.)
Levinson, Arthur, 23
Liberty ship, 244
license plates, 231–32
Lichtenberg, Georg Christoph, 127
Lichtenberg figures, 127
LifeVest, 412
lift, 247
light bulbs, 312–14
light emitting diodes, 313
lighting brightness, 312–13
Lincoln, Abraham, 9–10, 14, 221
Lindbergh, Charles A., 249
Lingzan, Liang, 118
Linotype machine, 50, 50 (ill.)
Linux operating system, 36
liquid measures, 96–97
liquid natural gas, 275
Lisa (personal computer), 37
Little Boy, 135
Littleton, Jesse T., 178
"live via satellite" television broadcast, 73–74
Livingston, Robert, 242
loadline, 244
local noon, 111
long term evolution (LTE), 78
long ton, 97
Louis XV, King of France, 384
Louis XVI, King of France, 384
Love Canal, 353, 353 (ill.)

low flow toilets, 363
low vision, 416
low-emissivity (low-e) glass, 178
Lowy, Douglas, 23

M

MacAdam, John Louden, 205
macadam roads, 205
Mach number, 249
machine gun, 132, 133 (ill.)
machines. *See also* tools
    breathalyzer, 126–27, 127 (ill.)
    cubic centimeters, 121
    cybernetics, 125
    donkey engine, 122
    fire hydrant nozzles, 125
    four-stroke engine vs. two-stroke engine, 121
    holography, 122
    industrial robots, 123–24
    internal combustion engine, 120, 121 (ill.)
    Lichtenberg figures, 127
    power take-off, 122
    road roller, 125
    robots, 122–24, 124, 124 (ill.)
    SCRAM, 127
    service robots, 123–24
    simple, 120
    Unimate (robot), 124, 124 (ill.)
Magee, Carlton C., 239
MAGLEV trains, 259
Maker Faire, 8
maker movement, 8, 87
makers, 8
male tanks, 131
Mall of America, 199
malls, 199
Malthus, Thomas Robert, 316
Manhattan Project, 134–35
manhole covers, 210
MANIAC, 28
Mansfield, Peter, 407–8
manual sequencing, 377
manufacturing
    acid-free paper, 188
    ammonia, 184
    aqua regia, 183
    bakelite, 181
    bone china, 179
    bulletproof glass, 177
    carbon black, 185–86
    carbon-carbon composites (CCCs), 180
    cashmere, 190
    ceramics, 179
    chromated copper arsenite, 185
    colored glass, 176
    composite materials, 182–83
    cord of wood, 187

CorningWare, 179–80
corrugated cardboard, 189
creosote, 185
crown glass, 175
definition, 174
diapers, 188–89
dry ice, 184 (ill.), 184–85
dynamite, 186
essential oils, 186–87
fiberglass, 183
fine china, 179
fireworks, 186
float glass process, 175–76
glass, 175–78
glass blocks, 177
glass windows, 177
glazes, 179
heat-resistant glass, 178, 178 (ill.)
hydrogen peroxide, 184
Kevlar, 191
low-e glass, 178
microfibers, 191–92
natural fibers, 189
paper, 187 (ill.), 187–88
paper towels, 189
parchment paper, 188
plastic, 180–81
polyester, 191
porcelain, 179
Pyrex, 178
railroad ties, 187
rosin, 186
ruby, 180
sandpaper, 188
silk, 190, 190 (ill.)
stained glass windows, 176
Styrofoam™, 182, 182 (ill.)
sulfuric acid, 183
synthetic fibers, 190–91
synthetic plastic, 180 (ill.), 180–81
technology and, 174
Teflon, 181–82
telephone poles, 187
tempered glass, 176, 176 (ill.)
thermopane glass, 177
TNT, 186
waterproof/water-repellent fabrics, 191
woods, 187
March, 109
Marconi, Guglielmo, 61, 62 (ill.)
Marcus, Siegfried, 225
Marescaux, Jacques, 420
Marie Antoinette, 384
Mark I computer, 28, 41
mark twain, 242
Marrison, Warren, 117
Marsh, George Perkins, 317
Martin, Pierre-Émile Martin, 169

materials standards, 19
Mathews, H. Hume, 13
Mauchly, John William, 28
Maudslay, Henry, 125
Maxam, Allan, 371
Maxim, Hiram S., 132
May, 109
May, Gene, 249
McAdoo, William, 238–39
measurements. See weights and measurements
mechanical engineering, 1–2
mechanical watch, 117
medicine
    alerting devices, 418
    artificial kidney, 421
    assisted reproductive technology, 421–22
    assistive listening devices, 417–18
    assistive technologies, 416
    automated external defibrillators, 412
    blood flow to heart/brain, 407
    brain-computer interfaces, 419
    breast diseases, 408
    cancer, 406
    CAT/CT scans, 405 (ill.), 405–6
    cochlear implant, 416–17, 417 (ill.)
    continuous glucose monitoring, 415
    defibrillator, 411–12, 412 (ill.)
    diabetes, 414
    dialysis machines, 421
    disabilities, 416
    drug delivery systems, 413
    electrical muscle stimulation, 420
    endoscope, 408–9, 409 (ill.), 410
    fetal image-guided surgery, 423
    fetal surgery, 422
    fetoscopic surgery, 423
    fingerstick glucose tests, 415
    hearing aids, 416
    heart-lung machine, 412–13
    insulin pumps, 414
    iontophoresis, 414
    lasers, 413, 423
    LASIK, 423–24
    LifeVest, 412
    needlesticks/sharps protection, 410
    new technologies for low vision, 418
    nuclear magnetic resonance, 407–8
    nuclear magnetic resonance imaging, 407
    obstetrics, 408
    pacemaker, 411

patient-controlled analgesia, 414–15
    personalized medicine, 403–4
    pharmacogenomics, 404
    positron emission tomography (PET) imaging, 406–7
    prosthetic devices, 418–19, 419 (ill.)
    pumps, 414
    robotic surgery, 419–20
    role of technology in, 403
    tattoos, 423
    telehealth, 403
    telesurgery, 420
    test-tube baby, 422
    thermometers, 409–10
    tissue engineering, 404
    transdermal patches, 413–14
    ultrasound, 408
    in vitro fertilization, 421–22
    wearable technology, 414–15
    X-rays, 405
medium truck, 239
megatall buildings, 201
Meinwald, Jerrold, 24
Mellon Arena, 203
meltdown, 284–85
memory cards, 35
Mercalli, Giuseppe, 100
Mergenthaler, Ottmar, 50
Merzenich, Michael M., 22
metal, 359
metallurgy, 156
metals. See also mining
    alkali metals, 156
    alkaline Earth metals, 156
    alloy, 163–64
    aluminum, 158–60, 160 (ill.)
    aluminum foil, 161
    basic oxygen furnace, 169–70
    bauxite ore, 159–60
    Bayer process, 160–61
    brass, 165
    bronze, 164
    carbide compounds, 164–65
    coins, 167
    coltan, 162
    continuous casting steel, 170–71
    copper, 162–63, 163 (ill.), 164, 165
    COR-TEN steel, 171
    electric arc furnace, 169–70
    elements, 156
    ferrous metals, 167
    fool's gold, 166
    German silver, 167
    gold, 165–66
    gold leaf, 166, 166 (ill.)
    gold-producing countries, 166
    Iron Age, 168

iron ore deposits, 168
metallurgy, 156
noble, 157
nonferrous metals, 167
oxidation, 163
pewter, 167
pitchblende, 157
platinum, 158
precious, 157
pure silver, 167
rare earth elements, 162
silver, 167
slag, 168
solder, 164, 164 (ill.)
stainless steel, 170
stainless steel flatware, 170
steel, 167, 169–72
steel mass production, 168–69
sterling silver, 167
tin, 167
titanium, 161, 162
titanium dioxide, 161–62
transition elements, 156–57
troy ounce of gold, 166
uranium deposits, 157–58
white gold, 166
zinc, 165
meter, 93
methane fuel, 356
metric system, 95–98
metric ton, 97
Metropolis, Nicholas C., 28
Meucci, Antonio, 74
Micali, Silvio, 25
Michaux, Pierre, 240
Michelin, André, 229
Michelin, Édouard, 229
Michelin tire, 229–30
microarray, 382
microarray technology, 386
microbes, 372
microfibers, 191–92
midnight, 101
MiG designation, 257
Mikoyan, Artem I., 257
military time, 114–15
military vehicles
    culin device, 255
    Humvee, 255
    Red Baron, 255
    tank, 255
Miller, Howard, 216
mine barrage, 131
minerals. See also metals; mining
    asbestos, 151 (ill.), 151–52
    China as leading producer of,
        156
    cinnabar, 152
    color standarization scheme,
        150
    corundum, 150

cosmetics, 155
cubic zirconia, 155
cubic zirconium, 155
Cullinan Diamond, 154–55
diamonds, 152–55, 154 (ill.)
elements, 148
emerald, 151
galena, 151
mineralogy, 147
Mohs scale, 150
native elements, 149
physical traits, 149–50
quartz, 147–48
rock, 147
silica, 148, 149
silica gel, 149
silicate, 148
silicon, 148
silicone, 148, 149, 149 (ill.)
star sapphires, 151
stibnite, 152
tiger's eye, 151
miner's canary, 144
miner's safety lamp, 143 (ill.), 144
mining. See also coal; minerals
    basic steps of, 139
    deepest underground mine, 143
    history of, 139
    importance of, 139
    miner's canary, 144
    miner's safety lamp, 143 (ill.),
        144
    open-pit mines, 141 (ill.),
        141–42
    surface, 140–41, 142
    underground, 140, 142–44
Mississippi River bridge, 216
Mobile Telephone Service (MTS), 77
mobile telephones, 76–77
modified Julian date, 106–7
Mohs, Friedrich, 100, 150
Mohs scale, 150
Moisseiff, Leon, 216
molecular cloning, 369
molecular genetic tests, 387
Molina, Mario, 342
monarch butterfly, 396
Monday, 108
monkey wrench, 126
Monnartz, Philip, 170
Monsanto Chemical Company, 77
months, 109
Montreal Protocol, 343
moon, 108
Moon, William, 52
moon walk, 8
Mooney, John J., 330
Moore, Gordon, 31
Moore's Law, 31
Morris, Robert, Jr., 45
Morse, Samuel F. B., 52, 52 (ill.)

Morse Code, 52–53
motor vehicle exhaust, 330
motorcycle, 240
Mount Rushmore, 221
mountaintop mining, 140
mouse, 33 (ill.), 33–34
Mponeng Gold Mine, 143
Muir, John, 317
Müller, Paul, 352
Mullis, Kary B., 21, 371, 378
multigene disorders, 386
Mundy, Peter, 221
municipal solid waste, 355, 356, 358
Murdock, William, 313
Murray, Andrew, 383
Murray, Cherry A., 23
Mushet, Robert F., 168
Muslim calendar, 104
mustard gas, 129
Mutke, Hans Guido, 249
mycotoxin, 130
Myriad Genetics, 375
myrrh, 174

N

nail, 126
Nakamura, Shuji, 21, 22, 314
nanometer, 3
nanotechnology, 3
napalm, 128
Napier, John, 27
Napier's Bones, 27
National Academies, 14
National Academy of Engineering,
    15, 21–22
National Academy of Sciences, 14
National Consumer Electronics Re-
    cycling Partnership, 366
National DNA Index System (NDIS),
    392
National Environmental Policy Act,
    322
National Highway System, 206–7
National Inventors Hall of Fame, 13
National Medal of Science, 23–24
National Medal of Technology and
    Innovation, 22 (ill.), 22–23
National Pollution Discharge Elimi-
    nation System, 345
National Research Council, 14–15
National Technical Information Ser-
    vice (NTIS) Bibliographic Data-
    base, 18
native elements, 149
NATO Phonetic Alphabet, 50–51
natural fibers, 189
natural gas, 265, 274–76, 275 (ill.)
natural materials
    amber, 173
    ambergris, 174

frankincense, 174
fuller's earth, 172
myrrh, 174
obsidian, 173, 173 (ill.)
petrified wood, 173, 173 (ill.)
nautical mile, 94
Nautilus, USS, 246
Navajo Code Talkers, 54, 54 (ill.)
NDIS (National DNA Index System), 392
Neanderthals, 119
needlesticks, 410
Nelson, Gaylord, 319
net registered tonnage, 244
Neumann, John von, 28
New Century Global Center, 200
New Hope Power Partnership, 293
newborn screening, 387
Newman, Max, 28
newspaper, 362
Newton, Isaac, 15
Niagara Power Plant, 299
night, 102
Nighthawk (aircraft), 257
nitrogen oxide emissions, 333
Nixon, Richard M., 319
Noah's ark, 241–42
Nobel, Alfred, 20, 186
Nobel Prize, 20, 21
noble metals, 157
Nokia 9000i Communicator, 78
nonferrous metals, 167
nonpoint source water pollution, 345
nonrenewable energy sources, 264
nonstop transatlantic flight, 248–49
noon, 102
North Dakota crude oil, 267, 268–69
Nouguier, Émile, 219
November, 110
Noyce, Robert, 30 (ill.), 30–31
nuclear annihilation, 118–19
nuclear energy
    Chernobyl accident, 285 (ill.), 285–86
    Fukushima nuclear accident, 286 (ill.), 286–87
    government agencies, 287
    meltdown, 284–85
    non-renewable energy source, 278
    nuclear accidents, 283–87, 285 (ill.), 286 (ill.)
    nuclear power plant, 278, 279 (ill.), 279–82
    nuclear waste, 287–88
    Rasmussen report, 284
    Three Mile Island, 283
    uranium, 278, 280
    vitrification, 288
nuclear magnetic resonance, 407–8

nuclear magnetic resonance imaging, 407
nuclear power plant, 278, 279 (ill.)
nuclear reactors, 278–79
Nuclear Regulatory Commission, 287
nuclear waste, 287–88
nuclear winter, 136
nuclear-powered automobile, 227
nuclear-powered vessels, 246–47

O

Oasis of the Seas (ship), 246
Obama, Barack, 118
object-oriented computer language, 43
Obninsk nuclear facility, 279
observation wheels, 222 (ill.), 222–23
obsidian, 173, 173 (ill.)
obstetrics, 408
ocean, 300
ocean debris, 347–49
octane numbers, 271–72, 272 (ill.)
October, 109
O'Dwyer Variable Lethality Law Enforcement pistol, 134
offshore drilling, 269
offshore oil rig, 218
offshore wind industry, 296
Ohl, Russell, 291
Ohm, Georg Simon, 100
oil, as fossil fuel, 265, 266–67, 268–70
oil fields, 266–67
oil pollution, 345–46
oil spills, 346–47, 347 (ill.), 348
Okamoto, Miyoshi, 192
Olds, Ransom Eli, 226
OLED technology, 69–70, 70 (ill.)
onager, 127–28
One World Trade Center, 202
open-pit mines, 141 (ill.), 141–42
open-source software, 44
operating system, 36
operating system software, 35–36
Oppenheimer, J. Robert, 135
O'Reilly, Tim, 87
Orient Express, 261
Oroville dam, 220
Osborne, Adam, 38
Osika, V. V., 155
Otis, Elisha, 201
Otto, Nikolaus August, 120
Otto Hahn (ship), 247
O_2 Dome, 204 (ill.), 205
overburden, 140
oxidation, 163
Oyster Creek, New Jersey, nuclear power plant, 280
ozone, 341 (ill.), 341–43

P

pacemaker, 411
Page, Larry, 84, 84 (ill.)
PAGE (PolyAcrylamide Gel Electrophoresis), 383
palindromic square, 55–56
Panama Canal, 219
paper, 187 (ill.), 187–88, 361–62, 362 (ill.)
paper towels, 189
Papworth, Neil, 78
papyrus, 49
parachute jump, 252
parchment paper, 188
Parkes, Alexander, 180–81
parking meter, 239
Parsons, William S., 257
Pascal, Blaise, 27
PASCAL computer language, 43
passenger ship, 246
passive solar energy systems, 289 (ill.), 289–90
"patent applied," 10
"patent pending," 10
patents
    copyrights and trademarks vs., 12
    definition, 9
    drawings, 10
    Edison, Thomas, 11
    first, 10
    human genes, 375
    length of protection, 12
    Lincoln, Abraham, 9–10
    number of, 9
    "patent applied," 10
    "patent pending," 10
    provisional application, 10
    things that cannot be patented, 10–11
    trade secret, 11
    types of, 4–5
    utility, 4–5
patient-controlled analgesia, 414–15
PCBs (polychlorinated biphenyls), 352
Pearl, Judea, 25
Pearson, Gerald, 291
Pennsylvania crude oil, 269
penny, 126
Pentagon, 200
performance standards, 19
Perkinson, Henry, 77
Permian Basin, 267
perpetual calendar, 105
Perreaux, Louis-Guillaume, 240
Pershing missile, 135
Persistent Organic Pollutants, 351–52
personalized medicine, 403–4
PET (polyethylene terephthalate), 360
PET imaging, 406–7

Peters, C. Wilbur, 409
petrified wood, 173, 173 (ill.)
pewter, 167
pharmacogenomics, 404
Philip, Prince, Duke of Edinburgh, 391
Philip Merrill Environmental Center, 194
Phillips screw, 126
Philosophical Transactions, 17
phishing, 58
phonetic alphabet, 50–51
photovoltaic (PV) cells, 290–91
physical disability, 416
phytoremediation, 328
Pilkington, Alastair, 175
pipeline system, 275
pisé, 197–98
pitchblende, 157
Pitts, Simon, 22
pixel, 38–39
placer mining, 140
plane, 120, 120 (ill.)
planes. *See* aircraft
plant patents, 9
plasma display panels, 69
plastics, 180–81, 359–60
platinum, 158
Plimsoll, Samuel, 244
Plimsoll line, 244
PLU codes, 57
plug-in hybrid electric vehicles, 228
Plunkett, Roy J., 181–82
P.M., 114
podcasts, 88
Polaniec biomass power plant, 293
police radar, 238
PolyAcrylamide Gel Electrophoresis (PAGE), 383
polychlorinated biphenyls (PCBs), 352
polyester, 191
polyethylene terephthalate (PET), 360
polymerase chain reaction, 378
polyvinyl chloride (PVC) plastics, 352–53
Pong, 46, 46 (ill.)
pontoon bridges, 215
porcelain, 179
portable computers, 38
portable storage media, 34–35
positron emission tomography (PET imaging), 406–7
Post, Wiley, 249
potash, 10
potential energy, 263
Potrykus, Ingo, 395
power, 264
power grids, 302
power outages, 302

power plant, 301
power take-off, 122
precious metals, 157
preimplantation testing, 387
prenatal testing, 387
Price Look-Up (PLU) codes, 57
primary air pollutants, 329
primary energy sources, 264
privately funded research reports, 17–18
programmers, 41
programming languages, 41–42
prosthetic devices, 418–19, 419 (ill.)
proteomics, 374–75
protocol identifier, 82
provisional patent application, 10
publications
    De Aquaeductu Urbis Romae (On the Aqueducts of Rome) (Frontinus), 18
    De re Metallica (Bauer), 16, 16 (ill.)
    grey literature, 16–17
    Journal des Sçavans, 16
    National Technical Information Service (NTIS) Bibliographic Database, 18
    Philosophiae Naturalis Principia Mathematica (Newton), 15
    Philosophical Transactions, 17
    standards, 18–21
    technical reports, 17–18
    Treatise on the Astrolabe (Chaucer), 18
pulley, 120, 120 (ill.)
pumps, 414
punched cards, 40 (ill.), 40–41, 42
pure silver, 167
Purple cipher machine, 53–54
PVC (polyvinyl chloride) plastics, 352–53
Pyrex, 178

Q

QM2 (ship), 246
QR code, 57, 57 (ill.)
quad, 304
quartz, 147–48
quartz clock, 117
quartz watch, 117
Quebec Bridge, 213
Queen Mary (ship), 246

R

radar, 238
radio
    AM radio stations, 62–63
    first broadcasting station, 61–62

FM radio stations, 62
HD Radio® Technology, 63
invention of, 61
K and W in call letters, 62
radio frequency identification systems (RFID), 65 (ill.), 65–66
satellite digital, 63, 64
submerged submarine communication, 64
torpedo detection, 64–65
radio frequency identification systems (RFID), 65 (ill.), 65–66
radiological dispersal device, 137
radon, 337
railroad ties, 187
railroads. *See* trains
rain forest, 324
RAM (random-access memory), 32–33
rammed earth, 197–98
random-access memory (RAM), 32–33
rare earth elements, 162
Rasmussen, Norman, 284
Rasmussen report, 284
reactive materials, 349
read-only memory (ROM), 32–33
recombinant DNA, 378
recordings
    CD-R, 79
    CD-RW discs, 79
    compact discs (CDs), 78–79, 79–80
    digital video recording (DVR), 71–72
    DVDs, 79–80
    video cassette recorder (VCR), 71
    video tape recorder, 70–71, 71 (ill.)
recycling
    aluminum, 359, 359 (ill.), 360
    biodegradable plastics, 359–60
    curbside recycling program, 358
    discarded tires, 364
    eCycling, 365
    electronic products, 364 (ill.), 364–66
    green electronics, 368
    green product, 368
    grocery bags, 361
    high-density polyethylene (HDPE), 360
    landfill, 361
    low flow toilets, 363
    materials with highest recycling rates, 359
    metal, 359
    municipal solid waste, 358

National Consumer Electronics Recycling Partnership, 366
newspaper, 362
paper, 361–62, 362 (ill.)
polyethylene terephthalate (PET), 360
products made from, 360–61
Resource Conservation and Recovery Act, 358
solid waste, 358
Sustainable Materials Management Electronics Challenge, 365
symbol, 358
washing dishes by hand vs. dishwasher, 363–64
water, 362–64
WaterSense, 363
Red Baron, 255
red dog, 147
Redfield, William Charles, 14
Redier, Antoine, 117
Reed, Ezekiel, 126
Reemtsma, Keith, 393
Rees, William, 316
refining process, 269–70
reformulated gasoline, 271
renewable and alternative energy sources
active solar energy systems, 289–90
biomass energy, 292–93
cord of wood, 293
definition, 264
electricity, 290
fossil fuels vs., 288
fuel cell, 291–92
geothermal energy, 297–98
geothermal power plant, 297–98, 298 (ill.)
hydroelectric power, 300
hydropower, 298–300
ocean, 300
offshore wind industry, 296
passive solar energy systems, 289 (ill.), 289–90
photovoltaic (PV) cells, 290–91
role of, 288
solar cell, 290–91
solar domestic hot water systems, 290
solar energy, 288–90, 291
solar generating plant, 291
Three Gorges Dam, 299 (ill.), 299–300
wind energy, 294–97
wind farm, 294–95, 295 (ill.)
wind turbines, 294
wood, 293
wood-burning stove, 293
research reports, 17–18

residential sector, 265
resistive screens, 39
Resource Conservation and Recovery Act, 358
restriction endonuclease, 380
restriction fragment length polymorphisms, 381
retractable roof stadium, 203–4
Richter, Burton, 24
Richter, Charles, 100
Richthofen, Manfred von, 255
ricin, 129
rigid beam bridge, 211
Ritchie, C. G., 191
Rittenhouse, Nicholas, 188
Rittenhouse, William, 188
road roller, 125
roads
civil engineering, 205
cloverleaf interchange, 209
coast-to-coast highway, 208
diverging diamond interchange, 209, 209 (ill.)
English Channel tunnel, 210
Interstate Highway System, 205–6
Jersey barrier, 210
macadam roads, 205
manhole covers, 210
most lanes, 208
most miles, 208
National Highway System, 206–7
numbering of U.S. highways, 208
traffic light, 209–10
tunnels, 210–11
robotic surgery, 419–20
robots, 122–24, 124 (ill.)
rock, 147
Rockville Bridge, 213
Roebling, John A., 213
Roebling, Washington A., 213
Roentgen, William Conrad, 405
Rogers Centre, 203–4
roller coasters, 223
ROM (read-only memory), 32–33
Roman calendar, 104–5
Roman measurements, 92
Romanov, Tsar Nicholas II, 391, 391 (ill.)
Romanov family, 391 (ill.), 391–92
Roosevelt, Theodore, 221
Roper, Sylvester Howard, 240
rosin, 186
Rosing, Boris, 66
rotational time, 101
Rowland, F. Sherwood, 342
Rowland, Thomas Fitch, 218
ruby, 180
rumble seat, 231, 231 (ill.)

Russ, Dolores H., 22
Russ, Fritz J., 22
Russia time zones, 112
Russky Bridge, 213
Rutan, Dick, 249
R-value, 196–97
Ryan, John T., Sr., 144

S

Saarinen, Eero, 217
safety recalls, 234
Sallo, Denis de, 16
Samuel, Arthur, 46
San Gorgonio Pass Wind Farm, 295 (ill.)
sandpaper, 188
Sanford, John, 382
Sanger, Fred, 371, 373, 375, 377
Sanger technique, 377
sapphires, 151
satellite, 73–74
satellite broadband, 82
satellite digital radio, 63, 64
satellite dish, 68
Saturday, 108
Saturn, 108
Sauvestre, Stephen, 219
Savannah (ship), 247
SBN system, 58
Scaliger, Joseph Justus, 106, 106 (ill.)
Scaliger, Julius Caesar, 106
Schiller, John, 23
science, technology, 1
scientific method, 315–16
scientists, 1
SCRAM, 127
screw, 120, 120 (ill.)
screw threads, 125
Scribner, Belding, 421
scrubbers, 334, 334 (ill.)
sea time, 115
seaplane, 254
search engine, 83
seasons, 110
seatbelts, 235
seawall, 218
secondary air pollutants, 329
secondary energy sources, 264
Secure Remote Continuous Alcohol Monitoring (SCRAM), 127
Segway, 240, 240 (ill.)
sensors, 39
September, 109
September 11, 2001, terrorist attack DNA investigation, 391
Serpollet, Leon, 231
service robots, 123–24
shaft mines, 143
shale, 267

447

shale gas, 276–77
Shanklin, Jonathan, 343
Sharkey, Tina, 88
sharps, 410
Shaw, Mary, 23
Shippingport, Pennsylvania, nuclear power plant, 280
ships. *See* boats and ships
shopping center, 199
short ton, 97
shot tower, 131–32
Shugart, Alan, 34
side-impact air bags, 236–37
Siemens, Frederick, 169
Siemens, William, 169
Siemens-Martin process, 169
Sikorsky, Igor I., 254–55
Silent Spring (Carson), 317, 318, 319, 319 (ill.), 352
Silevitch, Michael B., 22
silica, 148, 149
silica gel, 149
silicate, 148
silicon, 148
silicon carbide, 165
silicon chip, 30–31, 31–32
silicone, 148, 149, 149 (ill.)
silk, 190, 190 (ill.)
silver, 167
simple machines, 120
single gene disorders, 386
SiriusXM Radio, 63–64
SixDegrees, 88
skyscrapers, 200–202
slag, 168
Slate, Thomas Benton, 185
slope mines, 143
Smalltalk, 43
smart bomb, 135
Smart car, 229, 229 (ill.)
Smart Fortwo, 229
Smart Grid, 302–3
smart meters, 303
smartphone, 77–78
Smeaton, John, 15
Smith, Delford, 254
Smith, Edward John, 245
Smith, Michael, 378
Smith, Robert Angus, 333
Smith, William "Uncle Billy," 266
smog, 331 (ill.), 331–32
snow tires, 230–31
Sobriero, Ascanio, 186
social networking, 88–89
societies
    American Association for the Advancement of Science, 14
    American Philosophical Society, 14
    Institute of Medicine, 14, 15

National Academy of Engineering, 15
National Academy of Sciences, 14
National Research Council, 14–15
Society of Civil Engineers, 15
Society of Civil Engineers, 15
software, 30
solar cell, 290–91
solar domestic hot water systems, 290
solar energy, 288–90, 291
solar generating plant, 291
solder, 164, 164 (ill.)
solid waste
    garbage, 354
    garbage incinerator, 355
    landfill gas, 357
    landfills, 356 (ill.), 356–57
    methane fuel, 356
    municipal solid waste, 355, 356
    recycling, 358
Solomon, Sean C., 24
Sommelier, German, 125
Sopwith Camel, 256, 256 (ill.)
Sound Transmission Class (STC), 197
source reduction activities, 350
South China Mall, 199
spam, 86
specifications, 20–21
Spirit of St. Louis (airplane), 249
spring, 110
Spruce Goose (airplane), 253–54, 254 (ill.)
SR 520 bridge, 214
Stack, John, 249
Stackelberg, M. V., 155
stained glass windows, 176
stainless steel, 170
stainless steel flatware, 170
Standard Book Numbering (SBN), 58
standard gauge railroad, 258
standards, 18–21
star sapphires, 151
starboard, 241
Starlink corn, 396
Statue of Liberty, 220 (ill.), 220–21
statute mile, 94
STC (Sound Transmission Class), 197
stealth technology, 257–58
steamboats, 9–10
steam-powered ships, 242–43, 243 (ill.)
steel, 167, 169–72
steel mass production, 168–69
steel-arch bridge, 213
Stephenson, Adam, 15

Stephenson, George, 144, 258
Steptoe, Patrick, 422
sterling silver, 167
stibnite, 152
stone-arch bridge, 213
Stonebraker, Michael, 25
Stookey, Donald, 179–80
Strauss, Joseph B., 216
Strebe, Diemut, 400
structures. *See also* buildings
    dams, 220
    Eiffel Tower, 219, 219 (ill.)
    Ferris wheel, 221–22
    Gateway Arch, 216–17, 217 (ill.)
    Great Pyramid at Giza, 216
    Las Vegas High Roller, 222, 222 (ill.)
    Mount Rushmore, 221
    observation wheels, 222 (ill.), 222–23
    offshore oil rig, 218
    Panama Canal, 219
    roller coasters, 223
    seawall, 218
    Statue of Liberty, 220 (ill.), 220–21
    Texas tower, 218
    Washington Monument, 217–18
Styrofoam™, 182, 182 (ill.)
submarines, 64
Sugababe, 400
sulfur dioxide emissions, 333, 334
sulfuric acid, 183
Sullivan, Eugene G., 178
summer, 110
sun, 108
Sunday, 108
sundial, 116
sunset, 102
Superfund Act, 353
Superfund sites, 353 (ill.), 353–54
supersonic flight, 249
supertall buildings, 201
surface acoustic wave screens, 39
surface mining, 140–41, 142
suspension bridge, 212, 212 (ill.)
suspension bridges, 214
sustainability, 366–68
sustainable materials management, 322
Sustainable Materials Management Electronics Challenge, 365
Swade, Doron, 27
Swinton, E. D., 130
Swiss Army knife, 128
synthetic fibers, 190–91
synthetic fuels, 277
synthetic plastic, 180 (ill.), 180–81
Système Internationale d'Unités (SI), 92
Szostak, Jack William, 383

T

Tang, Ching W., 70
Taniguchi, Norio, 3
tank, 255
tanks, 130 (ill.), 130–31
Tarquinius Priscus, 105
tattoos, 423
taxicab, 239–40
Taylor, William C., 178
technical reports, 17–18
technology
    definition, 1
    high, 4–5
    time periods, 5–7
Teflon, 181–82
telecommunications. *See also* Internet
    caller ID, 75
    cell phones, 77–78
    communications satellite, 73
    digital technology, 77
    fax machine, 75–76
    fiber-optic cable, 76
    4G LTE, 78
    geosynchronous orbit, 73
    "live via satellite" television broadcast, 73–74
    Mobile Telephone Service (MTS), 77
    mobile telephones, 76–77
    smartphone, 77–78
    telephone, 74–75
    text messages, 78
    virtual reality, 79 (ill.), 80
    Voice over Internet Protocol (VoIP), 75
telehealth, 403
telephone, 74–75
telephone poles, 187
telesurgery, 420
television
    analog vs. digital, 68–69
    Dolby noise reduction system, 70
    earth station, 68
    flat-panel vs. traditional screens, 69
    founding of, 66–67, 67 (ill.)
    OLED technology vs. LCD technology, 69–70
    rain's effect on, 67
    satellite dish, 68
    video cassette recorder (VCR), 71
    video tape recorder, 70
Telford, Thomas, 205
Telstar 1, 73, 74
tempered glass, 176, 176 (ill.)
10-codes, 54–55
terminator gene technology, 388

terminology and symbols, 19
test methods, 19
test-tube baby, 422
Texas crude oil, 267
Texas tower, 218
text messages, 78
TGV (Train à Grande Vitesse) Atlantique, 258
Thacker, Charles P., 25
theodolite, 100–101, 101 (ill.)
thermometers, 409–10
thermopane glass, 177
thermoplastics, 181
thermosetting plastics, 181
thermostat, 310, 310 (ill.), 311
Thirteen-Month calendar, 105
Thomas, Seth, 118
Thompson, John Taliaferro, 132
Thompson, LaMarcus, 223
Thomson, William, Lord Kelvin, 100
Thor, 108
Thoreau, Henry David, 317
3-D printing, 38, 38 (ill.), 400
Three Gorges Dam, 299 (ill.), 299–300
Three Mile Island, 283
three-point safety belt, 235
Thursday, 108
Tibbets, Paul W., Jr., 257
tiger's eye, 151
time
    alarm clock, 117 (ill.), 117–18
    A.M., 114
    B.P., 111
    calendar year, 103
    calendars, 104–6
    century, 103
    Chinese years and animals, 107
    clockwise, 116
    Coordinated Universal Time, 112
    Daylight Savings Time, 111
    days, 108
    digital watch, 118
    doomsday clock, 118–19
    floral clock, 116
    grandfather clock, 116
    International Date Line, 112
    January 1 as first day of new year, 103–4
    Julian Day Count, 106
    leap second, 112, 114
    leap year, 108
    local noon, 111
    longest and shortest years, 107–8
    measurement of, 101–2
    mechanical watch, 117
    military time, 114–15
    modified Julian date, 106–7
    months, 109

    P.M., 114
    quartz clock, 117
    quartz watch, 117
    sea, 115
    seasons, 110
    sundial, 116
    time zones, 112
    timekeeping, 102
    units of, 102–3
    U.S. National Institute of Standards and Technology (NIST), 114
    U.S. Time Standard signal, 114
    water clocks, 116
    week, 109
    World Calendar, 106
    Y2K bug, 108
time zones, 112
timekeeping, 102
tin, 167
tires, 229–31, 306, 306 (ill.), 364
tissue engineering, 404
Titanic, 245 (ill.), 245–46
titanium dioxide, 161–62
Tiu, 108
TNT, 186
toilets, 363
Tokyo Sky Tree, 202
Tomlinson, Ray, 85–86
tonnage, 243–44
tools. *See also* machines
    compass, 119–20
    earliest agricultural, 119
    jackhammer, 125
    monkey wrench, 126
    nail, 126
    Neanderthal, 119
    penny, 126
    Phillips screw, 126
    screw threads, 125
top-level domain (TLD) suffixes, 83
torpedo detection, 64–65
Torvalds, Linus, 36
touch screen, 39–40
Townes, Charles H., 21
toxic materials, 349
Toxic Release Inventory, 349–50
Toxic Substances Control Act, 353
traction control systems, 233
trade secret, 11
trademarks
    copyrights and patents vs., 12
    definition, 11
    generic product name, 11–12
    length of protection, 12
    symbol, 11
traffic light, 209–10
trains
    cable car, 261–62
    cabooses, 260, 260 (ill.)
    chartered railroad, 260

449

fastest, 258–59
funicular railway, 261
gandy dancer, 261
longest railway, 259–60
MAGLEV, 259
Orient Express, 261
standard gauge railroad, 258
velocipede, 260–61
transcontinental railroad, 260
transdermal patches, 413–14
transduction, 380
transformation, 378
transgenic animal transplants, 393
transition elements, 156–57
transmission lines, 302
transposon, 373–74
Trans-Siberian Railway, 259–60
Treatise on the Astrolabe (Chaucer), 18
Trevithick, Richard, 225
tropical forests, 324–25
tropical rain forest, 324
troy measurements, 98–99
troy ounce of gold, 166
trucks, 239
tubeless tires, 230
Tuesday, 108
tugboats, 244
Tulloch, T. G., 130
tungsten carbide, 165
tunnels, 210–11
turbines, 301
Turco, Richard P., 136
Turing, Alan M., 24, 24 (ill.), 28, 46
Turing Award, 24–25
The Turk, 47, 47 (ill.)
twilight, 101
Twitter, 89
two-stroke engine, 121
Tyndall, John, 337

U

ultrasound, 408
unconventional gas, 276
underground mining, 140, 142–44
underinflated tires, 306, 306 (ill.)
Uniform Resource Locator (URL), 82–83
Unimate (robot), 124, 124 (ill.)
Universal Product Code (UPC), 56–57, 65
Universal Serial Bus (USB), 34
Upatnieks, Juris, 122
UPC, 56–57, 65
uranium, 278, 280
uranium deposits, 157–58
URL, 82–83
U.S. Department of Defense Advanced Research Projects Agency (DARPA), 80–81

U.S. Department of Energy, 287
U.S. National Institute of Standards and Technology (NIST), 114
U.S. Pavilion, 204–5
U.S. Time Standard signal, 114
USB, 34
utility patents, 4–5, 9

V

vaccines, 385
Vail, Alfred, 52
Valiant, Leslie G., 25
Van Slyke, Steven, 70
variable number of tandem repeats (VNTR), 388–89
VASCAR (Visual Average Speed Computer and Recorder), 238
VBA, 200
vector, 379–80
vehicle identification number (VIN), 232
velocipede, 260–61
Venter, Craig, 374
Verneuil, Auguste Victor Louis, 180
Verrazano-Narrows Bridge, 214
Verrazzano, Giovanni da, 214
video cassette recorder (VCR), 71
video tape recorder, 70–71, 71 (ill.)
Vindeby, Denmark, 296
virtual reality, 79 (ill.), 80
viruses, 379–80
visible-spectrum LED, 314
vision loss, 416
Visual Average Speed Computer and Recorder (VASCAR), 238
vitrification, 288
VNTR (variable number of tandem repeats), 388–89
Voice over Internet Protocol (VoIP), 75
volatile organic compounds (VOCs), 336–37
Volta, Alessandro, 100
Voyager (aircraft), 249

W

Wackernagel, Mathis, 316
Walton, Charles, 66
Wanxian Bridge, 213
Warren, David, 253
Washington, George, 221
Washington Monument, 217–18
water clocks, 116
water pollution
  British Petroleum offshore oil spill, 348
  Clean Water Act of 1972, 344–45

freshwater depletion and shortage, 345
  gray water, 345
  National Pollution Discharge Elimination System, 345
  nonpoint source, 345
  ocean debris, 347–49
  oil pollution, 345–46
  oil spills, 346–47, 347 (ill.), 348
  types of, 343–44
water recycling, 362–64
water weight, 98
waterproof/water-repellent fabrics, 191
WaterSense, 363
watershed, 327
Watson, James, 373
Watson, Thomas, 74
Watt, James, 100, 227
Watts, William, 131
WBZ (Springfield, Massachusetts), 61
WDMK-FM (Detroit, Michigan), 63
weapons
  aflatoxin, 130
  anthrax, 129
  atomic bomb, 134–35, 135 (ill.)
  bazooka, 132
  Big Bertha, 134, 134 (ill.)
  biological warfare, 129–30
  Blackout Bomb, 136
  botulism, 129
  Bowie knife, 128
  Cannon King, 132, 134
  chemical warfare, 128–29
  Civil War, 131
  clostridium perfringens, 130
  Colt revolver, 131
  cruise missile, 136
  Fat Man, 135
  flail, 128
  improvised explosive devices (IEDs), 136–37, 137 (ill.)
  Little Boy, 135
  machine gun, 132, 133 (ill.)
  Manhattan Project, 134–35
  mine barrage, 131
  mustard gas, 129
  mycotoxin, 130
  napalm, 128
  nuclear winter, 136
  O'Dwyer Variable Lethality Law Enforcement pistol, 134
  onager, 127–28
  Pershing missile, 135
  radiological dispersal device, 137
  ricin, 129
  shot tower, 131–32
  smart bomb, 135
  Swiss Army knife, 128

tanks, 130 (ill.), 130–31
weapons of mass destruction (WMDs), 137
weapons of mass destruction (WMDs), 137
wearable technology, 414–15
Web 2.0, 87
wedge, 120, 120 (ill.)
Wednesday, 108
week, 109
weights and measurements. *See also* time
   area, 97–98
   avoirdupois measurements, 98–99
   benchmark, 100
   biblical, 91
   distance to the horizon, 100
   dry measures, 96–97
   Egyptian, 91
   Greek, 91–92
   instruments, 99
   kilogram, 94–95
   liquid measures, 96–97
   long ton, 97
   meter, 93
   metric system, 95–98
   metric ton, 97
   nautical mile, 94
   pound of gold vs. pound of feathers, 99, 99 (ill.)
   Roman, 92
   shekel, 91
   short ton, 97
   statute mile, 94
   Système Internationale d'Unités (SI), 92
   theodolite, 100–101, 101 (ill.)

troy measurements, 98–99
units of measurement named after individuals, 99–100
water weight, 98
yard, 93–94
Weinreich, Andrew, 88
Welch, George, 249
Werner, Abraham Gottlob, 150
wetlands, 325–27, 326 (ill.)
wheel, 120, 120 (ill.)
Whinfield, J. R., 191
white gold, 166
Whitfield, Willis, 13
Whitney, Eli, 19 (ill.), 19–20
Whitworth, Joseph, 125
Wiener, Norbert, 125
wikis, 87–88
Wilbrand, Joseph, 186
Willis Tower, 202
Wilmut, Ian, 371, 400
Wilson, Blake S., 22
Wilson, Woodrow, 14 (ill.), 15
wind energy, 294–97
wind farm, 294–95, 295 (ill.)
wind tunnel, 253
wind turbines, 294
winter, 110
wireless fidelity (WiFi), 81–82, 82 (ill.)
Wirth, Niklaus, 43
WMDs (weapons of mass destruction), 137
Woden, 108, 109 (ill.)
Wolf, Edward, 382
wood, 293
wood-burning stove, 293
woods, 187
Work, Henry Clay, 116

World Calendar, 105, 106
World Trade Center, 202
World Wide Web, 80–81
Wright, Orville, 247
Wright, Wilbur, 247
writing, 49–50
Wyatt, Isaiah, 181

X

xenotransplantation, 392–93
Xing, Yi, 118
XM Satellite Radio, 63, 64
X-rays, 405

Y

Yamanaka, Shinya, 398
yard, 93–94
Yeager, Chuck, 249
Yeager, Jeana, 249
Y2K bug, 108
Yucca Mountain, Nevada, 287–88
yurts, 198, 198 (ill.)

Z

Zeidler, Othmar, 352
zinc, 165
Zip® disks, 34
Zoll, Paul, 411
Zuckerberg, Mark, 89, 89 (ill.)
Zuider Zee, 211
Zworykin, Vladimir K., 66–67, 67 (ill.)
zygote intrafallopian transfer, 422